# NUMERICAL/CHRONOLOGICAL/ AUTHOR INDEX 1954 - 1978

An Index of Publications of the
American Astronautical Society

## PUBLICATIONS OF THE AMERICAN ASTRONAUTICAL SOCIETY

Following are the principal publications of the American Astronautical Society:

JOURNAL OF THE ASTRONAUTICAL SCIENCES (1957- )

Published quarterly and distributed by AAS Business Office, 6060 Duke Street, Alexandria, VA 22304

AAS NEWSLETTER (1962- )

Published bimonthly and distributed by AAS Business Office, 6060 Duke Street, Alexandria, VA 22304

ADVANCES IN THE ASTRONAUTICAL SCIENCES (1957- )

Proceedings of major AAS technical meetings. Published and distributed for the American Astronautical Society by Univelt, Inc., P.O. Box 28130, San Diego, CA 92128

SCIENCE AND TECHNOLOGY SERIES (1964- )

Supplement to Advances in the Astronautical Sciences. Proceedings and monographs, most of them based on AAS technical meetings. Published and distributed for the American Astronautical Society by Univelt, Inc., P.O. Box 28130, San Diego, CA 92128

AAS HISTORY SERIES (1977- )

Supplement to Advances in the Astronautical Sciences. Selected works in the field of aerospace history under the editorship of Dr. Eugene M. Emme, NASA Historian. Published and distributed for the American Astronautical Society by Univelt, Inc., P.O. Box 28130, San Diego, CA 92128

AAS MICROFICHE SERIES (1968- )

Supplement to Advances in the Astronautical Sciences. Consists principally of technical papers not included in the hard-copy volumes. Published and distributed for the American Astronautical Society by Univelt, Inc., P.O. Box 28130, San Diego, CA 92128

Subscriptions to the Journal and the AAS Newsletter should be ordered from the AAS Business Office. Back issues of the Journal and all books and microfiche should be ordered from Univelt, Inc.

Front Cover Illustration

Space Shuttle Payload Deployment -
Courtesy of the National Aeronautics and Space Administration

Frontispiece

Retrieval of a Small Free-Flying Satellite by Use of Manned Maneuvering Unit from Space Shuttle -
Courtesy of NASA Johnson Space Center (NASA S 75 15043B)

Thanks are due Diana Law, of the Lockheed-California Company, Rosalie Brazee and Jeannette M. Martell both of Univelt, Inc., for indexing and/or preparing copy of these indexes.

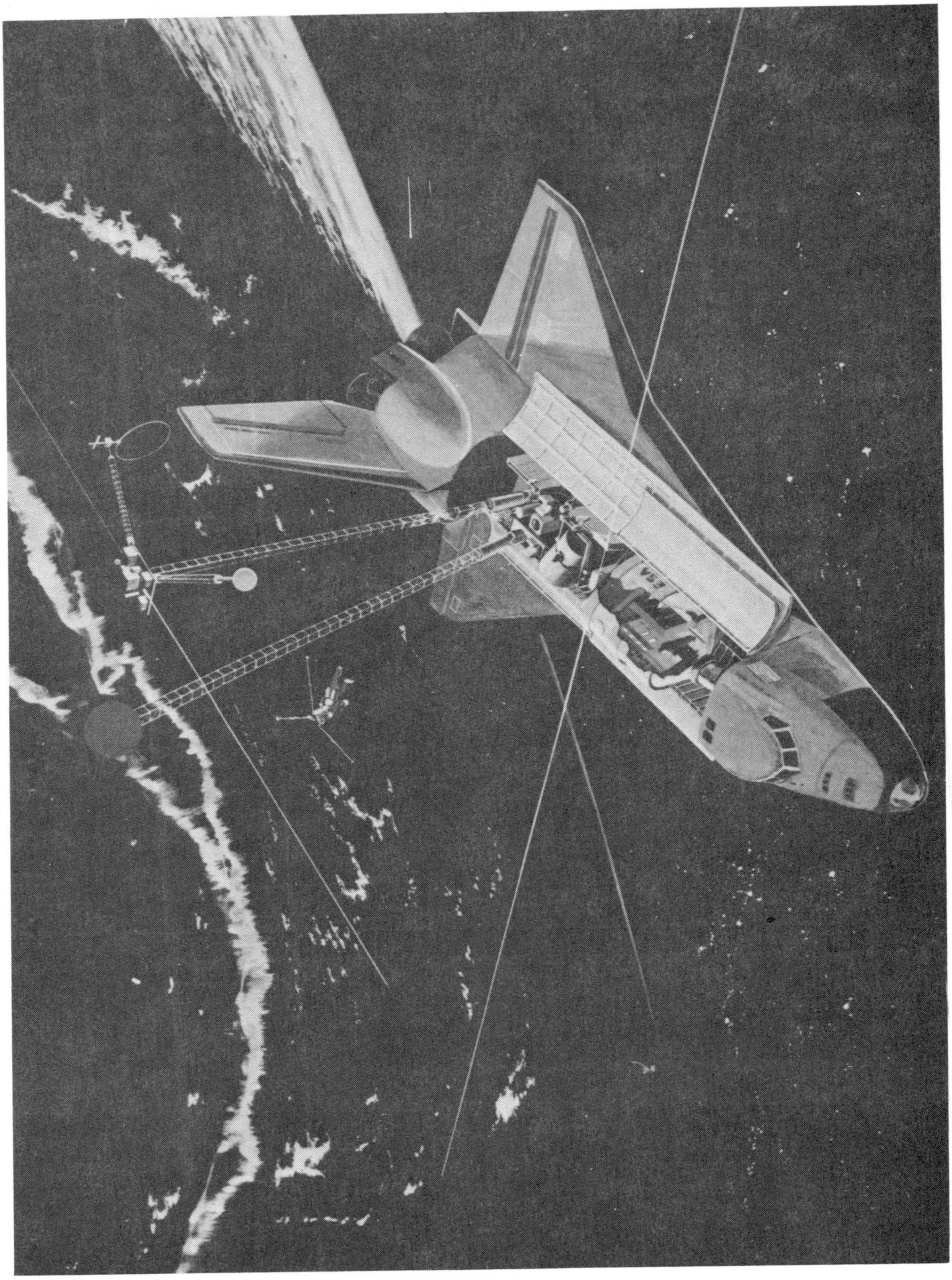

AN AMERICAN ASTRONAUTICAL SOCIETY PUBLICATION

# NUMERICAL/ CHRONOLOGICAL/ AUTHOR INDEX 1954 - 1978

Horace Jacobs
Robert H. Jacobs

## ADVANCES IN THE ASTRONAUTICAL SCIENCES
## SCIENCE AND TECHNOLOGY SERIES
and other AAS publications

Researched and published by Univelt, Inc.,
P.O. Box 28130, San Diego, California 92128

UNIVELT, INCORPORATED
P. O. Box 28130
San Diego, California 92128

ISBN 0-87703-102-9 (Hard Cover)
ISBN 0-87703-103-7 (Soft Cover)

Printed and Bound in the U.S.A.

## FOREWORD

This index volume presents a numerical/chronological index and an author index of all available articles and papers published by, for, or in conjunction with the American Astronautical Society from 1954 through 1978 with some citations from early 1979 added.

The Numerical/Chronological Index consists of three sections:

1. An index by AAS number in sequential order covering all papers that have appeared in proceedings volumes or their supplements. Included in this index are (a) some International Astronautical Federation papers, by IAF number assigned by the AAS for identification purposes only, and (b) some Deutsche Gesellschaft für Luft-und Raumfahrt papers by DGLR number--papers presented at a joint AAS/DGLR meeting. These IAF and DGLR papers appear at the end of the AAS papers for a given year. The papers are listed in numerical order under the volume or volumes in which they were published. In each case, the technical meeting at which the papers were presented is identified. Each number is followed by the title of the papers and the author(s). This is essentially a Citation Index in chronological/numerical order.

2. A chronological index of the contents of the Astronautical Sciences Review (1959-1962) by volume and issue, with titles, authors, and pages cited.

3. A chronological index of the contents of The Journal of the Astronautical Sciences (1954-1978) by volume and issue, with titles, authors, and pages cited.

The Author Index consists of an alphabetical list of all authors whose papers were published by the American Astronautical Society from 1954 through 1978. In each case, the author's name and initials only are given. Names are followed by paper number and/or journal volume and issue number and pages. It is necessary then to refer to the Numerical/ Chronological Index for the complete citation.

Since the Numerical/Chronological Index was prepared annually some minor variations in abbreviations and format occur.

The following appendices appear in this volume:

Appendix I.....Meetings of the American Astronautical Society (1954-1979)

Appendix II....AAS Meetings in Conjunction with the American Association for the Advancement of Science

Appendix III...AAS Goddard Memorial Symposia

Appendix IV....Books Published by or for the American Astronautical Society

A Subject Index will appear as a companion volume.

We consider this volume to be a most valuable reference tool in the field of astronautics and related disciplines in that it includes almost every available technical paper published by or for the American Astronautical Society in the first quarter century of the Space Age.

The Editors

# INDEX COVERAGE

THIS INDEX COVERS THE PERIOD 1954 THROUGH 1978 AND SOME CITATIONS FROM EARLY 1979. IT INCLUDES ALL AVAILABLE ARTICLES AND PAPERS THAT HAVE APPEARED IN THE FOLLOWING PUBLICATIONS OF THE AMERICAN ASTRONAUTICAL SOCIETY:

*THE JOURNAL OF THE ASTRONAUTICAL SCIENCES 1954-1978*

*ASTRONAUTICAL SCIENCES REVIEW 1959-1962*

*ADVANCES IN THE ASTRONAUTICAL SCIENCES 1957-June 1979*
*(Volumes 1-37 plus Volumes 38 and 39 which appeared in 1979)*

*SCIENCE AND TECHNOLOGY 1964-1978 (Volumes 1-46)*

*AAS HISTORY SERIES 1977 (Volume 1)*

*AAS MICROFICHE SERIES 1968-1978*
*(Volumes 1-29 plus Volume 30 which appeared in 1979)*

*SPECIAL VOLUMES PUBLISHED FOR OR IN CONJUNCTION WITH THE AAS*

IT IS A COMPLETE INDEX OF ALL AVAILABLE ARTICLES AND PAPERS PUBLISHED BY OR FOR THE AMERICAN ASTRONAUTICAL SOCIETY. ALMOST ALL CITATIONS CORRESPOND TO MATERIAL STILL AVAILABLE. A SUBJECT INDEX WILL APPEAR AS A COMPANION VOLUME.

# GLOSSARY

| | |
|---|---|
| AAAS | American Association for the Advancement of Science |
| AAS | American Astronautical Society |
| AAS History | *AAS History Series* |
| Adv | *Advances in the Astronautical Sciences* |
| AIAA | American Institute for Aeronautics and Astronautics |
| ASR | *Astronautical Sciences Review* |
| Bonn Proc. | Bonn Proceedings *(Utilization of Space Shuttle and Spacelab, 1976)* |
| DGLR | Deutsche Gesellschaft für Luft- und Raumfahrt |
| IAA | International Academy of Astronautics |
| IAF | International Astronautical Federation |
| JAS | *The Journal of the Astronautical Sciences*<br>*Astronautics*<br>*The Journal of Astronautics* |
| Mic | *AAS Microfiche Series* |
| Micro | *AAS Microfiche Series* |
| Sp v | Special Volume (see Appendix IV) |
| S&T | *Science and Technology* |

CONTENTS

CONTENTS (Cont'd)

# NUMERICAL/CHRONOLOGICAL INDEX OF AAS TECHNICAL PAPERS 1954 - 1978

The following pages provide a Numerical/Chronological Index of AAS Technical papers published from 1954 to 1978. They are organized by the technical meeting at which the papers were presented. Papers are identified by AAS + year + number. Also some papers, presented early in 1979, are included. Nearly all the papers indexed appear in proceedings volumes, all of which are available either in hard copy or microfiche from the publishers for the American Astronautical Society, Univelt, Inc., P. O. Box 28130, San Diego, California 92128.

In addition to AAS papers, a number of papers presented at annual meetings of the International Astronautical Federation/International Academy of Astronautics but published by the AAS are listed with an IAF prefix for identification purposes. Likewise some papers published at a meeting of the Deutsche Gesellschaft fur Luft- und Raumfahrt for which the AAS was a cosponsor are listed with a DGLR prefix. These special citations follow the AAS citations for the year in which the papers were presented.

Each citation includes the paper number, (AAS, IAF or DGLR), the full title and author(s). The heading under which the papers are listed identifies the volume(s) in which they appear and the technical event at which they were presented.

# AAS TECHNICAL PAPERS
# 1954 - 1957

# NUMERICAL/CHRONOLOGICAL INDEX OF AAS TECHNICAL PAPERS

FIRST ANNUAL MEETING
(Dec. 3, 1954, New York)

AAS 54-001 Logistics of Orbital Supply Systems, W. von Braun (*JAS v2-1, 1955, pp. 33-35)

SECOND ANNUAL MEETING
(Dec. 1, 1955, New York)

AAS 55-001 Economic Aspects of the Unmanned Satellite Vehicle, S. F. Singer (Not available)

AAS 55-002 to 55-014 Not Assigned

AAS 55-015 Fundamental Basis of Space Flight, W.O.Davis (JAS v3-1, 1956, pp. 9-10)

AAS 55-016 Satellite Ascent Guidance Requirements, J.Jensen (JAS v3-1, 1956, pp. 1-3)

AAS 55-017 Astronautics in Other Countries, F.C.Durant III, (JAS v2-4, 1955, p. 153)

AAS 55-018 Electrical Propulsion System for Space Ships,(Abstract) E. Stuhlinger (JAS v2-4, 1955, pp. 149-152; v3-1, 1956, pp. 11-14; v3-2, 1956, pp. 33-36)

AAS 55-019 Control and Flight Mechanical Problems of Orbital Rockets, J.J.Klein (Not Available)

AAS 55-020 Thermal Control in a Space Vehicle, P.E.Sandorff, J.S.Prigg, Jr., (JAS v3-1, 1956, pp. 4-8)

AAS 55-021 Acceleration Force and the Space Pilot, J.J.Raffone (JAS v2-3, 1955, pp. 100-104)

AAS 55-022 International Cooperation in the Field of Astronautics, A.G.Haley

VOL. 1, ADVANCES IN THE ASTRONAUTICAL SCIENCES, 1957
THIRD ANNUAL MEETING
(Dec. 6-7, 1956, New York)

AAS 56-001 Education in the Astronautical Sciences, C.S. Draper

AAS 56-002 The Satellite and Meteorology, H. Wexler

---

* JAS refers to the journal <u>Astronautics</u>, later designated the <u>Journal of Astronautics</u> and now titled the <u>Journal of the Astronautical Sciences</u>.

AAS 56-003 to 56-005 Not available

AAS 56-006 Orbit Lifetimes of Earth Satellites, H.B. Ketchum

AAS 56-007 The Problems of Goddard and Related Problems of Rocket Flight, F.D.Faulkner

AAS 56-008 Not available

AAS 56-009 Space Exploration - The New Challenge to the Electronics Industry, H.E. Prew

AAS 56-010 Not available

AAS 56-011 What the Future Holds for the Earth Satellite, R.P. Haviland

AAS 56-012 Not assigned

AAS 56-013 Meteoric Abrasion Studies Proposed for Vanguard, M.J. Swetnick

AAS 56-014 Radio Tracking of IGY Satellite, R.J. Easton

AAS 56-015 Project Moon Watch - Optical Tracking of IGY Satellite, A.N. Spitz

AAS 56-016 Long Term Human Confinement in Space Equivalent Vehicles, C.A.Dempsey, F.D.Van Wart, J.H.Duddy, J.K.Hockenberry

AAS 56-017 Accelerative Forces Associated with Leaving and Reentering Earth's Gravitational Field, R.R. Hessberg, Jr.

AAS 56-018 Mechanoreceptors, Gravireceptors, H. Strughold

AAS 56-019 Plants as a Means of Balancing a Closed Ecological System, J.G. Gaume

AAS 56-020 Environmental Problems Connected with Space Ship Occupancy, W.T. Ingram

AAS 56-021 Space Vehicles as Tools for Research in Relativity, S.F. Singer

AAS 56-022 Experimental Aspects of General Relativity Theory, L. Witten

AAS 56-023 Survey of Relativity, P. Bergmann

AAS 56-024 Principal Directions in Current Research on Gravitation, B.S.DeWitt

AAS 56-025 Cosmic Ray Instrumentation for the IGY Program, R.C. Haymes

VOL. 2, ADVANCES IN THE ASTRONAUTICAL SCIENCES, 1958
FOURTH ANNUAL MEETING
(Jan. 29-31, 1958, New York)

AAS 57-001 A survey of Propulsion and Space Dynamics, D.B. Cross, J. Jensen

AAS 57-002 Plasma Motors, W.H. Bostick

AAS 57-003 Beryllium: Promising Metal of the Space Age, R.E. Foster, L.A. Riedinger

1957

AAS 57-004 Returning Alive from Space, F.Riddell, R.W. Detra

AAS 57-005 Supersonic Flow about Blunt Bodies of Revolution, F.G.Gravalos

AAS 57-006 Optimum Thrust Programing Along Arbitrarily Inclined Rectilinear Paths, A. Miele, C.R. Caboti

AAS 57-007 Optimum Burning Program as Related to Aerodynamic Heating for a Missile Traversing the Earth's Atmosphere, (Abstract), A. Miele

AAS 57-008 Not available

AAS 57-009 Fundamentals of Inertial Guidance and Navigation, W.E. Frye

AAS 57-010 Determination of a Unique Attitude for an Earth Satellite, W.R. Davis

AAS 57-011 Not available

AAS 57-012 Inertial Aids for Space Travels, (Abstract), C. Mundo

AAS 57-013 Inertial Navigation and Space Flight, (Abstract), P.J. Newman

AAS 57-014 Steering of an Ascent Rocket for Maximum Cutoff Velocity, W.H. Foy, Jr.

AAS 57-015 Advances in Astro Biology, (Abstract), H. Strughold

AAS 57-016 The Selection and Training of a Bio-Satellite Crew, D.W. Conover, E.G.Aiken, C.M. Whitlock

AAS 57-017 Orientation of Research Needs Associated with Envircnment of Closed Spaces, W.T. Ingram

AAS 57-018 Ionization Dosage from X- and Beta-Rays in Flight through Auroral Displays, H.J. Schaefer

AAS 57-019 Five Days Aloft: Research in Human Travel on the Endless Frontier of Time, C.A.Dempsey, F.D.Van Wart, L.Eisen, J.G.Roth N.K.Morrison, C.Myers

AAS 57-020 Synthesis of Human-Automatic Control Systems for High Performance Vehicles, W.L. Morris, R.C. Kaehler

AAS 57-021 Isolation and Confinement in Space Flight, G.E.Ruff, E.C.Levy, V.H.Thaler

AAS 57-022 Design of an Algal Culture Chamber Adaptable to a Space Ship Cabin, J.G. Gaume

AAS 57-023 A Universal Radio Astronomy System for Radio Telescopes, Space Vehicle Tracking and Scatter Propagation Studies, G.J. Doundoulakis

AAS 57-024 RF Propagation in Interplanetary Communications (Abstract), A.M. Peterson

AAS 57-025 Television Telemetry for Missile Test Programs, D. Hochman, J.P. Taylor

AAS 57-026 Data Links in Space Exploration-Their Nature, Applications and Limitations, H.E. Prew

AAS 57-027 Not available

AAS 57-028 Some Observations on Russian Electronic Techniques, C.L. Rouaalt

AAS 57-029 Approaches to the Development of Space Vehicles, G.W. Hoover

AAS 57-030 Space Cabin Design, A.M. Mayo

AAS 57-031 Hydrogen Peroxide as a Source for Oxygen, Water, Heat, and Power for Space Travel, N.S. Davis

AAS 57-032 A Recoverable Emulsion Package for Extra Atmospheric Study of Cosmic Rays, R.A. Webster

AAS 57-033 Prediction of Cratering by Meteor Impacts, M. Kornhauser

AAS 57-034 Not available

AAS 57-035 On the Generation of Temperatures to $30{,}000^\circ$K, P.E. Glaser

AAS 57-036 Considerations about Visibility of Satellites for the Unaided Eye, I. Schmidt

AAS 57-037 Potential Application of Nuclear Energy to Space (Abstract) C.J. Wang, C.W. Anthony

AAS 57-038 Not available

AAS 57-039 Minimum Time Interplanetary Orbits, D.M. Cole

AAS 57-040 Interplanetary Applications of Automatic Navigation, E.V. Stearns

AAS 57-041 Not available

AAS 57-042 Plastic Balloons for Planetary Research, M.D. Ross

AAS 57-043 Project Skyball: A Proposed Basic Trainer for Space Flight, N.V. Petersen

AAS 57-044 A Preliminary Determination of the Position of Some First- and Second-Magnitude Stars on Mars' Celestial Sphere, F.R. West, Jr.

AAS 57-045 On Plasma Propulsion, Y.A. Yoler

AAS 57-046 The Use of Planetary Atmospheres for Propulsion, S.T. Demetriades, C. Kretchmer

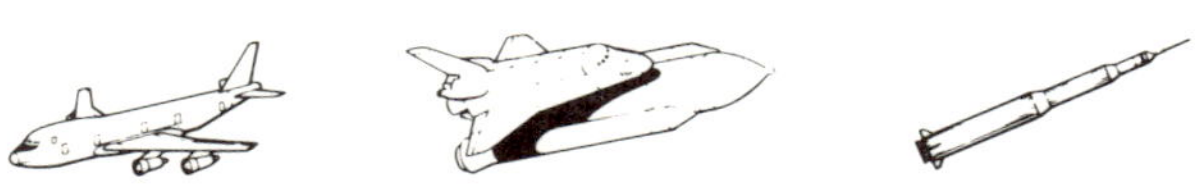

# AAS TECHNICAL PAPERS
# 1958 - 1959

VOL. 3, ADVANCES IN THE ASTRONAUTICAL SCIENCES, 1958
FIRST WESTERN REGIONAL MEETING
(Aug. 18-19, 1958, Palo Alto, Calif.)

AAS 58-001 Problems of Round Trip Manned Flight Outside the Earth's Atmosphere, (Abstract), E.P. Hiatt 1958

AAS 58-002 Time-Activity-Environment Analysis: A Way of Planning for Human Space Flight, J.E. Mangelsdorf

AAS 58-003 Manned Cabin Surface Temperatures and Heat Transfer Rates for a Reentry Hypersonic Glide Vehicle, T.E. Mauritsen

AAS 58-004 Man's Operational Environment in Space, G.W. Hoover

AAS 58-005 Satellite Propulsion Systems, W.R. Corliss, M.A. Zipkin

AAS 58-006 Electromagnetic "Pinch Effect: as a Space Propulsion System, (Abstract), A.E. Kunen

AAS 58-007 Laboratory Experiments in Hydromagnetic Propulsion, J.Gauger, V.Vali, T.E.Turner

AAS 58-008 Some Design Aspects of an Interplanetary Exploratory Vehicle, H.K. Hebeler, R.D. White

AAS 58-009 Structural Aspects of Earth Glide Re-entry Vehicles, M.T. Braun, E.G. Czarnecki

AAS 58-010 Techniques in the Prediction of Space Vehicle Reliability, J.S. Youtcheff, E.D. Karmiol

AAS 58-011 Space-to-Space Propagation Phenomena, C.N. Lee

AAS 58-012 Preliminary Analysis of Satellite Observation, (Abstract), O.K.Garriott

AAS 58-013 Satellite Faraday Effect Observations, (Abstract), C.R. Moe, J.L. Bennion

AAS 58-014 Space Communication Limits of Low Noise Receivers, (Abstract), C.T. McCoy

AAS 58-015 Flame Attenuation of RF Signals, R.J. Schmelzer

AAS 58-016 Not available

AAS 58-017 Next: Maneuverable Satellites, C.C. Kelber

AAS 58-018 Microthrust Maneuver Capabilities, A. Dobrowalski

AAS 58-019 Aspects of Transfer between Satellite Orbits, G.S.Gedeon

AAS 58-020 Optimal Transfer between Coplanar Terminals in a Gravitational Field, L.G. Vargo

AAS 58-021 Collision Trajectories in the Three-Body Problem, E.T. Benedikt

AAS 58-022 Temperature Equilibria in Space Vehicles, R.A.Cornog

AAS 58-023 Interplanetary Dust Distribution and Erosion Effects, D.B. Beard

AAS 58-024 A New Instrument for Measuring Density and Temperature at Satellite Altitude, H.A.Dessler, W.B.Hanson, M.Hertzberg, D.D.McKibbin, R.C.Wrigley

AAS 58-025 A Simulator for the Gravitational Field, R.E. Gaskell

AAS 58-026 Problems of Ground Simulation of Long-Range Space Flight Environmental Conditions, T.C. Helvey

AAS 58-027 Around the Moon in 80 Hours, D.M. Cole

AAS 58-028 Solar and Lunar Perturbations of a Satellite Orbit, V. Gradecak

AAS 58-029 Minimum Weight and Optimum Flight Path of Low-Acceleration Space Vehicles, H.F. Michielsen

AAS 58-030 Practical Considerations for Astronautical Guidance and Control, K.P. Gow, R.E. Roberson

AAS 58-031 Comparison Two-Dimensional and Three-Dimensional Analysis of Earth-Moon Flight, R.J. Gunkel, G.C. Goldbaum

AAS 58-032 The Effect of Aerodynamic Forces on Satellite Attitude, D.B. DeBra

AAS 58-033 Problems of Reentry into the Earth's Atmosphere, A.C. Robinson

AAS 58-034 Elliptical Orbit Lifetimes, J. Breakwell, L. Koehler

AAS 58-035 The Effect of Drag on Elliptic Orbits, H.E. Williams

AAS 58-036 Encke's Method and Variation of Parameters as Applied to Re-entry Trajectories, R.M.L. Baker, Jr.

AAS 58-037 Two-Dimensional Librations of a Dumbbell-Shaped Satellite in a Uniform Gravitational Field, T.A.J. Stocker, R.F. Vachino

AAS 58-038 Impact Protection for the Human Structure, M. Kornhauser

AAS 58-038-1 Biomedical Considerations of Weightlessness, J.E. Ward

AAS 58-038-2 Space Flight Symposium: Man on Moon Soonest, S.B.Batdorf (See AAS 58-038-6 to 12 for panel papers)

AAS 58-038-3 Status of Astronautical Development and Future Research Needs, F.M.Perkins

AAS 58-038-4 The Role of the Missile Industry in the Field of Astronautics, L.E.Root

AAS 58-038-5 Space Flight Legislation, R.B.Bryan

AAS 58-038-6 Flight Mechanics of a Lunar Vehicle, L.G.Walters

AAS 58-038-7 Environmental Control for the Man in Space, A.M.Mayo

AAS 58-038-8 Guidance and Control, W.E.Frye

AAS 58-038-9 Communication Problems, E.A.Blasi

AAS 58-038-10 Lunar Landing and Earth Recovery, A.J.Eggers,Jr.

AAS 58-038-11 Propulsion and Structures, Y.E.Lee

AAS 58-038-12 Biosatellites and Early Research, D.G.Simons

VOL. 4, ADVANCES IN THE ASTRONAUTICAL SCIENCES, 1959
FIFTH ANNUAL MEETING
(Dec. 27-31, 1958, Washington, D.C.)

AAS 58-039 Orbit Decay and Prediction of Motion of Artificial Satellites, H.F. Michielsen

AAS 58-040 Perturbations of Selenoid Satellites, E.T. Benedikt, P.S. Cohen

AAS 58-041 Satellite Rotation, R.N. Bracewell

AAS 58-042 Corpuscular Radiation in Outer Space, J.A. Van Allen

AAS 58-043 Artificial Modification of the Earth's Radiation Belt, S.F. Singer

AAS 58-044 Satellite Rendevous Operations, N.V. Petersen

AAS 58-045 Heat Transfer in High-Speed Slip Flow, R.A. Oman, R.A. Scheuing

AAS 58-046 Some Aspects of the Electrical Properties of the Upper Atmosphere, P.A. Goldberg

AAS 58-047 Discontinuity Surfaces in Magneto-Fluid Dynamics, L.G. Napolitano

AAS 58-048 A Study of Entry into Earth's Atmosphere, D.B. Cross

AAS 58-049 Horizon-Scanning for Atmospheric Re-entry, E.J. McCartney

AAS 58-050 On Gravitation, V. Gradecak

AAS 58-051 The High-Temperature Research Challenge in Space Vehicle Design, P.E. Glaser

AAS 58-052 Solar Turbo-Powerplant Design, D.B. Mackay

AAS 58-053 Lubrication Problems in Space Vehicles, M.M. Freundlich, A. Robertson

AAS 58-054 Assembly of a Multimanned Satellite, R.A. Byers, S.B. Kramer

AAS 58-055 Space Vehicle Attitude Problems, H. Perkel

AAS 58-056 Simplified Space Guidance Systems Analysis, C.A. Brown, R. Fleisig

AAS 58-057 Guidance Analysis for Ballistic Interplanetary Trajectories, T.A. Magness, J.B. McGuire, O.K. Smith

AAS 58-058 Proposed Method of Navigation on the Earth by Means of an Earth Satellite, R. Etcheverry, R. Brooks, J. Russell

AAS 58-059 Long-Range Detection by Star Occultation, H. Dubner

AAS 58-060 A Television Camera for Space Applications, J. Staniszewski, M.H. Mesner, P.H. Werenfels

AAS 58-061 The Man-Machine System in Space Vehicles, G.W. Hoover

AAS 58-062 Protection of Man against Transient Exposure to High Heat Loads, P. Webb

AAS 58-063 Physiological Telemetry, M. McLennan

AAS 58-064 Biodynamics of Launch and Reentry Profiles, E.P.Hiatt

AAS 58-065 Phychological Research in the Environment of Outer Space, F. Rohles, Jr.

AAS 58-066 Not available

AAS 58-067 A Comparison of Artificial Environments Used in Sealed Cabins during Flights into the Stratosphere, M.C. Ross

AAS 58-068 Survival of Terrestrial Microorganisms under Simulated Martian Environmental Conditions (Abstract) J.D. Fulton

AAS 58-069 Not available

AAS 58-070 Closed Ecological Subsystems, Some Practical Application Problems in Manned Space Flight, B.W. Pinc

AAS 58-071 Exploratory Research on the Theoretical Considerations of Waste Water Cycles in a Closed Ecological System, W.T. Ingram

AAS 58-072 Not available

AAS 58-073 Rocket Exploration of the High Atmosphere for the IGY, J.W. Townsend

AAS 58-074 Rocket Astronomy, H. Friedman

AAS 58-075 Radio Observations of Earth Satellites, G.W. Swenson, Jr.

AAS 58-076 The Variable Acceleration of Satellites, L.G. Jacchia

AAS 58-077 Some Results from the Pioneer III Flight, W.H. Pickering

AAS 58-078 Space Exploration and Human Welfare, H.L. Dryden

VOL. 5, ADVANCES IN THE ASTRONAUTICAL SCIENCES, 1960
SECOND WESTERN NATIONAL MEETING
(Aug. 4-5, 1959, Los Angeles)

AAS 59-001 Human Factors in Probing Outer Space, H. Imus

AAS 59-002 Shielding and Nuclear Propulsion, R. Trapp, E.B. Konecci

AAS 59-003 The Reaction Control Simulator - Flight Trainer for Manned Orbital Vehicles, W. Kuchnegger

AAS 59-004 Human Factors in Spacecraft Technology, R.N. DeCallies

AAS 59-005 Fuel Requirements for Crude Interplanetary Guidance, J.V. Breakwell

AAS 59-006 Guidance Requirements for Interplanetary Flight, F.H. Kierstead

AAS 59-007 Satellite Orbit Control System, M.C. Fitzgibbon

1959

AAS 59-008 Scientific Significance of Deep Space Exploration, E.T. Benedikt

AAS 59-009 Dynamic Response Surface for Nonlinear Equations of Powered Flight, R.E. Ross

AAS 59-010 Design Optimization of Thermal Radiators for Space Vehicles, R.A.Cornog

AAS 59-011 Solar Photovoltaic Power Sources for Space Vehicles, J.D. Sailor

AAS 59-012 The Problem of Electrical Neutrality in Ion Beams, E. Fisher

AAS 59-013 The Magnetic Pinch Engine for Space Flight, A.E.Kunen, W. McIlroy

AAS 59-014 Round-Trip Trajectories for Mars Observation, P.G. Johnson

AAS 59-015 A Systematic Approach to a Matrix of Space Research Vehicles, O.W. Nicks

AAS 59-016 The Influence of Launch Conditions on the Friendly Rendezvous of Astrovehicles, N.V. Petersen, R.S. Swanson

AAS 59-017 Effect of Re-entry Technique on the Design of a Space Vehicle, J. Stalony-Dobrzanski

AAS 59-018 Global Communication Using Passive or Active Satellite Relay Stations, F. Vilbig

AAS 59-019 Requirements of Data Transmission on Solar System Reconnaissance Vehicles, C.H. Child

AAS 59-020 Measuring Amplitude and Phase, R. Goodman

AAS 59-021 Design Techniques for Space Television, A. Viterbi

AAS 59-022 The Origin of the Gravitational Force, F.E.E. Alzofon

AAS 59-023 The Lunar Gravitational Potential, I. Alexandrov

AAS 59-024 An Analytical Method for Determining the Orbital Characteristics of a Nonlifting Body which Passes through the Lower Atmosphere of a Planet, G.S.Gedeon

AAS 59-025 Lunar and Terrestrial Atmospheres, R. Jastrow

AAS 59-026 Shock Tube Experiments Simulating Entry into Planetary Atmospheres, K.K. Chan, R.W. Rutowski

AAS 59-027 Keynote Address, (Abstract), W. von Braun

AAS 59-028 Panel Discussion: Probing Space, H.E. Newell, Jr.

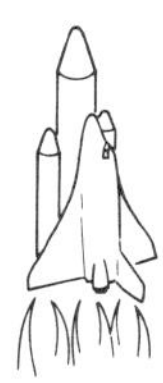

# AAS TECHNICAL PAPERS
# 1960 - 1961

VOL. 6, ADVANCES IN THE ASTRONAUTICAL SCIENCES, 1961
SIXTH ANNUAL MEETING
(Jan. 18-21, 1960, New York)

1960

| | |
|---|---|
| AAS 60-000 | Panel Discussion: Man in Space, When?, D.D. Flickinger |
| AAS 60-001 | A Comparison of FM Demodulation Methods, E.J. Baghdady |
| AAS 60-002 | Transmission Efficiency between Simple Antennas on the Basis of Orientation and Polarization, J.B. Rankin |
| AAS 60-003 | Not assigned |
| AAS 60-004 | An Earth-Oriented Communication Satellite of the Passive Type, W. Gillespie |
| AAS 60-005 | Project 217, Signal Corps Communication Satellite, S.P. Brown |
| AAS 60-006 | Space Missions for Ion Propulsion Systems, E. Stuhlinger |
| AAS 60-007 | A Survey of Ion Sources for Electrical Propulsion, K. Foreman |
| AAS 60-008 | Application of the Nuclear Rocket Engine for Space Propulsion, C.H. Trent,H.S. McQueen |
| AAS 60-009 | Nuclear Rocket Engine Control Problems in the Upper Stage Interplanetary Vehicle, B.P. Helgeson |
| AAS 60-010 | Straight-Line Flight to Mars, T.C. Tsu |
| AAS 60-011 | Guidance for Interplanetary Landings, E.V. Stearns |
| AAS 60-012 | An Astrovehicle Rendezvous-Guidance Concept, N.V. Petersen, L.R. Hoover |
| AAS 60-013 | On the Simplification of the Attitude Equations of a Satellite, C.T. West, R. Goodstein |
| AAS 60-014 | Some Display-Control Aspects of Manual Attitude Control in Space, M.L. Ritchie, L.F. Hanes, T.E. Hainsworth |
| AAS 60-015 | Path Control for Satellite Rendezvous, R.E. Roberson |
| AAS 60-016 | Not assigned |
| AAS 60-017 | Analysis of a Time Varying Gain Inertial Loop, E. Levinson |
| AAS 60-018 | Trajectory Prediction of Lunik III, J.E. Michaels |
| AAS 60-019 | Celestial Moving Target Indicator, H. Dubner |
| AAS 60-020 | Artificial Cooling of Mammals in a Hot Environment, V.O. Erk, R.E. Bundy, A. Arendas |
| AAS 60-021 | Methods of Predicting Radiation Dosages in Space Flight, A.F. Bond, M.G. DelDuca, A.D. Babinsky |
| AAS 60-022 | Basic Research Efforts in Astrobiology, R.S. Young |
| AAS 60-023 | Ecological Criteria for a Space Rocket Relay Station, R.E. Ross |

| | |
|---|---|
| AAS 60-024 | The Tunnel Display Concept, D. Ellis, L. Steinman, F. Ludwig |
| AAS 60-025 | The Use of Vegetable Cultures as the Photosynthetic Component of Isolated Ecological Cycles for Space Travel, L.G. Rich, W.M. Ingram, B.D. Berger |
| AAS 60-026 | Navigation and Energy Display Requirements for Pilot-Controlled Satelloid Flight, R.C. Kaehler, S. Romano |
| AAS 60-027 | Aerobic Biological Degradation of Human Waste in Closed Systems, R. Bogan, D.D. Chapman, L.H. Ericsson |
| AAS 60-028 | Anaerobic Waste Processing for Closed Ecological Systems, R.D. Gafford, H.L. Pote, D.E. Richardson |
| AAS 60-029 | The Oblatory Perturbations of Satellite Orbits, N.S. Hall, H.F. Galowica |
| AAS 60-030 | Velocity Increments Required to Reduce Target Miss on Coasting Trajectories, J. Lorell |
| AAS 60-031 | Determination of the Characteristics of Rapidly Decaying Orbits, G.S. Gedeon |
| AAS 60-032 | Rapid and Precise Method of Numerical Integration of the Motion of Planetary Bodies and Space Vehicles in the Solar System, F.W.A. Knoll |
| AAS 60-033 | Application of Lunar Theory to the Motion of Satellites, P. Lanzano |
| AAS 60-034 | The Influence of Gravity on the Angular Motions of an Earth Satellite, W.C. Triplett, B.F. Doolin |
| AAS 60-035 | Minor Circle Flight for Boost Glide or Orbital Reentry Vehicles, W.H.T. Loh |
| AAS 60-036 | Circularization of Elliptic Orbits, B.H. Gundel, D.B. DeBra |
| AAS 60-037 | Disintegration Barriers to Extremely High Speed Space Travel, E.T. Benedikt |
| AAS 60-038 | Not assigned |
| AAS 60-039 | Heat Rejection from Space Vehicles, D.P. Ross, E. Ray, H.C. Haller |
| AAS 60-040 | Optimizing the Design of a Solar Power Supply System, C.C. Osgood, S.H. Winkler |
| AAS 60-041 | Multistage Rocket Staging Optimization, R.L. Chase |
| AAS 60-042 | Reliability for Spacecraft Electronics, R.N. DeCallies |
| AAS 60-043 | Nuclear Power Plants for Space Vehicles, S. Naymark |
| AAS 60-044 | A Lithium Vaporization System for Hypersonic Cooling, H.K. Hebeler |
| AAS 60-045 | Not assigned |
| AAS 60-046 | A Nomographic Method of Material Selection for Ablating Shell Structures, J. Frisch |
| AAS 60-047 | Not assigned |

AAS 60-048 Orbital Patterns for Satellite Systems, L.G. Vargo

AAS 60-049 The Feasibility of Propelling Vehicles by Contained Nuclear Explosions, D.M. Cole

AAS 60-050 One-Way Reconnaissance to Mars, J.V. Hughes, G.N. Nomicos

AAS 60-051 Visual Tracking of Space Vehicles, H.C. Shepler

AAS 60-052 The Operational Support of Space Vehicle Missions, H.S. Dordick

AAS 60-053 The Radiation Belts of Planets Mars and Venus, S.F. Singer

AAS 60-054 An Estimate of the Nuclear Radiation at the Lunar Surface, J.A. Barton

AAS 60-055 The Application of Integral Theorems in Magnetohydrodynamics, L. Zadoff, J. Reinheimer, M. Bernstein

AAS 60-056 The Effect of Nonuniform Magnetic Fields on Internal Flows of Conducting Fluids, A. Sherman

AAS 60-057 Basic Requirements for the Exploration of Jupiter and Its Moons, R.G. Voss, W.G. Straley

AAS 60-058 Comparative Costs of Nondestructive Methods of Transporting Material to a Lunar Base, R.A.Cornog

AAS 60-059 Kinematics of a Planetary Ballistic Probe, M. Yachter

AAS 60-060 Human Factors in Long-Range Aerospace Flight, F.D. Van Wart

AAS 60-061 Test of Theory of Gravitational Light Deflection, R.L. Lillistrand

VOL. 7, ADVANCES IN THE ASTRONAUTICAL SCIENCES, 1961
THIRD WEST COAST ANNUAL MEETING
(Aug. 8-11, 1960, Seattle, Wash.)

AAS 60-062 A Conceptual Design for a Manned Mars Vehicle, P. Bono

AAS 60-063 Scientific Goals of Radar Astronomy, V.R. Eshleman, A.L. Peterson

AAS 60-064 Analysis of Minimal Data Orbit Determination by a Single Radar Site, E.W. Radany

AAS 60-065 Optimization of Three Dimensional Boost-Turn Performance, H.J. Kelley, H.K. Hinz

AAS 60-066 Determination of Rocket Engine ISP from Inverse Acceleration, W.R. Davis

AAS 60-067 Kinematics of Orbital Motions, G.S. Gedeon

AAS 60-068 Foamed-In-Place Structures for Space Vehicles and Stations, C.C. Osgood

AAS 60-069 Transfer between Noncoplanar Orbits with Minimum Velocity Requirements, W.E. Brunk

AAS 60-070 Motions of a Dumbbell Satellite, H.R. Rosner

AAS 60-071 Rendezvous with Satellites in Elliptical Orbits of Low Eccentricity, P.W. Soule

AAS 60-072 Not assigned

AAS 60-073 Ion Engine Development I, Diffusion Type Ion Sources, R.J. Sunderland, J.R. Radbill, R.D. Gilpin

AAS 60-074 Escape from an Oblate Planet, M.L. Anthony, G.E. Fosdick, L.M. Perko

AAS 60-075 Use of Perigee Propulsion for Orbital Launch of Nuclear Rockets, P.G. Johnson, F.E. Rom

AAS 60-076 The Spacing of Corrective Thrusts in Interplanetary Navigation, J.V. Breakwell

AAS 60-077 Sun Tracker with Moon and Planet Tracking Modifications, R.Wuensch, D. Wilson

AAS 60-078 Preliminary Considerations for Attitude Control of Space Vehicles, R.L. Stafford

AAS 60-079 Space Vehicle Attitude Dynamics During Velocity Corrections, J. Marino

AAS 60-080 Cryogenic Propellant Storage Parameters on Moon, C.C. Love, Jr.

AAS 60-081 Considerations Affecting the Design of a Mars Probe, W.A. Webb, J.A. Harder

AAS 60-082 A Life Support System for Near-Earth or Circumlunar Space Vehicle, G.B. Whisenhunt, Jr.

AAS 60-083 Habitability and the Maintenance of Human Performance in Long-Duration Space Missions, B.B. Adams, J.T. Celentano

AAS 60-084 Trends in Human Performance Augmentation, A.T. Kornfield

AAS 60-085 Some Considerations for Extra-Vehicular Protection Systems, E.J. Merrick (Not available)

AAS 60-086 Direct Re-entry at Escape Velocity, B.A. Galman

AAS 60-087 A Parametric Examination of Reentry Vehicle Size and Shape for Return at Escape Velocity, E.F. Styer

AAS 60-088 Reaction Controls for Reentry Vehicles, G.W. Freeman

AAS 60-089 Not assigned

AAS 60-090 Operation of an Electric Ramjet in a Planetary Atmosphere, E.P. French

AAS 60-091 Motion Picture Simulation of Unmanned Lunar Spacecraft Approaches to the Moon, R.L. Peterson

AAS 60-092 Stochastic Processes Applied to the Dynamics of Rocket Flight, R.E. Ross

SPECIAL VOLUME: WEIGHTLESSNESS - PHYSICAL PHENOMENA AND BIOLOGICAL EFFECTS, ed. E.T. BENEDIKT, 1961 (AAS Special Volume No. 1) *
ZERO GRAVITY RESEARCH SYMPOSIUM
(July 1, 1960, Hawthorne, Calif. )

AAS 60-093 General Behavior of a Liquid in a Zero or Near-Zero Gravity Environment, E.T. Benedikt

AAS 60-094 Behavior of Liquids in Free Fall, W.C. Reynolds

AAS 60-095 On the Mathematical Aspects of Benedikt's Representation of the Equations of Epihydrodynamics, M.Z. Krzywoblocki

AAS 60-096 Liquid Behavior Investigations under Zero and Low-Gravity Conditions, E.W. Schwartz

AAS 60-097 Human Performance and Behavior During Zero Gravity, E.L. Brown

AAS 60-098 An Experimental Study of Boiling in Reduced and Zero Gravity Fields, C.M. Usiskin, R. Siegel

AAS 60-099 An Experimental Study of the Transition from Nucleate to Film Boiling under Zero Gravity Conditions, H.F. Steinle

AAS 60-100 Orbital Force Field Boiling and Condensing Experiments, J.G. Reitz, R.L. Cummings, P.E. Grevstad

AAS 60-101 The Effect of Gravity upon Nucleate Boiling Heat Transfer, M. Adelberg, K. Forster

AAS 60-102 Current WADD Weightless Research, J.C. Simons

AAS 60-103 Zero Gravity Simulation, R.B. Levine

**AAS 60-103a Experimental Production of a Zero or Near-Zero Gravity, E.T.Benedikt, R.Lepper**

SPECIAL VOLUME: LUNAR EXPLORATION AND SPACECRAFT SYSTEMS, eds. R. FLEISIG, ET AL., 1962 (AAS Special Volume No. 2) *
LUNAR FLIGHT SYMPOSIUM
(Dec. 27, 1960, New York)

AAS 60-104 Scientific Objectives of Lunar Exploration, R. Jastrow

AAS 60-105 Distribution of Interplanetary Dust in Cislunar Space, S.F. Singer

AAS 60-106 Comparison of Special Perturbation Methods in Celestial Mechanics with Application to Lunar Orbits, S. Pines, M. Payne, H. Wolf

AAS 60-107 Radiation Shielding of Lunar Spacecraft, T.G. Barnes, E.M. Finkelman. A.L. Barazotti

AAS 60-108 A Family of Radioisotope-Fueled Auxiliary Power Systems for Lunar Exploration, R.J. Wilson

---

* Designated AAS Special Volume (Sp.v1, Sp.v2) for convenience

| | |
|---|---|
| AAS 60-109 | Extending the Range of Radar Beacon Tracking for Lunar Probes, N.S. Greenberg |
| AAS 60-110 | Horizon Trackers for Lunar Guidance and Control Systems, K.G. Kuhn, E.W. Stark |
| AAS 60-111 | Lunar Soft-Landing Guidance Sensor, A. Barabush |

VOL. 8, ADVANCES IN THE ASTRONAUTICAL SCIENCES, 1963
SEVENTH ANNUAL MEETING
(Jan 16-18, 1961, Dallas, Texas)

| | |
|---|---|
| AAS 61-001 | Interplanetary Spacecraft Communications Systems, P.F. Glaser |
| AAS 61-002 | Courier Satellite Communication System, E. Imboldi, D. Hershberg |
| AAS 61-003 | Wave Propagation through a Hot Plasma, F.J. Tischer |
| AAS 61-004 | Selection of a Television System for Lunar Exploration, W.S. Pope, P.J. Rousculp |
| AAS 61-005 | Latest Results on Project Echo, H.S. Black |
| AAS 61-006 | The Scientific Results of the Pioneer V Space Probe, J.W. Lindner |
| AAS 61-007 | Interplanetary Correlation of Geologic Time, E.M. Shoemaker, P.J. Hackman |
| AAS 61-008 | On the Nature of Comets and Some New Ideas Relating to their Origin, (Abstract), D.H. Robey |
| AAS 61-009 | A Study of the Martian Upper Atmosphere and Ionosphere, G. Yanow |
| AAS 61-010 | Geomagnetic Activity and Major Solar Flares, B. Bell |
| AAS 61-011 | Newer Concepts and Recent Developments in Bioastronautics, C.F. Gell |
| AAS 61-012 | Adverse Effects Due to Space Vehicle Rotation, G.T. Beauchamp |
| AAS 61-013 | Requirements for Artificial Gravity During Prolonged Space Flight-Impact on Vehicular Design and Operation, A.M. Mayo |
| AAS 61-014 | Monitoring of Astronaut Brain Function by Means of Telemetered Electroencephalogram with Automatic Analysis, R.M. Morell |
| AAS 61-015 | A Regenerative Life Support System for Long-Term Space Flight, J.J. Konikoff |
| AAS 61-016 | Comparative Feasibility and Logistics of Oxygen Recovery Systems, A. Bialecki |
| AAS 61-017 | Drag Modulation and Celestial Mechanics, P.H.Rose J.E. Hayes |

AAS 61-018 Heat Transfer around Blunt Bodies with Surfaces of Finite Catalytic Efficiency in Frozen Dissociated Hypersonic Flow, P.M. Chung, A.D. Anderson

AAS 61-019 Hypersonic Viscid-Inviscid Interaction Solutions for Perfect Gas and Equilibrium Real Air Boundary Layer Flow, W.M.Mann,Jr., R.C.Bradley,Jr.

AAS 61-020 Escape of Manned Vehicles from Lunar Mission Ascent Trajectories, W. Carter, P.C. Kramer

AAS 61-021 Booster Recovery Concepts, (Abstract), C.A. Lysdale

AAS 61-022 Geophysical Exploration with Satellites, (Abstract), R. Jastrow

1961

AAS 61-023 Tiros I Meteorological Satellite and Some Early Results, L.F. Hubert

AAS 61-024 Dynamics of the Upper Atmosphere as Revealed and Its Energy Unbalance, (Abstract), W.W. Kellogg

AAS 61-025 The Odd Harmonics of the Earth's Gravitational Field, H.F. Michielsen

AAS 61-026 Estimation of Longitudinal Variations in the Earth's Gravitational Field from Satellite Observations, W.M. Kaula

AAS 61-027 Analysis and Standardization of Astrodynamic Constants, R.M.L. Baker, M.W. Makemson, G.B. Westrom

AAS 61-028 An "Invariant" Property of Satellite Motion in a Dissipative Medium, B.C. Murray, A.B. Lees

AAS 61-029 Unified Theory and Differential Correction for Nearly-Parabolic and Nearly-Rectillinear Orbits, C.T. Van Sant, J.L. Arsenault

AAS 61-030 Extremal Rocket Trajectories in Position and Time Dependent Force Fields, G. Leitmann

AAS 61-031 The Orbital Motion of Pellet Clouds, (Abstract), S. Ross (For complete paper, see Journal of the Astronautical Sciences, Vol. 8, No. 3, 1961, pp 79-83

AAS 61-032 Costing Methodology and the Program Costs for a Manned Lunar Landing and Return Mission-Project Mallar, T.E. Dolan, W.G. Long, R.H. Lundberg

AAS 61-033 Military Interests in Space Systems, H.C. Sebring

AAS 61-034 Protection Against Solar Flare Protons, T. Foelsche

AAS 61-035 Not assigned

AAS 61-036 Manned Satellite Escape Systems, H.L. Bloom, J.H. Quillinan

AAS 61-037 Tracking Accuracies for Lunar Missions, F.A. Speer

AAS 61-038 Gimballess Inertial Navigation in Lunar or Planetary Guidance, D.F. McAllister

AAS 61-039 Guidance Concepts for Lunar Landings, P.J. de Fries

AAS 61-040 A Terminal Guidance Law which Achieves Collision Based on Coriolis Balance Techniques, M.L. Nason

AAS 61-041 Guidance of a Space Vehicle to a Desired Point on the Earth's Surface, E.C. Foudriat, J.W. Young, J.A. White

AAS 61-042 Cosmology and the Structure of Fundamental Particles, L. Motz

AAS 61-043 The Clock Paradox in Vertical Free Fall, (Abstract), E.T. Benedikt

AAS 61-044 Not assigned

AAS 61-045 Radio Astronomical Measurements Pertinent to Cosmology, A.E. Lilley

AAS 61-046 The Earth's Dust Belt, F.L. Whipple

AAS 61-047 Experiments for the Unmanned Scientific Exploration of the Moon, N.W. Cunningham

AAS 61-048 Atmosphere Near the Moon, S.F. Singer

AAS 61-049 Recent Cinematographic Work in Lunar Photography from Pic-du-Midi, Z. Kopal

AAS 61-050 Remote Programing Problems for Scientific Experiments on the Moon, C.C. Mercier

AAS 61-051 Radar Measurements of the Lunar Surface, G. Pettengill

AAS 61-052 The Social Implications of Space Activities: Why be Concerned? (Abstract), J.C. Oppenheimer

AAS 61-053 Some Cold War Effects of Space Activities, (Abstract), J. Nehnevajsa

AAS 61-054 Why Space Law? (Abstract), A.B. Nadel

AAS 61-055 Some Implications of Living in Today's Space Community, (Abstract), J. Fink

AAS 61-056 Astronomy Education for the Astronautical Sciences, R. Fleischer

AAS 61-057 Social and Political Implications of the Ultimate Human Society, D.M. Cole

AAS 61-058 Four Vector Kinematics of Interstellar Travel, G.H. Brigman

AAS 61-059 Not assigned

INTERACTIONS OF SPACE VEHICLES WITH AN IONIZED ATMOSPHERE, ed. S.F.SINGER (International Series of Monographs in Aeronautics and Astronautics, Division IX Symposia, Volume 18, Pergamon, 1965) *

(First Goddard Memorial Symposium, March 17, 1961, Washington, D.C.)

AAS 61-060 The Satellite Ionization Phenomenon as Studied by CW-reflection and Pulse Radar Techniques, J.D.Kraus

AAS 61-061 The Detection of Artificial Satellites by their Influence on the Ionosphere, T.G.Hame, W.D.Stuart

AAS 61-062 Plasma Sheath Effects on Rocket Antennas, W.Pfister, J.C.Ulwick

AAS 61-063 Particle Distribution and Motion in a Field of Force, E.J.Opik

AAS 61-064 An *in situ* Probe System for the Measurement of Ionospheric Parameters, R.T.Bettinger

AAS 61-065 Plasma Sheath and Screening around a Stationary Charged Sphere and a Rapidly Moving Charged Body, E.H.Walker

AAS 61-066 An HF-Radar Search for the Effects of Earth Satellites upon the Ionosphere, T.A.Croft, O.G.Villard

AAS 61-067 On the Formation of Satellite Electron Sheaths Resulting from Secondary Emission and Photoemission, D.B.Medved

AAS 61-068 Plasma Compression Effects Produced by Space Vehicles in a Magnetoionic Medium, S.F.Singer, E.H.Walker

AAS 61-069 Coherent Scattering of a Metallic Body in the Presence of an Ionized Shell, L. Peters, W.G.Swarner

AAS 61-070 Disturbance Due to a Satellite in a Plasma Medium and Its Effect on Radar Return, K.M.Chen

AAS 61-071 A Communication on the Analysis of Single-frequency Radio Transmission from Satellites, T.G.Hame, B.C.Potts

AAS 61-072 Free Molecular Heat Transfer in the Ionosphere, L.M.Gilbert, S.M.Scala

AAS 61-072a Reflection of Radio Waves Incident on Boundary Surfaces with Application to the Theory of the Kraus Effect, E.H.Walker

---

* In this index this volume is designated AAS Special Volume No. 4 (Sp.v4) for convenience

VOL. 9, ADVANCES IN THE ASTRONAUTICAL SCIENCES, 1963
FOURTH WESTERN REGIONAL MEETING
(Aug. 1-3, 1961, San Francisco)

| | |
|---|---|
| AAS 61-073 | An Approximate Analytical Determination of Long-Term Orbital Elements in the Gravitational Field of an Oblate Planet, M.L. Anthony |
| AAS 61-074 | An Integrated Approach to the Determination and Selection of Lunar Trajectories, W.C. Nelson |
| AAS 61-075 | Velocity Vector Direction Changes for Coplanar Orbital Transfer, S.P. Altman, J.S. Pistiner |
| AAS 61-076 | Vectors and Dyadics: The Natural Tool of Space Vehicle Mechanics, D.B. DeBra |
| AAS 61-077 | Arms Control and Space Weapons, C.C. Abt |
| AAS 61-078 | Satellite Networks for Global Coverage F.W. Gobetz |
| AAS 61-079 | Prediction of Eclipses of Artificial Satellites, L.G. Stoddard |
| AAS 61-080 | The Meteorite Danger on Earth, M. Kornhauser |
| AAS 61-081 | Guidance and Control Considerations for Guided Interceptor Vehicles in Space, M.H. Reiss |
| AAS 61-082 | The Space Booster Gap and the Panama Theory of Space Flight, D.M. Cole |
| AAS 61-083 | A Model of the Electromagnetic Environment in Cislunar Space, W.A. Livingston, W. Hart |
| AAS 61-084 | Regarding the Predictability of Flares, G.F. Anderson |
| AAS 61-085 | Solar Corpuscular and Radio Emission, D.H. Robey |
| AAS 61-086 | The Effects of Outer-Space Environment Important to the Simulation of Space Vehicles, E. Hart, W. Livingston |
| AAS 61-087 | Life Support Parameters in the Space Environment, H.H. Bovee, G.M. Christensen, I. Zommers, J.M. Thompson |
| AAS 61-088 | Ballistic Re-entry at Small Angles of Inclination, W.H.T. Loh |
| AAS 61-089 | Attitude Stabilization by Means of Solar Radiation Pressure, R.L. Sohn |
| AAS 61-090 | Electromagnetic Docking Operations in Space, E.T. Benedikt |
| AAS 61-091 | A Rate Damping System for Augmentation of the Aerodynamic Stability of a Satellite, P.O. Jarvinen, A.J. Merlini |
| AAS 61-092 | The Application of Statistical Filter Theory to Optimal Trajectory Determination On-Board a Circumlunar Vehicle, G.L. Smith, S.F. Schmidt |
| AAS 61-093 | Optimal Filtering and Linear Prediction Applied to an On-Board Navigation System for the Circumlunar Mission, J.D. McLean, S.F. Schmidt |

AAS 61-094 Interplanetary Navigation by Noise Correlation, D.M. Anderson

AAS 61-095 Self-Contained System for Interplanetary Navigation, R.L. Lillestrand, J.E. Carroll

AAS 61-096 Horizon Seeker Errors Arising out of the Statistical Properties of the Horizon, D. Bishop

AAS 61-097 A Space Vehicle Design for Unmanned Rendezvous, R.J. Johnson, G.R. Arthur

AAS 61-098 The Ultimate Communications Satellite -- Now, J.D. Sailor

AAS 61-099 Reliability -- An Approach to Predicting Human Errors and Parts Malfunctions in Space Operations, (Abstract), M.I. Kaufmann, R.A. Kaufman

AAS 61-100 Not assigned

AAS 61-101 Booster Thrust Requirements for Manned Space Missions, D. MacDonald, R.F. Burry, W.S. Eierman

AAS 61-102 Motion of Orbiting Vehicle under Application of Tangential Microthrust, M.L. Anthony, C.J. Maday

AAS 61-103 Continuous Thrust Rendezvous Transfers -- Approximately Mass Extremized, G.A. Bakalyar

AAS 61-104 The Use of High- and Low-Thrust Propulsion in Combination for Space Missions, T.N. Edelbaum

AAS 61-105 Equations of Motion of an Arbitrary Body of Variable Mass, M. Dryer

AAS 61-106 Linear Accuracy Analysis of Free-Flight Motion about a Spherical Homogeneous Body, W.A. McReynolds, D.B. Cross

VOL. 10, ADVANCES IN THE ASTRONAUTICAL SCIENCES, MANNED LUNAR FLIGHT, 1963
SYMPOSIUM ON MANNED LUNAR FLIGHT (Dec. 29, 1961, Denver)

AAS 61-107 Propulsion Capability for Lunar Missions, Y.C. Lee, B. Karen

AAS 61-108 Lunar Spacecraft Designs, G.R. Arthur

AAS 61-109 Lunar Navigation and Guidance, W. Elfers

AAS 61-110 Preliminary Investigation of Lunar Surface Communications, J.P. Ferrara, M. Chomet

AAS 61-111 Lunar Landing Problems, W. Duke

AAS 61-112 Reentry from Lunar Missions, P.H. Rose

AAS 61-113 Rendezvous for Lunar Missions, C.L. Kober

AAS 61-114 Neurophysiological Aspects of Space Flight, W.D. Winters

AAS 61-115 Hemodynamics Evaluation at Null Gravity, N. Pace

AAS 61-116 Lunar Topography, E. Whitaker

AAS 61-117 Lunar Basing, J. DeNike

AAS 61-118 Lunar Exploration Vehicles and Equipment, J. Froehlich, A. Hazard

# AAS TECHNICAL PAPERS
# 1962 - 1963

VOL. 11, ADVANCES IN THE ASTRONAUTICAL SCIENCES , 1963
EIGHTH ANNUAL MEETING
(Jan. 16-18, 1962, Washington, D.C.)

AAS 62-001 First Results in the NASA Topside Sounder Satellite Program, J.E. Jackson

AAS 62-002 The Theory of Satellite Orbits, I.I.Shapiro

AAS 62-003 Not assigned

AAS 62-004 Formation of the Solar System, A.G.W. Cameron

AAS 62-005 Ruby Optical Masers for In-Flight Measurement of Hypersonic Flow Field Parameters, G.K. Wessel, H.C. Rothenberg, B. Zei

AAS 62-006 Attitude Stability and Motions of Passive Gravity-Oriented Satellites, D.B.Debra 196

AAS 62-007 Attitude Control of Satellites Using Integrating Gyroscopes, J.E.De Isle, E.G.Ogletree, B.M.Hildebrant

AAS 62-008 Passive Magnetic Attitude Control for Earth Satellites, R.E. Fischell

AAS 62-009 Analysis of Guidance Techniques for Achieving Orbital Rendezvous, P.G. Felleman

AAS 62-010 An Automatic Flight Path Control System for the Terminal Phase of Orbital Rendezvous, R.L. Stapleford

AAS 62-011 New Concepts in the Monitoring of Astronaut Brain Function, (Abstract), R.M. Morrell

AAS 62-012 The Effects of Man on the Weight of a Space Vehicle, H.C. Sanders

AAS 62-013 Display and Control in Manned Space Vehicles, A. Shulman

AAS 62-014 Practicality and Selection of Space Cabin Environment, D.L. Foster, M.D. Ross

AAS 62-015 A Current Survey of Space Radiation Hazards, R. Madey

AAS 62-016 Space Scan, A Radio Broadcast System of Space Navigation, A. Tatz

AAS 62-017 Electromagnetic Radiations from Satellites and Space Vehicles, (Abstract), K.N. Satyendra

AAS 62-018 Propagation of Electromagnetic Waves through Reentry Induced Plasmas, B. Rosenbaum, P.J. Friel

AAS 62-019 A truly Global Satellite Communications System, H. Trotter

AAS 62-020 International Planning for Communications Satellites, P.A. Farley

AAS 62-021 The Geology of the Lunar Base, J. Green

AAS 62-022 Proton Dose Rates in Manned Space Vehicles, (Abstract), R. Madley

AAS 62-023 The Crater Contribution to Surface Roughness of the Moon, M. Kornhauser

AAS 62-024 Doppler Orbits for Satellites of Other Planets, (Abstract), A.J. Deutsch

AAS 62-025 Not available

AAS 62-026 An Analysis of Lunar Injection Parameters and Their Effect upon the Characteristics and Entry into the Earth's Atmosphere, (Abstract), J. Swider, R. Taylor

AAS 62-027 A Higher Order Theory of Ballistic Reentry, W.H.T. Loh

AAS 62-028 Determination of Saturn-Dyna-Soar Optimum Trajectories Using a Combination Analog-Digital Computational Procedure, R.L. Chase

AAS 62-029 Interplanetary Trajectories, F.G. Gravalos

AAS 62-030 Optimum Round Trip Trajectories for Mars and Venus, J.C. Walker

AAS 62-031 The Need for Astronomical Telescopes in Tracking Space Probes, J.C. Walker

AAS 62-032 A Possible Military Application of a Cis-Martian Asteroid, (Abstract), D.M. Cole

AAS 62-033 A Probability for Designing Military Satellite Systems, J.A. Joseph

AAS 62-034 Not available

AAS 62-035 Military Applications of Space Systems, C. Mitchell

AAS 62-036 Abort During Manned Ascent into Space, R.L. Gervais, M.C. Johnson

AAS 62-037 Radiation Considerations for Lunar Missions, J. Abel

AAS 62-038 Heat Transfer to Porous Nose Caps and Leading Edges of Hypersonic Vehicles and Reentry Aids, H.G. Lorsch

AAS 62-039 Aerodynamic Heating for Lunar Reentry Vehicles, B.H. Little,

AAS 62-040 Reentry/Recovery Design Considerations of a Lunar Return Data Vehicle, (Abstract), D.M. Waltz

TORQUES AND ATTITUDE SENSING IN EARTH SATELLITES, ed. S.F. SINGER (Mathematics and Mechanics, Vol. 7, Academic Press, New York, 1964)*
(Second Goddard Memorial Symposium, Mar. 15-16, 1962, Washington, D.C.)

AAS 62-041 Generalized Gravity Gradient Torques, R.E. Roberson

AAS 62-042 Rotational Motion of Passive Space Vehicles, G.S. Reiter, W.T. Thomson

AAS 62-043 Torques and Attitude Problems for Synchronous Satellites, D.D. Williams

---

* Designated AAS Special Volume No. 5 (Sp.v5) for convenience.

AAS 62-044 Torque and Attitude Problems Experience on the Tiros Satellite, B.J. Conrath

AAS 62-045 On the Motions of Explorer XI around Its Center of Mass, G. Colombo

AAS 62-046 An Investigation of Observed Torques Acting on Explorer XI, R.J. Naumann

AAS 62-047 Horizon Sensing for Attitude Determination, B.K. Lunde

AAS 62-048 Not assigned

AAS 62-049 Aerodynamic and Radiation Disturbance Torques on Satellite Having Complex Geometry, W.J. Evans

AAS 62-050 Forces and Torques due to Coulomb Interaction with the Magnetosphere, S.F. Singer

AAS 62-051 Exploitation of Magnetic Torques on Satellites, R.H. Wilson, Jr

AAS 62-052 Passive Gravity-Gradient Stabilization for Earth Satellites, R.E. Fischell

AAS 62-053 Satellite Angular Momentum Removal Using the Earth's Magnetic Field, R.J. McElvain

AAS 62-054 Speculations Concerning the Non-observation of West Ford, R.A.Lyttleton, S.F.Singer

AAS 62-055 The Application of Gyrostabilizers to Orbiting Vehicles, J.E.DeLisle, E.G.Ogletree, B.M.Hildebrant

AAS 62-056 Magnetic Attitude Control of the Tiros Satellite, E.Hecht, W.P.Manger

AAS 62-057 Horizon Sensing in the Infrared: Theoretical Considerations of Spectral Radiance, D.Q.Wark, J.Alishouse,

AAS 62-058 to -079 Not Assigned

VOL. 12, ADVANCES IN THE ASTRONAUTICAL SCIENCES, SCIENTIFIC SATELLITES, 1963
SYMPOSIUM ON SCIENTIFIC SATELLITES - MISSION AND DESIGN
(Dec. 27, 1962, Philadelphia)

AAS 62-080 Scientific Satellites - A Perspective, B.A. Kossiakoff

AAS 62-081 S-6, An Aeronomy Satellite, R. Horowitz

AAS 62-082 Topside Sounding of the Ionosphere, J.H. Chapman

AAS 62-083 The NASA Topside Sounder Program, L.J. Blumle, R.J. Fitzenreiter, J.E. Jackson

AAS 62-084 Geophysical Research with Satellites Injuns I, II, and III, B.J. O'Brien

AAS 62-085 The Mission of the Orbiting Geophysical Observatories, G.J. Gleghorn

AAS 62-086 The Engineering Design of the Orbiting Geophysical Observatories, G.J. Gleghorn

AAS 62-087 The Mission of the Advanced Orbiting Solar Observatory, J.C. Lindsay

AAS 62-088 One Approach to the Engineering Design of Orbiting Solar Observatory, A.J. Cervenka

AAS 62-089 The Mission of the Orbiting Astronomical Observatory, R.R. Ziemer, J.E. Kupperian, Jr.

AAS 62-090 The Engineering Design of the Orbiting Astronomical Observatory, W.H. Scott

VOL. 13, ADVANCES IN THE ASTRONAUTICAL SCIENCES, INTERPLANETARY MISSIONS, 1963
NINTH ANNUAL MEETING
(Jan 15-17, 1963, Los Angeles)

AAS 63-001 A Role for Nuclear Rockets, R.B. Dillaway, A.G. Negro

AAS 63-002 Not available

AAS 63-003 Booster Recovery by Paraglider Methods, R.F. Brodsky, J.D. McNerney

AAS 63-004 Stability and Control of Manned Rotating Space Stations, J.G. Shearin, P.R. Kurzhals

AAS 63-005 Single-Stage Launching of Large Preassembled Space Station Structures, B.H. Neuffler

AAS 63-006 Radiation Doses in Interplanetary Flight, T. Foelsche

AAS 63-007 A Systematic Approach to the Study of Nonstop Interplanetary Round Trips, S. Ross

AAS 63-008 A Systematic Approach to the Study of Stopover Interplanetary Round Trips, R.W. Gillespie

AAS 63-009 Application of a Combined Optical and R-F Communications System for Nuclear Powered Interplanetary Space Vehicles, (Abstract), L.B. Wilker, V.B. Morris, Jr.

AAS 63-010 Distance Judgment and Lateral Displacement Preception Studies Under a Simulated Vision Environment of Space, R. McKinner, C.R. Adams, M.D. Arnoult

AAS 63-011 Celestial Navigation in Space, R.P. Haviland, C.M. House

AAS 63-012 Unfurlable Metal Structures for Space Exploration, J.D. MacNaughton

AAS 63-013 The Dissociative Relaxation of Carbon Dioxide Behind Normal Shock Waves, J.T. Howe, J.R. Viegas

AAS 63-014 Aerodynamic Force Coefficients and Equilibrium Surface Temperatures in Free Molecule Flow, K.S. Irwin

AAS 63-015 Effects of Shape of Total Radiative and Convective Heat Inputs at Hyperbolic Entry Speeds, P.J. Bobbitt

AAS 63-016 Energy Considerations for Precision Landing from Hyper-Orbital Approach Trajectories, R.E. Fortier

AAS 63-017 Influence of Shape on Aerodynamic Damping of Oscillatory Motion During Planet Atmosphere Entry and Measurement of Pitch Damping at Large Oscillation Amplitudes, B. Dayman, Jr., J.M. Brayshaw, D.A. Nelson, T.L. Babineaux

AAS 63-018/-019 Not available

AAS 63-020 Terrain and Soil of Mars, R.W. Johnson

AAS 63-021 Not available

AAS 63-022 An Analysis of a Fiber Method of Illumination for an Algal Life Support System, R.S. Novosad, R.S. Dezur

AAS 63-023 Mariner II, R.J. Parks

AAS 63-024 Mission Analysis of Fast Manned Flights to Venus and Mars, K.A. Ehricke 1963

AAS 63-025 Optimum Earth to Mars Round-Trip Trajectories Utilizing a Low-Thrust Power-Limited Propulsion System, C.G. Sauer, Jr., W.G. Melbourne

AAS 63-026 Mars Surface Manned Missions by Circumsolar Rendezvous, (Abstract), E. Fonseca

AAS 63-027 Analysis of Planetary Flyby Missions, R.V. Ragsac, R.R. Titus

AAS 63-028 Astronautic Applications of the Asteroids, (Abstract), D.M. Cole

AAS 63-029 On Meteoroids and Penetration, F.L. Whipple

AAS 63-030 A New Approach to Hypervelocity Impact Theory, S.W. Yuan, C.N. Scully

AAS 63-031 Natural Time Measurements in Uniformly Accelerated Frames of Reference, J.E. Romain

AAS 63-032 The Solar System and Extraterrestrial Life, R. Berger

AAS 63-033 Solar Particles and Interplanetary Fields, C.S. Warwick

AAS 63-034 Radio Astronomy of the Solar System, (Abstract) V. Radhakrishnan

AAS 63-035 to AAS 63-039 Not assigned

VOL. 14, ADVANCES IN THE ASTRONAUTICAL SCIENCES, PHYSICAL AND BIOLOGICAL PHENOMENA IN A WEIGHTLESS STATE, 1963
SECOND SYMPOSIUM ON PHYSICAL AND BIOLOGICAL PHENOMENA UNDER ZERO G CONDITIONS (Jan. 18, 1963, Los Angeles)

AAS 63-040 Generalization of Equations of Epihydrodynamics in Benedikt's Formulation of Magnetohydrodynamics, M.Z.V. Krzywoblocki

AAS 63-041 Capillary Stability in an Inverted Rectangular Channel, P. Concus

AAS 63-042 Effect of Gravity Upon Nucleate Boiling, M. Adelberg, R. Jetter

AAS 63-043 Equilibrium Configuration of the Liquid-Vapor Interface in a Rotating Container Under Zero g Conditions, O.D.DiMaggio

AAS 63-044 The Zero-G Flow Loop: Steady-Flow, Zero-Gravity Simulation for Investigation of Two-Phase Phenomena, J.T. Congelliere, W. Unterberg, E.B. Zwick

AAS 63-045 Dynamic Techniques for Extending Zero g Duration, J.W. Cummings

AAS 63-046 Orientation of Fluids in Zero g Through Surface Tension Effects, R. Jetter

AAS 63-047 Liquid Behavior in Zero Gravity, G.B. Wood

AAS 63-048 Review of Zero-G Studies Performed at General Dynamics/Astronautics, H.F. Steinle

AAS 63-049 NSL Optical Epihydrogoniometer, R. Lepper

AAS 63-050 Propellant Orientation in Zero Gravity with Electric Fields, J.B. Blackmon

AAS 63-051 Not available

AAS 63-052 Nucleate Transition and Film Boiling Heat Transfer at Zero Gravity, J.A. Clark, H. Merte, Jr.

AAS 63-053 Effect of Gravity Upon Boiling, R.W. Graham

AAS 63-054 Fundamentals of Low Gravity Phenomena Relevant to Fluid System Design, P.J. Berenson

AAS 63-055 Effects of Zero-G Conditions on Gyro Unit Operation, C.S. Draper, W.G. Denhard, E.J. Hall

AAS 63-056 Accelerometer Operation in a Low-G Environment, C.S. Draper, M.S. Sapuppo, L. Hughes

AAS 63-057 Electroencephalographic Observations of Human Zero g Flights, C.W.Sem-Jacobsen

AAS 63-058 A Study of the Effects of Zero Gravity on Cell Physiology, R. McKinney, P.O.B. Montgomery, C.F. Gell

AAS 63-059 Physiological Response to X-15 Flight Profile, H.R. Bratt

AAS 63-060 Predictions on the Biological Effects of Weightlessness, C.C. Cain

AAS 63-061 Bioengineering for the Zero G Environment, J.T. Celentano, H.S.Alexander, R.A.Baer, D.P.Balkin, S.P.Johnson, J.B.Reynolds

AAS 63-062 to AAS 63-064 Not assigned

VOL. 15, ADVANCES IN THE ASTRONAUTICAL SCIENCES, EXPLORATION OR MARS, 1963
(June 6-7, 1963, Denver, Colorado)

| | |
|---|---|
| AAS 63-065 | Influence of the Martian Ephemeris on Interplanetery Trajectories, R.M.L. Baker |
| AAS 63-066 | Role of Large Thrust Nuclear Power in Mars Flight, K. Boyer |
| AAS 63-067 | Electric Propulsion Systems for Mars Missions, W.E. Moeckel |
| AAS 63-068 | NOVA Launch Vehicle Design Studies, A. Kalitinsky |
| AAS 63-069 | Launch Facility Requirements for Mars/NOVA Vehicle, H. Keyser |
| AAS 63-070 | Vehicle Design for Earth Orbit to Mars Orbit and Return, H.O. Ruppe |
| AAS 63-071 | Vehicle Design for Mars Landing and Return to Mars Orbit, D.M. Hammock, B.G. Jackson |
| AAS 63-072 | Mars - A Target for Advanced Propulsion, H.B. Finger |
| AAS 63-073 | Research on Aerophysics of Re-entry at Super-Circular Speeds, P.H. Rose |
| AAS 63-074 | A Systems Approach to Vehicle Design for Earth Re-entry from an Interplanetary Mission, F.P. Dixon, L.D. Stimpson |
| AAS 63-075 | Not available |
| AAS 63-076 | Human Requirements for the Mars Mission, E.B. Konecci |
| AAS 63-077 | Fully Regenerative Life Support Systems for Mars Missions, R.D.Gafford |
| AAS 63-078 | Psychological Aspects of Extended Manned Space Flight, J.M. Christensen |
| AAS 63-079 | Some Metabolic Aspects of Extended Space Flight, D. Young |
| AAS 63-080 | Insuring the National Space Program, E.M. Zuckert |
| AAS 63-081 | Instrumentation and Communication Aspects of Mars Mission, E. Rechtin, R. Titsworth, W. Victor |
| AAS 63-082 | (A) Non-conventional Communication Devices for the Mars Mission, and (B) Use of Radar Techniques to Select Mars Landing Site, K.M. Siegel |
| AAS 63-083 | Power Supply Aspects of the Mars Mission, J.R. Wetch, R. Balent |
| AAS 63-084 | The History of the Concepts about Mars, W. Ley |
| AAS 63-085 | Preliminary Design of Mars Basing, P.H. Bliss, R.W. Hall |
| AAS 63-086 | Use of Extraterrestrial Resources for Mars Basing, E.A.Steinhoff |

AAS 63-087 Vehicle Designs for Exploration of Mars, A.M. Lippisch

AAS 63-088 The Ecological Profile of Mars: Bioastronautical Aspect, H. Strughold

AAS 63-089 Optical Studies of the Surface and Atmosphere of Mars, G. de Vaucouleurs, R. Norton

AAS 63-090 Spectroscopic Studies of the Surface and Atmosphere of Mars, G. Kuiper

AAS 63-091 Evidence of the Existence of Life on Mars, W.M. Sinton

AAS 63-092 Biological Experiments for a Manned Mars Mission, C. Sagan

AAS 63-093 Astrophysical Experiments for Mars Mission, G.J. MacDonald

AAS 63-094 Remarks on the Meteorology of Mars, S.L. Hess

AAS 63-095 Symposium Synopsis: the When and What of the Mars Trip, W.O. Roberts

AAS 63-096 to AAS 63-099 Not assigned

VOL. 16 I and II, ADVANCES IN THE ASTRONAUTICAL SCIENCES, SPACE RENDEZVOUS, RESCUE, AND RECOVERY, 1963 *
SYMPOSIUM ON SPACE RENDEZVOUS, RESCUE, AND RECOVERY (Sept. 10-12, 1963 Edwards AFB, Calif.)

AAS 63-100 Guidance Equations for Automatic Docking, L.A. Irish

AAS 63-101 Error Analysis, Apogee Biasing and Reference System Consideration for Satellite Rendezvous, A.E. Wudell, G.H. Smith, E.G. Slaughter

AAS 63-102 An Exact Closed-Form Solution to the Rendezvous Problem, A.E. Schwarzbein, R.H. Gersten, G.E. Smith

AAS 63-103 Rendezvous Energy Requirements for Collision and Pursuit Course Guidance, E. Harrison

AAS 63-104 Techniques for Rendezvous and Docking, D.L. Farless, G. Caggiano

AAS 63-105 $T_1$ Compensation Guidance Law for Ascent of Manned Space Vehicles, S. Chomak

AAS 63-106 Space Rescue System, H.L. Bloom, J.H. Quillinan

AAS 63-107 Emergency Escape from a Space Station, R.D. Heitchue

AAS 63-108 Crew Transfer in Orbiting Spacecraft, N. Belasco

AAS 63-109 Single and Multi-Crew Inflatable Re-entry Vehicles, J.E. Crawford

AAS 63-110 Investigation of the Effectiveness of a Non-Re-entry Mode of Escape as Applied to the Rescue Mission of a Disabled Space Station, T.P. Lubinski

AAS 63-111 Design and Similitude Testing of Dynamically Scaled Paraglider Model, B.W. Raff

* All papers appear in Vol. 16I except those marked with an asterick which appear in Vol. 16II.

AAS 63-112 Inflatable Parawing Deployment Studies Using a Dynamically and Elastically Scaled Model, H.N. Kelly, J.F. McNulty

AAS 63-113 Flight Characteristics of the Paraglider Spacecraft Recovery System, C.R. Faulder

AAS 63-114 The Design and Environmental Structural Testing of a Re-entry Paraglider, F. Warren, J.F. Keville

AAS 63-115 Flexible Wing Re-entry and Landing Systems, J. D. Sutliff

AAS 63-116 Manned Control of Space Vehicle Docking, J.E. Montgomery

AAS 63-117 Rendezvous Terminal Phase with a Combination Manual-Automatic Control-Display System, M.L.Ritchie, E.L. Warren

AAS 63-118 Control Criteria for Manual Orbital Rendezvous, D.C. Born

AAS 63-119 Visual Rendezvous Along a Fixed Line by Nulling the Apparent Drift of the Star Background, D. Novak

AAS 63-120 Recall and Return of a Manned Vehicle from Satellite Orbit: Reductions in Recovery Force Requirements Achieved with Maneuver Capability and Unique Recovery Area Geometry, E.J. Boyle, Jr.

AAS 63-121 An Energy Management System for Vehicle Re-entering from Earth Orbit, R.W. Austin, J.M. Ryken

AAS 63-122 An Approach and Landing Technique for Low L/D Manned Re-entry Vehicles, J.P. Dunn

AAS 63-123 On-Board Navigation and Landing Site Selection for Project Gemini, J.R. Burton, M.J. Schrader

AAS 63-124 Some Considerations of the Performance of a Maneuverable Lifting-Body Entry Vehicle, C.A. Syvertson, B.L. Swenson, J.L. Anderson, G.C. Kenyon

AAS 63-125 Space Rendezvous Propulsion System Requirements, R.L. Phen

AAS 63-126 A Proposed Dual Purpose Airlock and Docking Probe, L. Carlyle, F.H. Duskin, K. Englar

AAS 63-127 In-Flight Operational Evaluation of a Space System, S.P. Altman

AAS 63-128 Biologistic Tasks for Rendezvous Vehicles, N.M. Burns

AAS 63-129 Manned Maintenance of Space Vehicles, W.E. Powe, W.E. Bailey

AAS 63-130 A Manned Docking Simulator with Five Degrees of Freedom, A.C. Little

AAS 63-131 Visual Rendezvous Simulation, T.P. Neuberger

AAS 63-132 Simulation of Manned Rendezvous Under Adverse Dynamic Conditions, J.B. Roberts

AAS 63-133 A Full Scale Six-Degree-of-Freedom Orbital Docking Simulator, G.H. Smith, W.L. DeRocher, Jr.

AAS 63-134 Simulation Techniques for Astronaut Training, J. Prodan

*AAS 63-135 Visual Display Simulation of Lunar Landing Under Remote Control by a Human Operator, H.F. Meissinger, F.W. Belsky

*AAS 63-136 Spacecraft Structural Design Criteria Recovery Phase, F.C.S. Hung

AAS 63-137 Recovery of Lunar Vehicles--Geometric Restraint on Trajectories, D.M. Caldwell

AAS 63-138 Estimating Size and Mass of Artificial Satellites from Drag and Magnitude Data, T.W.Petrie

AAS 63-139 Economics of Booster Recovery, T.P. Sapp

AAS 63-140 Redstone Booster Recovery, J.P. Dei, C.C. Mair

*AAS 63-141 Recovery Control for Contingency Return from Lunar Flight, J. Zvara

AAS 63-142 Orbital Rendezvous Considerations for a Mars Mission, J.D. Bird, D.P. Thomas, R.L. Collins

AAS 63-143 Slow-Fly-By Approach to Satellite Inspection, J.S. Gansler, J. Carus

AAS 63-144 A Comparison of Methods for Determining Optimum Trajectories for Rendezvous, Landing and Orbit Transfer, F.D. Faulkner

AAS 63-145 The Orbital Characteristics of Vostok III and IV During the Apparent Near Rendezvous Flight of August 12-15, 1962, A.L. Jones, T.W. Petrie

AAS 63-146 Rendezvous Possibilities for the Impulse of Optimum Two Impulse Transfer, D.F. Bender

AAS 63-147 Arrested Rendezvous - A New Concept, D.C. Weiss

*AAS 63-148 Design of a Satellite-Based Radar Seeker, S.K. Kamen, M.S. Tanenbaum

AAS 63-149 An Optical Aid for Manual Acquisition and Tracking of Target Satellite During a Rendezvous Mission, F.E.Baird R.A. Schindler, R.N. Smith

AAS 63-150 Overshoot Orbits Reduce Sensor Requirements with Minimum Fuel Penalty, G. Henf

AAS 63-151 Not available

AAS 63-152 Systems Studies of Orbital Rendezvous, J.L. Hudson

AAS 63-153 Argument for a Universal Rendezvous Docking and Coupling Mechanism, J.W. James

AAS 63-154 Design Requirements for Orbital Attachment, Connection and Docking System, W.C. Buzzard

---

* In Vol. 16 II. All other papers appear in Vol. 16 I.

AAS 63-155 Coupling and Assembly of Orbiting Pressurized Modules, J. Zoszak

AAS 63-156 Aerodynamic Interference in Space Operations, W.M. Howard

AAS 63-157 Dynamic Behavior of Large Flexible Rotating Space Stations During Docking Maneuvers, F.W. Hopper

AAS 63-158 Concepts and Techniques for Attitude Control, Dynamic Balance and Stabilization of a Manned Rotating Space Station, J.E. Drake, D.S. Jackson

VOL. 17, ADVANCES IN THE ASTRONAUTICAL SCIENCES, BIOASTRONAUTICS - FUNDAMENTAL AND PRACTICAL PROBLEMS, 1963
(AAS/AAAS Meeting, Dec. 30, 1963, Cleveland, Ohio)

AAS 63-159 The Necessity for Biological Experimentation in Space, L.D. Carlson

AAS 63-160 Acceleration, Vibration and Impact, N.P. Clarke

AAS 63-161 The Pathophysiology of Disuse and Some Problems of Prolonged Weightlessness, M. McCally

AAS 63-162 The Biological Rhythm Problem and Its Bearing on Space Biology, F.A. Brown

AAS 63-163 Extraterrestrial Life, A.R. Brown

AAS 63-164 The Role of Man in Space, H.C. Urey

AAS 63-165 Selecting the Man for Space Exploration, D. Flickinger

AAS 63-166 Not available

AAS 63-167 Not available

AAS 63-168 The Space Sciences and Capitol Hill, C.A. Mosher

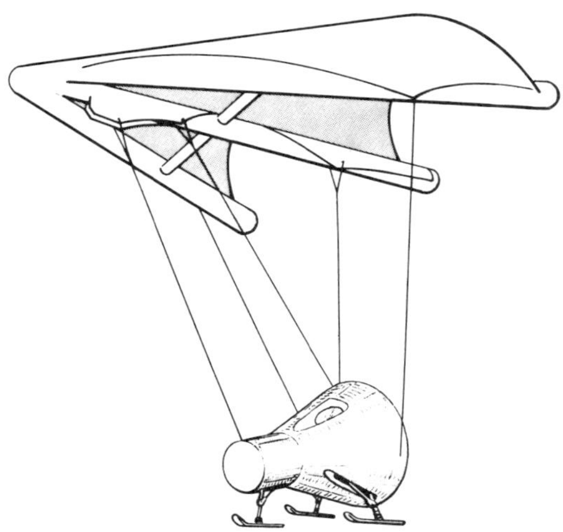

# AAS TECHNICAL PAPERS
# 1964 - 1965

VOL. 18, ADVANCES IN THE ASTRONAUTICAL SCIENCES, LUNAR FLIGHT PROGRAMS, 1964
TENTH ANNUAL MEETING
(May 4-7, 1964, New York)

AAS 64-001 The Manned Space Flight Program, G.E. Mueller

AAS 64-002 New Evidence Concerning the Lunar Surface, T. Gold

AAS 64-003 Inferences Concerning the Lunar Surface - From Radar, Radio, and IR Measurements, D.O. Muhleman

AAS 64-004 Not available

AAS 64-005 Unmanned Lunar Probes and the Manned Lunar Mission, W. Boyle

AAS 64-006 Ranger TV Subsystem, B.P. Miller

AAS 64-007 Lunar Orbiter: Its Mission and Capability, I. Taback

AAS 64-008 Surveyor Lander Mission and Capability, M. Beilock

AAS 64-009 The Apollo Mission, A.B. Shepard 1964

AAS 64-010 Launch Operations Support in Saturn/Apollo, K.H. Debus

AAS 64-011 Apollo Flight Operations, C.C. Kraft, Jr.

AAS 64-012 Space Medicine Experience in Manned Space Flight and Applications to Apollo, C.A. Berry

AAS 64-013 Technical Development Status of the S-IC Stage Saturn V Launch Vehicle, H.C. Runkel

AAS 64-014 Technical Development Status of the S-IVB Stage: Saturn V Launch Vehicle, T.D. Smith

AAS 64-015 Technical Development Status of the Apollo Command and Service Modules, N.J. Ryker

AAS 64-016 Technical Development Status of the Apollo Lunar Excursion Module, T.J. Kelly

AAS 64-017 Technical Development Status of Apollo Guidance and Navigation, N.E. Sears

AAS 64-018 Manned Lunar Scientific Operations, P. Lowman, D.A. Beattie

AAS 64-019 Extended Lunar Exploration, T.C. Evans

AAS 64-020 Advanced Lunar Transportation Systems, T.R. Salter

AAS 64-021 Apollo Logistic Support Systems, H. Schaefer, L.S. Yarbrough

VOL. 1, SCIENCE AND TECHNOLOGY, 1964
MANNED SPACE RELIABILITY SYMPOSIUM
(June 9, 1964, Anaheim, Calif.)

AAS 64-022 Integrating Crew Performance into Space Vehicle System Design for Optimum Reliability, W.P. Chase

AAS 64-023 The Applicability of Environmental Testing for the Reliability Demanded by a Manned Space Vehicle - with Particular Reference to Space Simulation, H.W. Lekuck

AAS 64-024 The Multidisciplinary Nature of Space Systems Reliability Assessment, G.E. Monroe

AAS 64-025 Not available

AAS 64-026 Environments of the Orbiting Space Station and the Logistic Supply Vehicle, D.F. McKenzie

AAS 64-027 Monte Carlo Simulation for Reliability and Maintainability, A. Economos

AAS 64-028 On Murphy's Law, W.C. Williams

VOL. 2, SCIENCE AND TECHNOLOGY,
TOWARDS DEEPER SPACE PENETRATION, 1964
(AAS/AAAS Meeting, Dec. 29, 1964, Montreal)

AAS 64-029 Lunar and Planetary Environments, J.M. Eggleston, D.E. Evans, J.W. Harris

AAS 64-030 Some Theoretical Considerations Arising from Guidance Analysis, C.G. Pfeiffer

AAS 64-031 Detection of Life Related Compounds on Planetary Surfaces by Gas Chromatograph - Mass Spectrometry Techniques, N.E. Bently, G.E. Giffin, D.G. Whitten, W.F. Wilhite

AAS 64-032 Alouette's Successors, C.D. Florida

AAS 64-033 Flight Mechanics Considerations for Manned Landing Missions to Mars, G. Markus, D.N. Lascody

AAS 64-034 The Design of the Canadian Communication Satellite Ground Station, O.L. Britney

VOL. 19, ADVANCES IN THE ASTRONAUTICAL SCIENCES, UNMANNED EXPLORATION OF THE SOLAR SYSTEM, 1965
(Feb. 8-10, 1965, Denver, Colorado)

AAS 65-001 Some Unsolved Problems of the Solar Atmosphere, R.G. Athay

AAS 65-002 Solar Radio Bursts, J. Warwick

AAS 65-003 Interplanetary Plasma, E.N. Parker

AAS 65-004 Meteorites and Asteroids, J.A. Wood

AAS 65-005 Comets, F.L. Whipple

AAS 65-006 Zodiacal Light and Interplanetary Dust, E.P. Ney

AAS 65-007 The Promise of Solar System Exploration, D.R. Ostrander

AAS 65-008 Not available

AAS 65-009 Probe Experiments Concerning the Atmosphere and Surface of Mars, G. de Vaucouleurs

AAS 65-010 A Martian Horror Story: Requirements vs. Capabilities for the Photographic Exploration of Mars, B.C. Murray

AAS 65-011 High Resolution Spectra of Planetary Atmospheres, H. Spinrad

AAS 65-012 Circulation of Planetary Atmospheres, Y. Mintz

AAS 65-013 Atmosphere of Venus, P. Thaddeus

AAS 65-014 Criteria for the Detection of Biological Systems, E.A. Shneour

AAS 65-015 The Goal of Experimental Biology in Space, R.G. Lindberg

AAS 65-016 The NASA Biosatellite Program, D.W. Jenkins

AAS 65-017 The Objectives and Technology of Spacecraft Sterilization, L.B. Hall, J.R. Miles, C.W. Bruch, P. Tarver

AAS 65-018 Limitations in Designing Life Detection Experiments, G.A. Soffen

AAS 65-019 Implant Transducers and Biological Rhythms -- on Earth and in Space, W. Ko, O. Lindan, J.B. Reswick 1965

AAS 65-020 Detection of Microorganisms on the Planet Mars, W. Vishniac, D.E. Buckendahl

AAS 65-021 Scientific Objectives and Instrumentation for a Solar Probe, J.P. Martin

AAS 65-022 Deep Space Probes: Sensors and Systems, H.A. Lassen, R.A. Park

AAS 65-023 Electric Power for the Scientific Exploration of the Solar System, A.H. Smith

AAS 65-024 Use of Entry Vehicle Responses to Define the Properties of the Mars Atmosphere, A. Seiff, D.E. Reese, Jr.

AAS 65-025 Some Considerations in Establishing the Feasibility of Data Compression in Unmanned Spacecraft, D.G. Bourke, P. Jewell

AAS 65-026 Low Data Rate Telemetry, R.M. Goldstein, W.B. Kendall

AAS 65-027 A Critical Data Recorder for Spacecraft, D.N. Tompkins

AAS 65-028 Infrared Instrumentation in Space Exploration, F.B. Gray

AAS 65-029 Orbital Boundary Value Problems, R.H. Battin

AAS 65-030 Trajectories for Unmanned Interplanetary Missions, R.L. Sohn, D.N. Lascody, J.P. O'Malley

AAS 65-031 The Use of Close Venusian Passages During Solar Probe Missions, F.G. Casal, S. Ross

AAS 65-032 Monte Carlo Simulation of an Adaptive Policy for Multiple Impulse Correction of the Trajectory of a Spacecraft, D.W. Curkendall

AAS 65-033 Recent Mariner Spacecraft - Design and Flight, D. Schneiderman, G.A. Reiff, J.N. James

AAS 65-034 Conceptual Design Studies of an Advanced Mariner Mars Flyby/Lander System, W.P. Neacy

AAS 65-035 Preliminary Investigation of Unmanned Missions to the Asteroid Belt with Growth Potential for Jupiter Flyby, D.J. Shapland

AAS 65-036 Advance Pioneer - Synthesis of System Concepts for a Mission to 10 AU, J.H. Chestek

AAS 65-037 Titan III for Unmanned Space Exploration, R.B. Demoret

AAS 65-038 Description and Status of Saturn IB Program, F.L. Williams

AAS 65-039 Return from Mars, P.K. Eckman

AAS 65-040 French Hybrid Rocket Propulsion, A. Moutet

AAS 65-041 A Review of Biological Material in Meteorites, H.C. Urey

AAS 65-042 The Military Value of Space Exploration, A.H. Flax

AAS 65-043 Spadework for Exploration of the Solar System, H.E. Newell

AAS 65-044 Welcoming Address, M.S. Peters

AAS 65-045 Objectives of Unmanned Exploration of the Solar System, M.W. Hunter II

AAS 65-046 The Scientific Goals of the Space Exploration Program G. Munch

AAS 65-047 Biological Objectives of Unmanned Exploration of the Solar System, G.V. Levin

AAS 65-048 Space Exploration in the Seventies, D.M. Cole

AAS 65-049 The ISIS Satellite, J. Almond

AAS 65-050 The French Scientific Satellite Program, R. Hamelin

AAS 65-051 Recent Advances in the Italian Space Program, F. Fiorio

VOL. 4, SCIENCE AND TECHNOLOGY, SCIENTIFIC EXPERIMENTS FOR MANNED ORBITAL FLIGHT, 1965
(Third Goddard Memorial Symposium, Mar. 18-19, 1965, Washington, D.C.)

AAS 65-052 Infrared Spectral Returns and Imagery of the Earth from Space and their Application to Geologic Problems, S.J. Gawarecki, R.J.P. Lyon, W. Nordberg

AAS 65-053 Applications of Remote Sensing in Agriculture and Forestry, R.N. Colwell, J.R. Shay

AAS 65-054 Some Applications of Radar Return Data to the Study of Terrestrial and Oceanic Phenomena, W.J. Pierson, Jr., B.B. Scheps, D.S. Simonett

| | |
|---|---|
| AAS 65-055 | Data Systems Requirements for Geographic Research, W.L. Garrison, R. Alexander, W. Bailey, M.F. Dacey, D.F. Marble |
| AAS 65-056 | Hyperaltitude Photography and its Applications, P.D. Lowman, Jr., T.L. Chang |
| AAS 65-057 | Aerial Observations of the Sea Surface off the Atlantic Coast of the United States, K.O. Emery |
| AAS 65-058 | The Use of Terrestrial Test Sites and Aircraft Flights in Preparing for Remote Sensing from Earth Orbital Spacecraft, P.C. Badgley, L. Childs, W.A.Fischer |
| AAS 65-059 | The Research Potential of Manned Earth Orbiting Spacecraft in the Field of Meteorology, S.F. Singer |
| AAS 65-060 | Research for the Development of an Earth Atlas Based on Viewing the Earth from Space with Various Remote Sensors and the Application for Global Planning, J.E. Gillis, Jr., R.A. Leestma |
| AAS 65-061 | Scientific Experiments on Gemini and Apollo Earth Orbital Flights as Precursors to Apollo Extension and ORL Earth Orbital Missions, J. Trombka, J.R. Gill, L. Lewyn |
| AAS 65-062 | The Study of Basic Physical/Biological Phenomena under Zero-G conditions in Earth Orbital Spacecraft, J. Trombka, J.M. Teem, H.R. Smith |
| AAS 65-063 | Critical Areas for Biomedical Research on Future Manned Orbital Spacecraft, W.M. Helvey |
| AAS 65-064 | The Overall Spectrum of Astronomical Research Possibilities Utilizing Manned Earth-Orbital Spacecraft, W. Tifft, J.R. Gill, L. Roberts |
| AAS 65-065 | The Possible Application of Remote Geochemistry in Planetary Exploration, A.R. Barringer |
| AAS 65-066 | Gravity Gradient Sensors and Their Applications for Manned Orbital Spacecraft, L.G.D. Thompson, R.O. Bock, P.H. Savet |
| AAS 65-067 | Communication and Terrestrial Navigation Applications from Earth Orbiting Spacecraft, E. Ehrlich |
| AAS 65-068 | National and International Cooperation in Scientific Experiments on Future Earth-Orbiting Spacecraft, H.L. Dryden |
| AAS 65-069 | Spectral Reflectance and Planetary Reconnaissance, D.M. Gates, H.J. Keegan, V.R. Weidner |
| AAS 65-070 | Not assigned |

VOL. 5, SCIENCE AND TECHNOLOGY, PHYSIOLOGICAL AND PERFORMANCE DETERMINANTS IN MANNED SPACE SYSTEMS, 1965
(April 14-15, 1965, Northridge, Calif.)

| | |
|---|---|
| AAS 65-071 | Not available |
| AAS 65-072 | Prediction of Human Thermal Comfort in Oxygen-Nitrogen Atmospheres, P.J. Berenson |
| AAS 65-073 | The Influence of Man on Spacecraft Design, A.H. Urmer |
| AAS 65-074 | Effect of Simulation and Environment on Man's Performance in Space, R.E. Geller |
| AAS 65-075 | An Integrated Approach to Evaluating the Performance Capabilities and Physiological State of Spacecraft Crews, R.S. Lincoln, J.E. Mangelsdorf |
| AAS 65-076 | Human Performance Capabilities in Spacecraft Control, L.G. Summers |
| AAS 65-077 | Not available |
| AAS 65-078 | Foods and Food Systems as Influenced by Zero Gravity Space Flight, A. Richards, I. Streimer, B. Wendrow |
| AAS 65-079 | Weightlessness vs Artificial Gravity, G.N.Hoover |
| AAS 65-080 | Determining Flight Crew Capability for a Manned Orbital Research Laboratory, J.D. Brower |
| AAS 65-081 | Body Heat Storage in Full Pressure Suits, E.C. Wortz, L.E. Browne, E.J. Prescott, D.K. Edwards |
| AAS 65-082 | Two-Man Space Station Simulation, J. Prodan |
| AAS 65-083 | Some Considerations Regarding the Definition of Man's Role in Future Space Missions, E.R. Jones |
| AAS 65-084 | Studies Concerning Low Friction Simulation, W.R. Pierson, A.T. Pessa |
| AAS 65-085 | Testing the Effects of Extended Space Missions on Piloting Skill, C.R. Kelley |
| AAS 65-086 | Not available |
| AAS 65-087 | Telescope Field-of-View Requirements for Star Recognition, R.W. Allen, M.L. Hershberger |
| AAS 65-088 | Not assigned |
| AAS 65-089 | Not assigned |

VOL. 20 I & II, ADVANCES IN THE ASTRONAUTICAL SCIENCES, POST APOLLO EXPLORATION, 1966 *

ELEVENTH ANNUAL MEETING (May 3-6, 1965, Chicago)

| | |
|---|---|
| AAS 65-090 | The Importance of Manned Exploration in Space, S.A. Korff |
| AAS 65-091 | Exploring in Extreme Terrestrial Environments, B.C. Bishop |
| AAS 65-092 | Performance Capabilities of Man on Earth and in Space: A Study in Continuity, J.M. Christensen |
| AAS 65-093 | Exploration - Yesterday and Tomorrow, G.H. Stine |
| AAS 65-094 | The Rationale and Scope of Biomedical/Behavioral Experiments in Earth Orbit, P.R. Tobias, W.J. Burke |
| AAS 65-095 | A Program: Biological Research in a Manned Space Laboratory, H. Richter, F. Crandall, A. McGuire |
| AAS 65-096 | Representative Astronomical Experiments Associated with Manned Earth Orbital Programs, R. Frost, I. Finkelstein, L. Narodny |
| AAS 65-097 | Manned Earth Orbital Program in Earth Sensing, D.S. Lowe, R.B. MacDonald, I.J. Sattinger, R.K. Walter |
| AAS 65-098 | System Reliability Analysis of the Manned Orbiting Research Laboratory, N. Kallay |
| AAS 65-099 | Application of Saturn S-IVB Apollo Systems to Planetary Exploration, M.W. Root |
| AAS 65-100 | The Use of High-Energy Upper Stages for Space Research, J.B. Kendrick |
| AAS 65-101 | Titan III Capability for Ferry-Resupply, W.D. Cowan, W.L. DeRocher, Jr. |
| AAS 65-102 | Not Available |
| AAS 65-103 | Spacecraft and Engineering Aspects of AES Lunar Orbital Missions, C.C. Johnson, M.W. Krueger |
| AAS 65-104 | Lunar Ground Data Required for Interpretation of AES Orbital Experiments, J. Friedman, R.J.P.Lyon, D.A.Beattie, J. Downey |
| AAS 65-105 | Ultraviolet Investigations for Lunar Missions, W.R. Hemphill, W.A. Fischer, J.E. Dornbach |
| AAS 65-106 | Radio Frequency Reflectivity Experiments Proposed for AES Lunar Orbital Flights, A.R. Barringer |

* Papers AAS 65-090 through -133, -144 appear in Vol. 20 Part I; papers AAS 65-114 through -143 appear in Vol. 20, Part II.

AAS 65-107 A Proposed Multispectral Photography Experiment for AES Lunar Orbital Mission, J.F. Cronin, J.B. Adams, R.N. Colwell, W.G. Tifft

AAS 65-108 Not available

AAS 65-109 System Approach to Lunar Surface Exploration, C.J. Weatherred, R.E. Wong, W.E. Crosmer

AAS 65-110 Apollo Extension System (AES) Lunar Surface Missions, D.A. Beattie

AAS 65-111 Lunar Exploration Systems/Lem Truck-Shelter, E. Markow, T. Barnes

AAS 65-112 Scientific Exploration of the Moon Using a Roving Vehicle, O.L. Tiffany, J.A. Downey III, E.M. Zaitzeff

AAS 65-113 Recent Results of Lunar Eclipse Measurements Showing Hot Spots, R.W. Shorthill, J.M. Saari

AAS 65-114 A Mobile Lunar Laboratory System, J.H. Carr, S. Romano

AAS 65-115 Extended Lunar Exploration, C.W. Henderson

AAS 65-116 Scientific Missions for Extended Lunar Space Exploration, W.B. Foster, D.F. Mitchell

AAS 65-117 Lunar Exploration - What is the Proper Course, T.C. Evans

AAS 65-118 Not available

AAS 65-119 The Utilization of Lunar Resources for Propellant Manufacture, S.D. Rosenberg, G.A. Guter, F.E. Miller

AAS 65-120 Water Resources on the Moon: Experimental Studies, J.N. Weber, R.Roy, R.T. Greer

AAS 65-121 The Role of Solar Furnaces in the Processing of Extraterrestrial Resources, P.E. Glaser, A.E. Wechsler, J. Green

AAS 65-122 An Approach to Indigenous Lunar Power - The Geothermal Concept, C.F. Austin, J.K. Pringle

AAS 65-123 Apollo Extension Systems Opportunities for Advanced Space Applications, G.E. Mueller

AAS 65-124 Mission Requirements for the Unmanned Exploration of the Solar System, F. Narin

AAS 65-125 The Voyager Project, D.P. Burcham

AAS 65-126 Comet and Close Approach Asteroid Mission, G.O.Moore, R. Jensen

AAS 65-127 A Survey of Manned Planetary Mission Possibilities, J.N. Smith

AAS 65-128 Space Navigation, Capt. P.V.H. Weems

AAS 65-129 The Space Velocity Meter, N.G. Lozins, W.W. Davenport

AAS 65-130 The Effects of High Luminance Sources upon the Visibility of Point Sources, R.F. Haines

AAS 65-131 Manual Guidance for Interplanetary Flight, S.P. Altman

AAS 65-132 Simulator Studies of MOL Tracking Experiments, J.R. Terell, W.H. Tobey, G.W. Smith

AAS 65-133 Grazing Trajectories and Capture of a Ballistic Vehicle, C.J. Maday

AAS 65-134 The Performance Characteristics and Mission Capabilities of the Nuclear Rocket Engine, A.R. Chovit, C.D. Kylstra, R.K. Plebuch

AAS 65-135 System Performance and Program Cost, R.S. Novosad, M.E. Capehart

AAS 65-136 The Extended Stay Lunar Exploration Mission, P.A. Penzo, R.W. Johnson

AAS 65-137 Maximum Data Recovery Using the Space Velocity Meter, N.G. Lozins, W.W. Davenport

AAS 65-138 Exploration of Cosmic Space in the USSR, G.A. Skuridin

AAS 65-139 Results of Medical Investigations Made on the Spaceships Voskhod and Voskhod-2, O.G. Gazenko, A.A. Gyurdzhian

AAS 65-140 A Method of Autonomous Navigation for Spaceships Orbiting Near Planets, I.K. Bazhinov

AAS 65-141 Methods of Study of the Dynamics of Booster Rockets and Spacecraft, G.N. Mikhishev

AAS 65-142 Survey of Methods for the Numerical Solution of Variational Problems of Flight Dynamics, V.K. Isayev, V.V. Sonin

AAS 65-143 Analysis of Flight Trajectories to the Moon, Mars, and Venus, I.K. Bazhinov

AAS 65-144 Manned Aerospace Flight in the Future, E.B. Konecci

VOL. 6, SCIENCE AND TECHNOLOGY, SPACE ELECTRONICS SYMPOSIUM, 1965 (AAS/AES Meeting, May 25-27, 1965, Los Angeles)

AAS 65-145 Snapoodle Demonstration Design Study, E.L. Leventhal, D.F. Berganini

AAS 65-146 Power Conditioning and Distribution of Lunar Electric Power, G.W. Bills

AAS 65-147 The Buck-or-Boost Regulator, L.A. Ule

AAS 65-148 Fuel Cell-Inverter/Converter System Synthesis, R.M. McKechnie

AAS 65-149 Charged Colloid Generation Research, R.E. Hunter, S.H. Wineland

AAS 65-150 The Hyperformance Electric Rocket Concept, S.L. Eilenberg, E. Caplinger

AAS 65-151 State-of-the-Art of Electric Propulsion - 1965, D.J. Kerrisk

AAS 65-152 Biochemical Fuel Cell Development Test, G.V. Colombo, T.J. Hooper

AAS 65-153 A Comparison of USA and USSR Life Support Systems, T.B. Weber

AAS 65-154 Impedance Pneumography, A.F. Pacela

AAS 65-155 Design and Test of Atmosphere Supply System for a Manned Space Cabin Simulator, J.K. Jackson

AAS 65-156 Regenerative Carbon Dioxide Removal Subsystem, J.F. Reed, A.D. Babinsky

AAS 65-157 Considerations in the Design of Aerospace Vehicle Antennas, E.L. Bock, M.F. Gunderson, R.L. Thomas

AAS 65-158 High Alphabet Command Decoder for Spaceborne Vehicles, L.A. Busby

AAS 65-159 The 300-Inch Diffraction-Limited "Orbiting Eye", J.H. Bloomer

AAS 65-160 The Early Bird Communications System, A.T. Owens

AAS 65-161 Optimizing Thrust Vector Control for Short Powered Flight Maneuvers, D.F. McAllister, E.E. Schiring

AAS 65-162 On the Simplification of the Inverse Problem of Optimal Control Theory, M.D. Schwartz

AAS 65-163 Control of the Syncom Communication Satellite, D.D. Williams

AAS 65-164 Redundant Microelectronic 1023 Counter Shift Register, G. Huffman, J. Billings

AAS 65-165 Deep Space Applications of a Secor-Type System, J.H. Reid

AAS 65-166 Electrostatic Detection of Objects in Space, M.G. Kaufman

AAS 65-167 Apollo-Saturn V Prelaunch Checkout Operations, J.H. Bell

VOL. 7, SCIENCE AND TECHNOLOGY, THEODORE VON KARMAN MEMORIAL SEMINAR, 1966
(May 12, 1965, Los Angeles, Calif.)

AAS 65-168 Luncheon Message, S. Yorty

AAS 65-169 Keeping the Peace in Space, B.A. Schriever

AAS 65-170 Important Benefits from the Space Program, R.F. Freitag

AAS 65-171 Keeping a Date in Space, W. Von Braun

AAS 65-172 What are Taxpayers Buying in Space?, G.P. Miller

AAS 65-173 Space Law and the Future, A.G. Haley

AAS 65-174 Man Can Live in Space, W.K. Douglas,

AAS 65-175 Launching a City into Space, S. Thomas

VOL. 8, SCIENCE AND TECHNOLOGY, IMPACT OF SPACE EXPLORATION ON SOCIETY, 1966
(Aug. 18-20, 1965, San Francisco)

AAS 65-176 Historical and Philosophical Background of the Space Age, F.G. Allen

AAS 65-177 Implications of Space Exploration for National Security, B.A. Schriever

AAS 65-178 Research on Social Consequences of Space Activities, J.M. Goldsen

AAS 65-179 The Space Program -- A Social Enigma, J.C. Fletcher

AAS 65-180 Engineering in a Perspective of Change, L.M.K. Boelter

AAS 65-181 Education for the Space Age, L.A. DuBridge

AAS 65-182 Influence of the Space Effort on Secondary Education, R.M. Clowes

AAS 65-183 Influence of the Space Effort on Higher Education, L.J. Stiles

AAS 65-184 Education for the Space Age, W.L. Whitson

AAS 65-185 Technical Re-education for the Space Age, R.C. Flothow

AAS 65-186 The Impact of Space Technology on the Geophysical Sciences, S.F. Singer

AAS 65-187 Planetary Exploration and Biological Research, J. Lederberg

AAS 65-188 Remarks on Prof. Lederberg's Paper, H.P. Klein

AAS 65-189 Impact of the Space Program on Medical Research and Practice, S. Bondurant

AAS 65-190 Impact of Space Studies on Astronomy, N.J. Woolf

AAS 65-191 The Exploration of Space as a National Goal, G.P. Miller

AAS 65-192 The European Regional Space Program, A.W. Lines

AAS 65-193 Space Programs of the Soviet Union, F.J. Krieger

AAS 65-194 A Summary Report of NASA's International Space Programs, A.W.Frutkin

AAS 65-195 The Value of Space Activity for Developing Countries, V. Sarabhai

AAS 65-196 The United Nations and International Cooperation in Space, M.A. Vellodi

AAS 65-197 The Development of a Global Communications Satellite Syatem, S.H. Reiger

| | |
|---|---|
| AAS 65-198 | The Impact of Satellites on Meteorology, D.S. Johnson |
| AAS 65-199 | The Use of Artificial Satellites for Geodesy, W.M. Kaula |
| AAS 65-200 | Examining Satellite Techniques for Navigation Systems, L. Jaffe |
| AAS 65-201 | Space Science and Public Policy, B. Graham |
| AAS 65-202 | The Return on our Investment in Space Exploration, L.E. Root |
| AAS 65-203 | Space and the World Economy, J.B. Condliffe |
| AAS 65-204 | Advanced Space Technology and the Future Political Economy, J.H. Rubel |

SECOND AAS SYMPOSIUM ON INTERACTIONS OF SPACE VEHICLES WITH AN IONIZED ATMOSPHERE
(Nov. 26-27, 1965, Miami, Florida)

Papers not available

VOL. 9, SCIENCE AND TECHNOLOGY, RECENT DEVELOPMENTS IN SPACE FLIGHT MECHANICS, 1966
(AAS/AAAS Meeting, Dec. 29, 1965, Berkeley, Calif.)

| | |
|---|---|
| AAS 65-205 | Trajectory design for Mission Analysis, S.E. Ross |
| AAS 65-206 | The Hodographic Theory of Newtonian Mechanics, S.P. Altman |
| AAS 65-207 | Recent Developments in Analytical Celestial Mechanics, R.F. Arenstorf |
| AAS 65-208 | Optimization Problems in Powered Space Flight, T. Edelbaum |
| AAS 65-209 | Trajectory Determination from Observational Data, J.D. Anderson |
| AAS 65-210 | Principles and Developments in Passive Attitude Control, D.B. DeBra |
| AAS 65-211 | Flight Dynamics of Planetary Entry, R.C. Wingrove |

VOL. 3, SCIENCE AND TECHNOLOGY, ORBITAL HODOGRAPH ANALYSIS (A MONOGRAPH), S.P. Altman, 1965

| | |
|---|---|
| AAS 65-212 | Basic Dynamics of the Orbital Hodographs |
| AAS 65-213 | Orbital Hodograph Transformation Theory |
| AAS 65-214 | Ballistic Missile and Intercept Trajectories |
| AAS 65-215 | Orbital Transfer |
| AAS 65-216 | Orbital Rendezvous |
| AAS 65-217 | Lunar Landing |
| AAS 65-218 | Potential Research Areas and Applications: Orbital Hodograph Analysis |

# AAS TECHNICAL PAPERS
# 1966

VOL. 21, ADVANCES IN THE ASTRONAUTICAL SCIENCES, PRACTICAL SPACE APPLICATIONS, 1967
(Feb. 21-23, 1966, San Diego, Calif.)

AAS 66-001 A Retrospective Look at the Application Satellite Program, H.J. Goett

AAS 66-002 Commercial Satellite Communications, S.W. Spaulding

AAS 66-003 Space Meteorology, Past and Present, S. Sternberg

AAS 66-004 Status of the Navy Navigation Satellite System, R.B. Kershner

AAS 66-005 International Aspects of Space Applications, J.D. Iams

AAS 66-006 International Satellite Communications: A Case Study, R.R. Colino

AAS 66-007 Legal Aspects of the Use of Satellites in the Exploitation of Natural Resources, H.J. Taubenfeld

AAS 66-008 Commercial Aspects of the Exploitation of Natural Resources by Satellite, H.C. Beggs

AAS 66-009 Applications of Saturn-Apollo Systems, W.B. Taylor

AAS 66-010 Space Experiments Program for Practical Applications, W.B. Foster 1966

AAS 66-011 Practical Applications from Advanced Research and Technology, J.L. Sloop

AAS 66-012 Commercial COMSAT Applications, S.W. Spaulding

AAS 66-013 Space Applications for the Environmental Sciences and Operations, J.P. Kuettner

AAS 66-014 The Economic Considerations of Space Operations, S.L. Hislop

AAS 66-015 Economic Aspects of Operational Satellite Systems, R.S. Field

AAS 66-016 Economic Justification for Manner Space Systems, P.A. Castruccio

AAS 66-017 Joint Venture Approach to Practical Space Utilization, D.R. MacQuivey

AAS 66-018 Practical Use of Extraterrestial Resources, E.A. Steinhoff

AAS 66-019 Utilization of Space Environment for Therapeutic Purposes, K.A. Ehricke

AAS 66-020 Commercial Development of the Resources of Space, A.N. Stanton

AAS 66-021 Potential Uses of a Survey Satellite in Vegetation and Terrain Analysis, R.N. Colwell

AAS 66-022 Satellite Detection of Natural Resources, W.A. Fischer

AAS 66-023 The Role of Satellites in the World Weather Watch, L. Krawitz

AAS 66-024 Operational Telecasting by Spacecraft After 1975, R.B. Marsten

AAS 66-025 Navigational Satellite System of the 1970s, E. Keats

AAS 66-026 Space and the Free Society, K.G. Harr. Jr.

AAS 66-027 Near Future Applications, D.M. Jones

AAS 66-028 Summary Panel, B.O. Evans

AAS 66-029 Not assigned

STEPPING STONES TO MARS
(AIAA/AAS Meeting, March 28-30, 1966, Baltimore, Maryland)

Papers available through the AIAA

VOL. 10, SCIENCE AND TECHNOLOGY, SPACE AGE IN FISCAL YEAR 2001, 1967
(Fourth Goddard Memorial Symposium, Mar. 15-16, 1966, Washington, D.C.)

AAS 66-030 Propulsion, W.A. Kuhrt

AAS 66-031 Structures, Materials, Re-entry, M.C. Adams

AAS 66-032 Communication, Guidance, and Radioastronomy, D.J. Fink

AAS 66-033 Life Sciences - Advanced Concepts, L.D.Proctor

AAS 66-034 Cislunar Transportation, H.W. Paige

AAS 66-035 Solar Transportation, K.A. Ehricke

AAS 66-036 The Case Against Space Ports in FY2001, W.G. Purdy

AAS 66-037 Postscript - The Stars, C. Sagan

AAS 66-038 Government Research and Development, E.Q. Daddario

AAS 66-039 Space Sciences, H. Friedman

AAS 66-040 The Overall Economic Outlook, C.S. Sheldon II

AAS 66-041 Space Commerce, E.P. Wheaton

AAS 66-042 Evaluation of Systems Approach, C.J. Dorrenbacher

AAS 66-043 A Space Age Projection, E.C. Welsh

AAS 66-044 Planning the Technological (R)Evolution, J.E. Karth

AAS 66-045 Impact of Automation, G. Kozmetsky

AAS 66-046 Community Planning,E.H. Sprague

AAS 66-047 Cities of the Future, C. Doxiadis

AAS 66-048 Applications in Aerospace Technology, B.M. Kerr

AAS 66-049 Tomorrow's Education, E.B. Konecci

AAS 66-050 Challenge to Education, R.L. Bisplinghoff

*VOL. 22, ADVANCES IN THE ASTRONAUTICAL SCIENCES, THE SEARCH FOR EXTRATERRESTRIAL LIFE, 1967, AND SUPPLEMENT:

VOL. 1, **AAS MICROFICHE SERIES, 1968**
(May 23-25, 1966, Anaheim, Calif.)

| | |
|---|---|
| AAS 66-051 | Not available |
| AAS 66-052 | The Biological Significance of the Search for Extraterrestrial Life, N.H. Horowitz |
| AAS 66-053 | Not available |
| AAS 66-054 | Indications of Possible Biological Substances in Carbonaceous Meteorites, B. Nagy (Vol. 1, Microfiche) |
| AAS 66-055 | Not available |
| AAS 66-056 | Delivery Systems: Orbiting Spacecraft and Planetary Landing Capsules, D.P. Burcham (Vol. 1, Microfiche) |
| AAS 66-057 | Development Planning for a Planetary Landing Vehicle, L.I. Mirowitz, K.W. Arnold |
| AAS 66-058 | Progressive Mars Capsule Mission Capability, C. de Moraes, R. Scarborough |
| AAS 66-059 | Instrumentation Integration for Planetary Missions, D.N. Tompkins |
| AAS 66-060 | Engineering Implications of Landing Site Selection, W.E. Tengelsen, J.P. Duda |
| AAS 66-061 | Landed Operations of Capsule Systems, P.G. Thome, E.J. Merz |
| AAS 66-062 | Data Transmission Capabilities of Mars Probes and Landing Capsules, J.J. Mahoney |
| AAS 66-063 | Potentialities of the GC/MS System for the Detection of Organic Compounds in Extraterrestrial Materials, K. Biemann |
| AAS 66-064 | Detection of Optical Activity as a Sign of Life, J.W. Westley |
| AAS 66-065 | Not available |
| AAS 66-066 | Life Detection by Means of Metabolic Experiments, G.V. Levin, G. Perez |
| AAS 66-067 | Life Detection by Visual Imaging, G.A. Soffen, R.K. Sloan |
| AAS 66-068 | Not available |
| AAS 66-069 | The Automated Biological Laboratory, W. Hostetler |
| AAS 66-070 | Sample Acquisition for Life Detection Experiments, J.L. Stuart |
| AAS 66-071 | Processing Planetary Samples for Life Detection Experiments, R.D. Johnson |

---
* Unless otherwise indicated, papers appear in Vol. 22

AAS 66-072 Logic and Programming for an Automated Biological Laboratory, H. Cole, P.R. Tobias

AAS 66-073 Information Processing and Data Compression for Exobiology Missions, L.L. Sutro

AAS 66-074 to AAS 66-076 Not available

AAS 66-077 Problems in Inter-species Communication, J.C. Lilly (Vol. 1, Microfiche)

AAS 66-078 to AAS 66-084 Not available

AAS 66-085 Prospects in the Search for Extraterrestrial Civilizations, F.D. Drake (Vol. 1, Microfiche)

AAS 66-086 to AAS 66-090 Not assigned

*VOL. 11, SCIENCE AND TECHNOLOGY, SPACE FLIGHT MECHANICS SPECIALIST SYMPOSIUM, 1967.

AND SUPPLEMENT:

VOL. 2, **AAS MICROFICHE SERIES, 1968**
(July 6-8, 1966, Denver, Colorado)

AAS 66-091 A New Approach to the Determination of the Earth's Zonal Harmonics, H.F. Michielsen

AAS 66-092 Not available

AAS 66-093 A Modified Encke Special Perturbation Method, W.T. Kyner, M. Bennett (Summary in Vol 11; in full in Vol. 2 Microfiche)

AAS 66-094 Rigorous Error Bounds on Position and Velocity in Satellite Orbit Theories, J.V. Breakwell, J. Varners

AAS 66-095 Some Qualitative Characteristics of Near-Lunar Satellite Orbits, B.D. Tapley, G.H. Born

AAS 66-096 A Graphical Method for Predicting Satellite Lifetime Based on a Time-Varying Atmospheric Density Model, T.J. Richards

AAS 66-097 Motion of a Vehicle in the Transition Region of the Restricted Three Body Problem (Summary), K. Forster

AAS 66-098 Normalization at $L_4$, A. Deprit, J.Henrard, A.R.M. Rom

AAS 66-099 Application of Krylov-Bogoliubov Me,the Stellar Three Body Problem, P. Musen (Summary in Vol. 11; in full in Vol. 2 Microfiche)

AAS 66-100 New Periodic Solutions of the Three Body Problem Applicable to the Lunar Theory (Summary), R.F. Arenstorf

AAS 66-101 Some Periodic Solutions of Elliptical and Restricted Four-Body Problems, I.S. Bernstein, S.A. Musa (Summary in Vol.11; in full in Vol. 2 Microfiche)

* Unless otherwise indicated papers appear in Vol. 11

AAS 66-102 On Constructing Formal Integrals of a Hamiltonian System Near an Equilibrium Point, F.G. Gustavson (Summary in Vol. 11; in full in Vol. 2 Microfiche)

AAS 66-103 Not assigned

AAS 66-104 Orbit Determination for Stationary Satellites, R.E. Balsam, S.E. Dunin

AAS 66-105 Physical Constants as Determined From Radio Tracking of the Ranger Lunar Probes, W.L. Sjogren, D.W. Trask, C.J. Vegos, W.R. Wollenhaupt

AAS 66-106 A Complete Error Budget of a Ballistic Flight Orbit Prediction Including Force Model Error Effects, J.F. Jones

AAS 66-107 Stability Analysis of Minimum Variance Estimators, J.F. Jones, J.M. Blair

AAS 66-108 A Unique System of State Transition Partial Derivatives, W.W. Lemon, L.S. Diamant (Summary in Vol. 11; in full in Vol. 2 Microfiche)

AAS 66-109 Orbit Computation for Motion in a Non-Central Force Field, C. Sheffield (Summary in Vol. 11; in full in Vol. 2 Microfiche)

AAS 66-110 Orbit Determination of Two Spacecraft Through the Use of Earth-Based Multiple-Link Doppler Tracking Data, D.W.Curkendall

AAS 66-111 A Parameter Optimization Procedure for Multi-Stage Vehicles, B.A. Glassman, F.I. Quinlan, V.C. Zvonkovich

AAS 66-112 A Combination Numerical-Analytical Approach to Ascent Trajectory Optimization, R. Rosenbaum

AAS 66-113 A Launch Vehicle Sizing Technique, R.H. Bingham, F.D. Breuer

AAS 66-114 Asymptotic Matching in Power-Limited Interplanetary Transfers, J.V. Breakwell, H.E. Rauch

AAS 66-115 Time Optimal Attitude Control of a Spinning Vehicle, D.H. deDoes

AAS 66-116 The Automation of Optimization Problems, H.L. Ingram

AAS 66-117 Optimization Problems Caused by Interaction of Guidance and Propellant Systems, D.C. Chandler, H.J. Horn, R.S. Ryan

AAS 66-118 An Accelerated Gradient Method for Parameter Optimization With Nonlinear Constraints, H.J. Kelley, W.F. Denham, I.L. Johnson, P.O. Wheatley (Summary in Vol. 11; in full in Vol. 2 Microfiche)

AAS 66-119 Minimum-Fuel Transfers Between Coaxial Orbits, Both Coplanar and Noncoplanar, C.B. Winn

AAS 66-120 An Analytic Investigation of the Escape Problem Using Tangential Thrust, M.L. Anthony, F.T. Sasaki

AAS 66-121 The Exact Analytic Solution of the Lunar Landing Problem, J.A. Jungmann

AAS 66-122 Geometric Constraints of the Disorbit Problem, A. Busemann, N.X. Vinh, R.D.Culp (Summary in Vol. 11; in full in Vol. 2 Microfiche)

AAS 66-123 Optimum Deorbit Positioning for Minimum Impulse Reentry, G.S. Stern

AAS 66-124 Asteroid Encounters on Mars Missions, D.F. Bender

AAS 66-125 Venus Swingbys for Manned Mars Missions During the 1978-1986 Period, A.A. Vander Veen

AAS 66-126 Trajectory Design for Impulsive Earth-Mars-Earth Trajectories Launched in 1969 and 1971, P.H. Feitis (Summary in Vol. 11; in full in Vol. 2 Microfiche)

AAS 66-127 Escape From Elliptical Parking Orbits, H.S. London (Summary in Vol. 11; in full in Vol. 2 Microfiche)

AAS 66-128 Interplanetary Trajectory Analysis for Combined High and Low-Thrust Propulsion Systems, J.L. Horsewood

AAS 66-129 Propellant Requirements for Unmanned Lunar and Interplanetary Soft Landing Vehicles, J.J. Ribarich

AAS 66-130 An Approximate Method for Predicting Lunar Satellite Lifetimes and Its Application to a Lunar Orbit Mission Analysis, C. Uphoff

AAS 66-131 Lunar Surface Accessibility for Manned Missions, 1968-1988, W.D. Kinney, J.O. Capellari, Jr., J.S. Dudek

AAS 66-132 Station-Keeping in the Vicinity of Collinear Libration Points with an Application to a Lunar Communications Problem, R.W. Farquhar

AAS 66-133 Abort Requirements for Lunar Polar Orbit, R.H. Hillsley (Summary in Vol. 11; in full in Vol. 2 Microfiche)

AAS 66-134 Three-Impulse Transfer from Lunar Orbits, E.D. Webb

AAS 66-135 Determination of Orbital Rendezvous Requirements: A Second Order Improvement (Summary), P.J. Bonomo, L.B. Schlegel

AAS 66-136 to AAS 66-138 Not assigned

VOL. 12, SCIENCE AND TECHNOLOGY, MANAGEMENT OF AEROSPACE PROGRAMS, 1967 (Nov. 16-18, 1966, Columbia, Missouri)

AAS 66-139 Aerospace Management and the Space Council, H.G. Edler

AAS 66-140 The Role of OST in Program Selection, W.L. Hooper

AAS 66-141 Management Systems Planning in New Flight Programs, P.K. Hatt

AAS 66-142 Apollo Applications Program Experiment Selection Process, J.K. Hancock

AAS 66-143 Not available

AAS 66-144 Manned Interplanetary Program Planning, F.P. Dixon

AAS 66-145 Not available

AAS 66-146 Bidding Strategy, J. Bjorksten

AAS 66-147 Not available

AAS 66-148 Product Oriented Cost Analysis for Major Development Production Programs, A.F. Klick

AAS 66-149 An Empirical Method for Estimating Unmanned Spacecraft Program Costs, C.A. Stone, W.P. Finnegan

AAS 66-150 Program Budgeting and the Space Program, M.L. Weidenbaum

AAS 66-151 Trends in Incentive Contracting, J.N. Daniel, Jr.

AAS 66-152 Management Schemes for Apollo, S.C. Phillips

AAS 66-153 Paper on Fixed Price Contract Management; F-111 Program, G.F. Keeling

AAS 66-154 Titan III: An Experiment in CPIF Management, W. Purdy

AAS 66-155 Cost Models for Complex Programs, D.D. Wyatt

AAS 66-156 Program Trend Line Analysis, An Aid To General Management, T.E. Jenkins

AAS 66-157 Gemini Incentive Program Model, W.C. Schneider

AAS 66-158 Gemini Program Management, H.W. Oldeg

AAS 66-159 Interface Management of Aerospace Systems, J.M. Schmissrauter

AAS 66-160 Not available

AAS 66-161 Data Utilization in ADC Weapon Systems Management, L.L. Knudsen

AAS 66-162 Operational Phase of the Minuteman Program, E.H. Manuel

AAS 66-163 Not available

AAS 66-164 Not available

AAS 66-165 Implications of On-line, Real-time Systems for Managerial Decision Making, D.C. Carroll

AAS 66-166 Not available

AAS 66-167 Not available

AAS 66-168 Not available

AAS 66-169 Not available

AAS 66-170 to AAS 66-179 Not assigned

VOL. 13, SCIENCE AND TECHNOLOGY, PHYSICS OF THE MOON, 1967
(AAS/AAAS Meeting, Dec. 29, 1966, Washington, D.C.)

| | |
|---|---|
| AAS 66-180 | Orbiter Observations of the Lunar Surface, L.C. Rowan |
| AAS 66-181 | Not available |
| AAS 66-182 | A Review of Interpretations of Lunar Surface Feature Origins, J. Green |
| AAS 66-183 | Crater Statistics and Erosion, R.J. Collins, G.B. Smith |
| AAS 66-184 | Review of Lunar Infrared Observations, J.M. Saari, R.W. Shorthill |
| AAS 66-185 | Optical Properties and Infrared Emission of the Moon, J.J. Hopfield |
| AAS 66-186 | Radar Properties of the Moon, T. Hagfors |
| AAS 66-187 | Remarks on the New Lunar Photography, J.A. O'Keefe |
| AAS 66-188 | Present Knowledge about the Thermal History of the Moon, R.A. Phinney, D.L. Anderson |
| AAS 66-189 | The Geometric and Dynamical Figures of the Moon, W.M. Kaula |
| AAS 66-190 | Lunar Seismic Exploration, R.L. Kovach |
| AAS 66-191 | Lunar Gravity Investigation, J. Weber |
| AAS 66-192 | The Origin and Dynamical Evolution of the Moon, S.F. Singer |
| AAS 66-193 | Panel Discussion on the Lunar Surface, D.H. Menzel |
| AAS 66-194 | Panel Discussion on the Lunar Interior, G.J.F. MacDonald |

VOL. 14, SCIENCE AND TECHNOLOGY, INTERPRETATION OF LUNAR PROBE DATA, 1967
(Sept. 17, 1966, Huntington Beach)

| | |
|---|---|
| AAS 66-195 | Lunar Probe Data as Related to Hypervelocity Impact Mechanisms (Abstract), D. Gault |
| AAS 66-196 | Lunar Vesiculation Processes as Related to Luna and Surveyor Probe Data, E. Azamon |
| AAS 66-197 | A Review of Shock Processes Pertinent to Fragmentation and Lithification of the Lunar Terrain, N.M.Short |
| AAS 66-198 | Cohesive Properties of Lunar Soils as Indicated by Surveyor I, J. Ryan |
| AAS 66-199 | Mechanical, Optical, Thermal, and Electrical Properties of Surveyor I Landing Site, J.D.Halajian |
| AAS 66-200 | The Nature of the Surface of the Moon, Illustrated in Part by 3-D Slides Taken from Ranger Photographs, D.H.Menzel |
| AAS 66-201 | The Impact of Lunar Probe Data on Future Space Missions - Panel Discussion, S. Greenfield, C. Schultz, J. Edson, G. Kuiper, J.W. Rawls, J.L. Burridge, R. McFee |

AAS 66-202 Seleno-Geological Evidence from Surveyor I, K. Von Bulow

AAS 66-203 What Have the Probes Told Us?, J. McCall

AAS 66-204 Some Geological Aspects of the Lunar Crust: Án Interpretation of Photographs Taken by Ranger, Surveyor and Orbiter Probes, S. Miyamoto

AAS 66-205 Lunar Surface Characteristics, P. Moore

AAS 66-206 On the Inapplicability of Baldwin's Relation for the Determination of the Causes of Lunar Craters, G.S. Shteinberg

AAS 66-207 Bonus Experiments for Surveyor Spacecraft, J.B. Edson

**VOL. 9 AAS MICROFICHE SERIES, SOUTHEASTERN SYMPOSIUM ON MISSILES AND AEROSPACE VEHICLE SCIENCES, 1969 (SE Section AAS Meeting, December 5-7, 1966, Huntsville, Alabama)**

AAS 66-208 Terrestrial Environment Guideline Document, G.E. Daniels

AAS 66-209 Preliminary Spacecraft Concept for an Operational Space Environment Monitoring System, A.S. Hill

AAS 66-210 Probabilities of Solar Flare Occurrence, W.T. Roberts, G.S. West

AAS 66-211 Prediction of the Lunar Thermal Environment, F.B. Tatom, K.V. Ramakrishna

AAS 66-212 Random Number Solar Cycle Prediction Technique, H.C. Euler

AAS 66-213 Exploration of the Magnetosphere by Analysis of High Altitude Nuclear Weapon Test Data, D.T. Hodder, R.A. Fowler, J. Kert

AAS 66-214 Theory of a Refined Earth Figure Model, H.G.L. Krause

AAS 66-215 Not available

AAS 66-216 The Use of Vaporizing Underlayers for Transpiration Cooling of Re-entry Vehicles, D.G. Harden

AAS 66-217 Massive Blowing From Slender Bodies in Viscous Hypersonic Flow, G.R. Inger

AAS 66-218 Application of the Method of Integral Relations to the Nonequilibrium Hypersonic Planetary Entry Problem, H.J.Deacon,Jr., C.C.Oliver

AAS 66-219 A Simplified Method for Calculating Heat Transfer Bodies at an Angle of Attack, F.R. DeJarnette

AAS 66-220 Design Problems Associated with Complex Reentry Vehicles, E. Offenhartz, W. Zeh

AAS 66-221 Base Flow Characteristics and Thermal Environment of Launch Vehicles with Strap-on Solid Rocket Motors, G.A. Etemad, K.D. Korkan

AAS 66-222 A Technique in Predicting Multiple-Engine Exhaust Geometry and Thermodynamic Properties with Single Reference Engine, C.R. Mullen, R.L. Bender

AAS 66-223 The Influence of Nonuniform Flow on Drag Forces, J.F. Harp

AAS 66-224 Kick Stage Propellant-Dependent Weights Study, J.J. Traxler, C.W. Colquitt, Jr.

AAS 66-225 Determination of Normal Force and Pitching Moment Characteristics of Bodies of Revolution at Finite Angles of Attack, S.V. Paranjape, T.J. Thomas, J.D. Warmbrod

AAS 66-226 Base Pressure Calculation of a Stepped Axisymmetric Body in Supersonic Flow S.C. Zien

AAS 66-227 Numerical Solution of Compressible Flow in Ducts, D.U. von Rosenberg, D.L. Beauchamp

AAS 66-228 Monte Carlo Simulations of Molecular Flow Through Cylindrical Ducts, C. Fan

AAS 66-229 Optimization of Manned Lifting Re-entry Vehicles Using the Occof Technique, A. M. Goldman, T.A. Campbell

AAS 66-230 Feasibility Study of an Orbiting Energy Depot, D.V. Merrifield, W.G. Offik

AAS 66-231 A New Approach for Mapping the Field of Fluctuation by Electro-Magnetic Techniques for Aero-Space Flight, S.S. Hu, S.V. Paranjape

AAS 66-232 Not available

AAS 66-233 A Methodology for the Technical Evaluation and Rating of Space Science Experimental Payloads, M.K. Fox, G.F. Gibson

AAS 66-234 Information Storage and Retrieval for Spacecraft Parts, Materials and Processes, L. Allen, M. Cook

AAS 66-235 Mars Flight II Space Cabin Simulator, R.A. Ibison

AAS 66-236 Multiple Payload Missions, R.S. Novosad, M.E. Capehart

AAS 66-237 Solar Pressure Perturbations on the Orbit of a Flat Reflector, B.S. Perrine, Jr.

AAS 66-238 Not available

AAS 66-239 A Review of Some Existing Literature Concerning Discontinuous State Variables in the Calculus of Variations, R.E. Burns

AAS 66-240 Mass Optimal Solutions of Some Variable End Point Trajectory Problems, M.S. O'Mahony, C.D. Eskridge, A.G. Bennett

AAS 66-241 Effects of High-Temperature Devices on Close Solar Probe Mission Planning, E.M. Schall, W.E. Brasher

AAS 66-242 An Alternate Form of the Iterative Guidance Mode, D.J. Dobner

AAS 66-243 Optimum Two-Impulse Transfer Between Coplanar Circular Close Orbits, N.F. Eraslan

AAS 66-244 Calculation of a Spacecraft Ephemeris for Compatibility Studies of Orbits and Experiments Considered for AAP Missions, J.L. Diven

AAS 66-245 Influence of IR Frequencies on Tracking for Optical Communication, E.B. Moss

AAS 66-246 Signal to Noise Performance of SEC Vidicon at Reduced Readout Rates, R.E. Marinaccio

AAS 66-247 Refinement of the Experiment Integration Process, P.C. Green, R.F. Filipowsky

AAS 66-248 A Mission for Surveying and Mapping the Lunar Surface, S.N. Heaps

AAS 66-249 A Direct Method for Estimating Lagrange's Multipliers in Trajectory Optimization Problems, M.L. Thompson

AAS 66-250 An Adaptive Guidance Scheme for Orbital Rendezvous, C.D. Eskridge, M.S. O'Mahony, A.G. Bennett, A.W. Brasse

AAS 66-251 Monte Carlo Performance Analysis of Lunar Rendezvous Guidance, T.S. Bettwy, F.A. Evans, D.M. Kisner, W.C. Ruland, J.T. Wagner

AAS 66-252 The Use of a Differential Approximation as a Basis for the Design of a Near-Time Optimal Guidance Scheme, J.R. Sklaroff

AAS 66-253 Critical Flow Rate of Two-Phase Nitrogen, H.M. Campbell, Jr. /J.J. Overcamp

AAS 66-254 Analysis of Fluid Conditions in the Discharge Line of a Cryogenic Container Undergoing Self-Pressurized Draining, H.M. Campbell, Jr., R.R. Head

AAS 66-255 A Similarity Study of Gravity and Pressure Driven Discharge of Liquids from Propellant Tanks, E.R.F. Winter, R.J. Schoenhals

AAS 66-256 Minimization of Slosh Amplification at Orbital Injection of the S-IVB/Saturn V by an Optimum Thrust Termination Sequence, H.S. Curtis

AAS 66-257 Flow Losses in Flexible Hose, J.R. Langley, K.L. Riley, C.A. Whitehurst, B.S. Pressburg

AAS 66-258 Capillary Stability of Rotating Dielectric Liquid Cylinder in an Electric Field, L.M. Habip

AAS 66-259 Alternate Mission Modes for Manned Mars Exploration, R.R. Titus

**AAS 66-260/261 Not Available**

AAS 66-262 Three-Impulse Interplanetary Rendezvous Trajectories, J.R. Doll, F.W. Gobetz

AAS 66-263 Application of the Saturn V Launch Vehicle to Unmanned Scientific Exploration of the Planet Jupiter, P.R. Odom, A.S. Hill, B.G. Brown

AAS 66-264 A Planetary Geodetic Laser Survey System, R.A. Fowler, V. Costellano, E.L. Cohn

AAS 66-265 The Atmosphere of Mars: A Derivation of Engineering and Design Parameters, W.T. Roberts, G.S. West

AAS 66-266 Environment of Jupiter, C.L. Hasseltine

AAS 66-267 Use of an Earth-Analog in Lunar Mission Planning, W.J. Hinze, R. Erhlich, H.F. Bennett, D. Fletcher, E. Zaitzeff, O.L. Tiffar

AAS 66-268 Predicting the Luminance of the Lunar Surface, W.H. Rauser

AAS 66-269 Post Apollo Lunar Mission Planning, E. Zaitzeff, O.L. Tiffany, /J. Downey III

AAS 66-270 A Technique for On-Board Fast Generation of Space Guidance Signals for Long Range Wide-Angle Turn Powered Flights, and Its Advantages over Other Current Techniques, S.S. Hu

AAS 66-271 Optimal and Suboptimal Guidance and Control for Low Thrust Orbital Transfer, A.P. Sage, T.W. Ellis

AAS 66-272 The Control Laws for Two-Finite-Burns Minimum Fuel Rendezvous, I.E. Kliger

AAS 66-273 Rapid Convergence to Optimum Solutions Using a Min-H Strategy, R.G. Gottlieb

AAS 66-274 Extremal Control of an Undamped Oscillatory Plant with Control Amplitude and Rate Saturations, J.Y.S. Luh, J.S. Shafran

AAS 66-275 Theoretical Optimization of Staged Ejector Systems, J.L. Loth

AAS 66-276 Investigation of Direct Energy Conversion Utilizing a Superconducting Flux Barrier, J.S. Goodling, R.K. Irey, C.D. Schwebel

AAS 66-277 ICARUS: A Proposed 0.1 AU Solar Probe, S.L. Dickerson E.J. Steinhardt

AAS 66-278 Ground Based Evaluation of EVA, C.H. Murrish, G.W. Smith, J.R. Tewell

AAS 66-279 A Methodology to Analyze and Evaluate Critical Human Performance, M.A. Barone

AAS 66-280 Lateral Stability Augmentation of the X-15 Using Multi-channel Optimal Feedback Control, R.W. Bass, J.W. Stoner, F.J. Taylor

AAS 66-281 A Study of Sub-Optimal Control Laws, K.V. Grider

AAS 66-282 Stability Analysis Methods for Hybrid Control Systems, C.L. Phillips, J.S. Boland, R.K. Cavin, III, D.L. Chenoweth

AAS 66-283 On Optimization and the Guidance Problem for System with Transport Lags, H.R. Sebestka, L.G. Clark

AAS 66-284 Optimum Sensitivities Obtained Without Using Finites Difference, W.C.Hines

AAS 66-285 Multiple Exhaust Nozzles for Advanced Nuclear Rocket Engines, E.E. Duke, D.F. Vanica

AAS 66-286 Liquid Rocket Engine Combustion Time Delay Analysis, C.E. Driskill, C.E. Lundy

AAS 66-287 Turbopump Configuration Selection for an Optimized Nuclear Engine, W.E. Campbell

AAS 66-288 An Early Earth Orbital Propellant Transfer Mission, M.C. Ziemke, L.E. Toole, T.F. Stomps

AAS 66-289 A Study of New Methods of Space Cryogenic Storage Involving Recovery of Vent Gases, J.H. Bell, Jr.

AAS 66-290 An Analytical and Experimental Examination of the Effect of Cryogenic Propellant Venting on Orbital Vehicle Dynamic Behavior, A.B. Walburn

AAS 66-291 Discontinuous Flows and Free Streamline Solutions for Axisymmetric Bodies at Zero and Small Angles of Attack, H.G. Struck

AAS 66-292 Not available

AAS 66-293 The Response of Launch Vehicles to Nonstationary Random Transonic Buffeting Excitation, H.C. Peterson, C.R. Howard, and R.A. Philippus

AAS 66-294 Optimum Design of a Ring-Airfoil to be used as a Performance Improving Device in a Conical Diffuser, T. Yang, H.O. Jones

AAS 66-295 Analog Simulation of Saturn IB Stage Propulsion System Dynamic Characteristics, J.W. Lehner

AAS 66-296 Statistical Analysis of Thrust Vector Characteristics for Rocket Engine Clusters, F.L. Echols

AAS 66-297 Propulsion Requirements for Low Gravity Liquid Settling, M.E. Meadows

AAS 66-298 Status of Nuclear Propulsion for Space Flight, J.H. Beveridge

AAS 66-299 Rankine Cycle Powerplant Characteristics for Low Acceleration Propulsion Manned Mars Missions, B. Gitlow, J.W. Schmitt, R. Ragsac

AAS 66-300 Free Molecule Aerodynamic Coefficients, B.H. Shirley

AAS 66-301 The Calculation of Near Free Molecular Flux Distribution to Simple Bodies in Hypervelocity Flow, J.T. Miller

AAS 66-302 A Study of the Relationship Between the Incipient Cavitation Number and the Reynolds Number, D.W. Lane, H.W. Hsu

AAS 66-303 Aerodynamic Considerations for a Planetary Entry Probe, P.A. Giragosian, E.T. Myers

AAS 66-304 A Statistical Method for Computing Launch Vehicle Loads, J.C. Angehr

AAS 66-305 On Liquid Sloshing in a Cylindrical Tank with a Flexible Bottom, J. Siekmann, S.C. Chang

AAS 66-306 A Model to Approximate the Effect of Propellant Slosh on Vehicle Dynamics Under Zero Gravity Conditions, A.S. Abbott, J.P. Gille

AAS 66-307 Development and Optimization of Bonded Honeycomb Panel Welded Joints, J.A. Mallon, E.J. Beck

AAS 66-308 Plastic Deformation of Cylindrical Missils in Impact, N. Davids, D. Raftopoulos, C. Graberek, A. Ricchiazzi

AAS 66-309 Thermal-Fatigue Analysis of a Cryogenically Cooled Rocket-Nozzle, A.E. Carden, D.G. Harman, E.A. Franco-Ferreira

AAS 66-310 Discrete Methods for the Implementation of Control Networks Using Integrated Circuitry, G.B. Doane III, C.C. Carroll

AAS 66-311 The Electrostatic Gyro - A Tutorial Model, R.E. Paulsen

AAS 66-312 Strapdown System Application Studies Related to a Surface-to-Surface Missile, R.E. Alongi

AAS 66-313 A Strapped Down Orbital Attitude Reference System, R.L. Gordon, W.J. Carroll

AAS 66-314 Some Cross-Coupling Effects in a Strapdown Orbital Gyro-compass, W.J. Carroll

AAS 66-315 A Mathematical Model for the Study of Slosh Effects on Rendezvous Dynamics, J.H. Dobkins

AAS 66-316 Axisymmetric Dynamic Response of Spherical and Cylindrical Shells, J.T.S. Wang, C.W. Lin, W. Stadler

AAS 66-317 Stiffness Matrix of an Unsymmetric Nodal Ring Element for Matrix-Displacement Analysis of Rib-Stiffened Shells, R.L. Thoms

AAS 66-318 Forced Resonant Nonlinear Oscillations of a Liquid in a Cylindrical Tank, O.D. DiMaggio, R.N. Salzman

AAS 66-319 Not available

AAS 66-320 Classification of Elastic-Plastic Waves in Plane Strain, C.Y. Yang

AAS 66-321 Not available

AAS 66-322 Not available

AAS 66-323 Guidance System Stability as Effected by Vehicle Systems Interaction, J.J. Kelly, C.T. Hagstrom

AAS 66-324 Radar-Updated Inertial Navigation of a Continuously-Powered Space Vehicle, B.A. Kriegsman

AAS 66-325 Filtering Horizon-Sensor Measurements for Orbital Navigation, R.J. Fitzgerald

AAS 66-326 to 329 Not Available

VOL. 24 SCIENCE AND TECHNOLOGY, AEROSPACE RESEARCH AND DEVELOPMENT

(An international dedication to Dr. Adolf Baeumker on the occasion of his 75th birthday)
AAS paper numbers assigned for identification purposes.

AAS 66-330 Past and Future Aspects of Space Exploration, W. von Braun

AAS 66-331 Grossforschung und Einzelforscher (Basic Research and Individual Researchers), A. Betz

AAS 66-332 The Role of the Technical Societies in Advancement of Science and Technology in the Space Age, B.H. Goethert

AAS 66-333 The Influence of Space Technology on General Technological Progress and Impact on Standard of Living and Economy, E.A. Steinhoff

AAS 66-334 Die Windkanaele des Instituts fuer Aerodynamik der DFL Braunschweig (The Windtunnels of the DFL Aerodynamics Institute, Braunschweig), H. Schlichting

AAS 66-335 Ceramic Research at the Aerospace Research Laboratories, B.C. Weber

AAS 66-336 Basic Research in the Space Age, H.H. Kurzweg

AAS 66-337 Warum Strahltriebwerke? Vergangenheit, Gegenwart und Zukunft der Luftfahrt-Gasturbine (Why Jet Engines? Past, Present, and Future of the Aircraft Jet Engine), B. Eckert

AAS 66-338 New Concepts in the Physics of Solids: A Monograph, J.N. Plendl

AAS 66-339 Is there a Martian Biosphere?, H. Strughold

AAS 66-340 Gas-to-Gas Film Cooling, E.R.G.Eckert

AAS 66-341 Interaction of Strong Blast Waves on Slender Supersonic Bodies, A. Busemann, V.D. Blankenship

AAS 66-342 What is Flow Acoustics? A Heuristic Study, G.E. Knausenberger

AAS 66-343 Man-Made ELF Phenomena in the Ionosphere and Exosphere, H.A. Bomke

AAS 66-344 Analytical Investigations of Hypersonic Flow with Non-equilibrium Oxygen Dissociation around Blunt and Pointed Bodies, R. Hermann

AAS 66-345 Nebelbeseitigung ueber Landebahnen auf Flugplaetzen mit Hilfe eines mechanischen Entnebelungs-Verfahrens (Fog Dispersal over Airport Runways with the Aid of a Mechanical De-fogging Process), U. Schmieschek

AAS 66-346 Eine moegliche Erklaerung des rotating Stall( A Possible Explanation for Rotating Stall),H. Soehngen

AAS 66-347 Flow Modification through Space Charges, O.M. Stuetzer

AAS 66-348 Ueberschallverbrennung in thermogasdynamisch anschaulicher Darstellung (Supersonic Burning in Thermogasdynamic Graphic Presentation), O. Lutz

AAS 66-349 Compressible Flows with Closed Streamlines at Large Reynolds Numbers, K.G. Guderley

AAS 66-350 Compound Ceramics, E. Ryshkewitch

AAS 66-351 Aerodynamics of Wing-Propeller Interaction, H. Multhopp

AAS 66-352 Zur Entwicklungageschichte der deutschen Flugtriebwerke-Höhenprüfstände (To the History of the Development of German Altitude Test Stands for Aircraft Engines), H. Barth

# AAS TECHNICAL PAPERS
# 1967

VOL. 16, SCIENCE AND TECHNOLOGY SERIES
**VOYAGE TO THE PLANETS, 1968**
**(Fifth Goddard Memorial Symposium, Mar. 14-15, 1967, Washington, D.C.)**

AAS 67-001 The Mariners, W.H. Pickering

AAS 67-002 Galactic Jupiter Probes, J.F. Clarke, E.W. Hymowitz, J.H. Trainor

AAS 67-003 The Voyager Program, E.M. Cortright

AAS 67-004 Paving the Way for Manned Planetary Exploration, C.W. Mathews

AAS 67-005 Manned Expeditions to Mars and Venus, E.Z. Gray, F.P. Dixon

AAS 67-006 The Path to Voyager (abstract), C.S. Pittendrigh

AAS 67-007 Meanwhile, Back on Earth, C.S. Sheldon II

AAS 67-008 National Prestige, G.P. Miller

AAS 67-009 The Role of the Diplomat, R.F. Packard

AAS 67-010 Planetary Exploration Strategy, J.W. Findlay

AAS 67-011 Doorway to the Universe, G.T. Seaborg

AAS 67-012 Man and Space, A.C. Clarke

AAS 67-013 to AAS 67-024 Not assigned

VOL. 15, SCIENCE AND TECHNOLOGY SERIES, 1967 **1967**
FUTURE SPACE PROGRAM AND IMPACT ON RANGE AND NETWORK DEVELOPMENT
(March 22-24, 1967, Las Cruces, New Mexico)

AAS 67-025 The Challenge of Space and Range Development, H.G. Davisson

AAS 67-026 Future Manned Earth Orbital Missions, W.H. Clohessy

AAS 67-027 Future Lunar Surface Missions, H.Wagner, A. Schaffer

AAS 67-028 Future Manned Planetary Missions, R.L. Sohn

AAS 67-029 Future Interplanetary Unmanned Space Missions, M.W. Hunter II, R.L. Moll

AAS 67-030 The Commercial Utilization of Space, A.M. Mayo

AAS 67-031 A Practical View of Future Space Activity, R.H. Lea

AAS 67-032 Some Space Goals in the Near Future from the Viewpoint of a Planetary Astronomer, C.W. Tombaugh

AAS 67-033 Surveyor Lunar Landing, S.C. Shallon

AAS 67-034 Aerodynamic Aspects of Future Space Vehicles and Implications for Testing, F.S. Nyland

AAS 67-035 Potential Requirements Imposed Upon Ranges and Networks by Future Unmanned Reentry Vehicles, H.L. Bloom

AAS 67-036 Manned Entry and Recovery - Its Possible Effects on Range Requirements, J. Putegnat

AAS 67-037 Launch Operations Requirements of Future Space Programs, J.P. Claybourne

AAS 67-038 Aerospace Application of Gun Launched Projectiles and Rockets, C.H. Murphy, G.V. Bull

AAS 67-039 World-Wide Space Traffic, J.K. Sterrett

AAS 67-040 Can Aerospace Firms Diversify?, D.C. Coddington, J.S. Gilmore

AAS 67-041 Canceled

AAS 67-042 A High Accuracy, Early-Launch Tracking System, H.H. Germond

AAS 67-043 High Accuracy Trajectory Determination, C.L. Carroll, Jr.

AAS 67-044 Data Processing Developments, H.G. Jones

AAS 67-045 Dynamical Approach to Some Problems in Range Safety, R.C. Amme

AAS 67-046 Computer Applications to the Test Scheduling Problem at the Air Force Eastern Test Range, F.F. Fennema

AAS 67-047 Spaceport Meteorology, W.L. Webb

AAS 67-048 Challenges of Range Management, L.I. Davis

AAS 67-049 Trends in Ground Support Systems for Space Flight, O.M. Covington

AAS 67-050 Canceled

AAS 67-051 Aspects of Vehicular S-Band Telemetry Antennas, D.G. Henry

AAS 67-052 Instrumentation Ships - Present and Future, T.T. Chick, R.E. Davis

AAS 67-053 Aircraft as Range Instrumentation Stations, C.G. McMullen

AAS 67-054 Onboard Data Systems and Their Relation to Ground Networks, E.C. Wood, W.H. Neuschaefer

AAS 67-055 Future Radar Developments, R.P. Cheetham

AAS 67-056 Future Support Requirements for the Manned Space Program, D.E. Fielder

AAS 67-057 OSD View of the Defense Ranges, C.J. Kronauer

AAS 67-058 The Developing Aerospace Law, M. Menter

AAS 67-059 The Management of a Modern Integrated Missile Range, G.D. Clark

AAS 67-060 Tools of the Trade, N. Wagner

AAS 67-061 Role of IRIG in Range Development, W.S. Hines

AAS 67-062 Recommended Developments for Ranges and Networks (Panel), G.W. Morgenthaler

AAS 67-063 A Quick Look at the Federal Budgetary Process, T.G. Morris

AAS 67-064 A Satellite Control Center Concept for 1970-1980, W.H. Jordon, Jr.

AAS 67-065 to AAS 67-073 Not assigned

VOL. 23 ADVANCES IN THE ASTRONAUTICAL SCIENCES, 1968
VOL. 3 AAS ASTRONAUTICS SERIES, 1968
COMMERCIAL UTILIZATION OF SPACE - AAS THIRTEENTH ANNUAL MEETING

| | |
|---|---|
| AAS 67-074 | Commercial Utilization of Space - Overview, A.M.Mayo |
| AAS 67-075 | Keynote Address, G.H. Stoner |
| AAS 67-076 | The Challenge of the Commercial Utilization of Space, L.V. Berkner |
| AAS 67-077 | NASA Applications Technology, L. Jaffe |
| AAS 67-078 | Satellite Communications - A Renewed Challenge in Resource Utilization (Abstract)*, R.C. Kirby, T. deHaas |
| AAS 67-079 | Microelectronics (Abstract)*, C.L. Hogan, S.L. Levy |
| AAS 67-080 | The Man-Machine-Environmental Interface(Abstract)*, A.H. Schwichtenberg |
| AAS 67-081 | Departures - Structures and Materials (Abstract)*, L.A. Riedinger |
| AAS 67-082 | Vehicles and Propulsion - An Overview (Abstract)*, A.B. Greenberg |
| AAS 67-083 | Commercial Utilization of Space - An Introduction, W. von Braun |
| AAS 67-084 | Future Trends in Earth Orbital Transportation, R.E. Fisher |
| AAS 67-085 | Not available |
| AAS 67-086 | Not available |
| AAS 67-087 | Estimation of Future Progress, J.M. West |
| AAS 67-088 | Space Communications - A Major Candidate for Commercial Utilization of Space, H.K. Ziegler |
| AAS 67-089 | NASA Space Applications - An Overview, R.C. Seamans, Jr. |
| AAS 67-090 | Earth Applications - An Overview, C.F. Luce |
| AAS 67-091 | Satellites and Future Communications, Including Broadcast (Abstract)*, J.L. Hult |
| AAS 67-092 | Space Communications in the 21st Century (Abstract)*, F.A. Gicca |
| AAS 67-093 | Propagation Limitations to Earth-Space Radio Communications for Frequencies Below 10 $GH_z$ (Abstract)*, C.G. Little |
| AAS 67-094 | Windows to Space (from $10^{10}$ $H_z$ up) - (Abstract)*, R. Kompfner |
| AAS 67-095 | Direct Satellite Broadcast (Abstract)*, S. Gubin |
| AAS 67-096 | Developing a National Information Network for the Space Age (Abstract)*, R.W. Hodgers, Jr. |

* In abstract in Vol. 23, Advances in the Astronautical Sciences. In full in Vol. 3, AAS Microfiche Series

| | |
|---|---|
| AAS 67-097 | Information Centers in the Space Age (Abstract)*, C.W. Sherwin |
| AAS 67-098 | Information Handling (Abstract)*, P.C. Goldmark |
| AAS 67-099 | Retention and Data Display (Abstract)*, M.S. Day |
| AAS 67-100 | Remote Access to Information (Abstract)*, J. Becker |
| AAS 67-101 | The Future Potentials of Navigation Satellites (Abstract)*, E. Ehrlich |
| AAS 67-102 | Satellite Navigational Aids for Transportation of the 1970s (Abstract)*, D.W. Roese |
| AAS 67-103 | The Use of Satellites to Track Ghost Balloons (Abstract)*. V.E. Lally |
| AAS 67-104 | Environmental Prediction Using Orbital Sensors (Abstract)*, W.O. Davis, J.P.Kuettner |
| AAS 67-105 | Meteorological Satellites and the World Weather Watch (Abstract)*, C.A. Spohn |
| AAS 67-106 | Mapping of the Land from Space (Abstract)*, F.J. Doyle |
| AAS 67-107 | The Oceans, an Overview Regarding the Commercial Utilization of Space (Abstract)*, R.E. Stevenson |
| AAS 67-108 | Remote Sensing in Agriculture and Forestry (Abstract)*, A.B. Park |
| AAS 67-109 | Applications of Space Data to Water Resources (Abstract)*, P. Bock |
| AAS 67-110 | Mineral and Land Resources (Abstract)*, W.T. Pecora |
| AAS 67-111 | Tools for the Future - Remote Sensing (Abstract)*, M.R. Holter |
| AAS 67-112 | New Tools for the Inquisitive Mind - The Industrial Research Laboratory in Space, T.S. Burkhalter, S.S. White |
| AAS 67-113 | Design of a Commercial Space Station (Abstract)*, V.D. Kirkland, R.L. Gervais |
| AAS 67-114 | Not available |
| AAS 67-115 | New Uses of Nuclear Energy in Outer Space (Abstract)*, P.Harteck, S. Dondes |
| AAS 67-116 | Local Transportation in Commercial Space Operations (Abstract)*, H.M. Graham |
| AAS 67-117 | Space Optics and Astronomy (Abstract)*, D.N. Buell, M.L. Scot |
| AAS 67-118 | The Uses of Orbiting Reflector Satellites (Abstract)* A.G. Buckingham, H.M. Watson |
| AAS 67-119 | Not available |
| AAS 67-120 | World Health Maintenance (Abstract)*, E.B. Konecci |

* In abstract in Vol. 23, Advances in the Astronautical Sciences. In full in Vol. 3, AAS Microfiche Series

AAS 67-121 Biological "Fuel Cells" - from the Electric Fish to Man (Abstract)*, H. Strughold

AAS 67-122 The Use of Space as a Research Tool (Abstract)*, J.D. Gatts

AAS 67-123 The Utilization and Engineering of an Orbital Hospital (Abstract)*, O.H. Minney

AAS 67-124 Not available

AAS 67-125 Commercial Utilization of our Space Program in its Application of Space Medical Research (Abstract)*, L.D. Proctor

AAS 67-126 Hotels in Space, B. Hilton

AAS 67-127 Space Tourism, K.A. Ehricke

AAS 67-128 Space Manufacturing, R.F. Mettler

AAS 67-129 Future Spaceport Facilities Technology and Economics, K.H. Debus

AAS 67-130 Impact of Performance Improvements and Reuse on the Cost of Space Transportation (Abstract)*, C.J. Dorrenbacher, R.J. Gunkel

AAS 67-131 Unconventional Propulsion Systems (Abstract)*, R.J. Rosa

AAS 67-132 Considerations of Nuclear Power and Propulsion (Abstract)*, J.D. Lafleur, Jr., F.C. Schwenk

AAS 67-133 An Economical Approach to Space Transportation (Abstract)*, E.N. Hall

AAS 67-134 Commercial Space Transportation Possibilities (Abstract)*, M.W. Hunter II

AAS 67-135 The First 2000 Years of the Space Age (Abstract)*, P.A. Campbell

AAS 67-136 Present Values and Future Aspirations (Abstract)*, J.E. Coppage, J.R. Tabibian

AAS 67-137 A Critique on the Application of Systems Analysis to Social Problems (Abstract)*, I.R. Hoos

AAS 67-138 Commercial Utilization of Space: Some Aspects of a Possible Future (Abstract)*, K.B. Bley, R.D. Shaver

AAS 67-139 Commercial Utilization of Space: The Economic, Political, and Cultural Impact on the World of Tomorrow (Abstract)*. J.C. Casey, E. Gardner

AAS 67-140 Man and His Environment: A Challenge to Education, S.F. Singer

AAS 67-141 Man-Rating the Educational System (Abstract)*, R.C. Flothow

AAS 67-142 Impact of Commerce in Space on Education: Technology in Education (Abstract)*, N. Hackerman

---

* In abstract in Vol. 23, Advances in the Astronautical Sciences. In full in Vol. 3, AAS Microfiche Series

AAS 67-143 Space Science and the Universities (Abstract)[*], J.C. Allred

AAS 67-144 Not available

AAS 67-145 A World Federation of Learning (Abstract)[*], D.A. Cowan

AAS 67-146 Patterns of International Space Applications and their Extension, A.W. Frutkin

AAS 67-147 Not available

AAS 67-148 International Co-operation from the UN Viewpoint (Abstract)[*], K. Waldheim

AAS 67-149 International Co-operation from the Business Viewpoint (Abstract)[*], R.W. Porter

AAS 67-150 International Co-operation in Satellite Communications (Abstract)[*], J.A. Johnson

AAS 67-151 The Challenge to Private Enterprise, G.M. Bunker

AAS 67-152 Requirements for Management of Space Enterprises (Abstract)[*], G.W. Hoover

AAS 67-153 Management of Space Enterprises (Abstract)[*], G. Kozmetsky

AAS 67-154 The Impact of Communications Satellites on the Television Industry (Abstract)[*], L.L. Johnson

AAS 67-155 Financing Space Ventures (Abstract)[*], A.H. Frady, Jr.

AAS 67-156 Commercial Utilization of Space and the Law (Abstract)[*], P.G. Dembling

AAS 67-157 Law and the Utilization of Space (Abstract)[*], R.F. Packard

AAS 67-158 Educational Implications of Commercial Utilization of Space, H.G. Stever

AAS 67-159 Economic Need and Payoff, D.A. Hicks

AAS 67-160 Commercial Exploitation of Space Technology, G.E. Mueller

AAS 67-161 Development of Needed Technology, R.L. Bisplinghoff, J.L. Sloop

AAS 67-162 Realizing the Industrial Opportunity, A.N. Stanton

AAS 67-163 Insurance for Commercial Ventures in Outer Space (Abstract)[*], C.C. Pell, Jr.

AAS 67-164 Communications in Space - Hotline to the World, D.W. Coyle

AAS 67-165 to AAS 67-167 Not assigned

---

[*] In abstract in Vol. 23, Advances in the Astronautical Sciences. In full in Vol. 3, AAS Microfiche Series

VOL. 17 SCIENCE AND TECHNOLOGY SERIES, 1968
VOL. 5 AAS MICROFICHE SERIES, 1968 (supplement to Vol. 17)
USE OF SPACE SYSTEMS FOR PLANETARY GEOLOGY AND GEOPHYSICS
(May 25-27, 1967, Boston, Mass.)

AAS 67-168 Concepts in the Application of Technology to Space Systems, H.T. McAdams

AAS 67-169 Interpretation and Application of Photometric-Polarimetric Observations of Planetary Surfaces, W.G. Egan, K.M. Foreman, E.A. Nowatski

AAS 67-170 Orbital Surveys of the Earth, W.A.Fischer

AAS 67-171 Geologic Exploration and Mapping from Space (Abstract)* M. Abdel-Gawad

AAS 67-172 Mission Support for Lunar Geoscience, J.F. McLaughlin

AAS 67-173 Cryonics and the Grand Design, R.C.W. Ettinger

AAS 67-174 Interstellar Matter, F.M. Johnson

AAS 67-175 Utilization of the Deep Space Environment by Unmanned Interstellar Probes, R.L. Forward

AAS 67-176 The Atmosphere of the Sun and Other Stars, H.Y. Chiu

AAS 67-177 The Solar Electromagnetic Radiation Environment, H.H. Malitson

AAS 67-178 Gravisphere, Gravipause, H. Strughold, O.L. Ritter

AAS 67-179 The Lithosphere - The Solid State of Matter, R.M. Haag

AAS 67-180 Automatic Checkout of Space Systems, L.T. Mast

AAS 67-181 Atmospheres and Telescopes, L.H. Narodny

AAS 67-182 A Photometric Technique for Deriving Slopes from Lunar Orbiter Photography, J.J. Lambiotte, G.R. Taylor

AAS 67-183 Seismic Signatures, J.E. Hawkins

AAS 67-184 Technological and Operational Considerations Affecting Development of Reusuable Space Transportation Systems, S. Enzer

AAS 67-185 Electro-Propulsion with Colloids, S.P. Harris, M. Farber

AAS 67-186 Bibliography of Interstellar Travel and Communication, R.L. Forward

AAS 67-187 Asteroid Belt Investigation Using Small Spin-stabilized, Fly-by Probes, R.L. Forward, L.S. Pilcher, V.T. Norwood

AAS 67-188 Solar Electric Propulsion Mission to the Comets, R.A. Park, W. Bailey III, W.D. Dickerson, D.B. Smith

AAS 67-189 Role of Manned Orbiting Laboratories in the Exploration of the Planets, W.E. Stoney, Jr., C. Covington

AAS 67-190 Signals and Noise, H.R. Raemer

AAS 67-191 Satellite High Resolution Radar Mapping Techniques, J.J. Kovaly

AAS 67-192 Analysis and Detection of Non-Recurring Phenomena, S. Himmelstein

AAS 67-193 Bistatic Cross Section of the Sea for Beaufort 5 Sea, V.W. Pidgeon

AAS 67-194 Radar Land Clutter for Small Grazing Angles at X- and L-Band, V.W. Pidgeon

AAS 67-195 Time, Frequency, and Spatial Correlation of Radar Sea Return, V.W. Pidgeon

AAS 67-196 The Role of Product Assurance in Spacecraft Sterilization, H.C. Schwartz

AAS 67-197 Interplanetary Missions and Rendezvous, A.A. Aintablian

AAS 67-198 Gravity Thrust and Interplanetary Transportation Networks, M. Minovitch

AAS 67-199 Unmanned Exploration of the Inner Planets, Comets, the Outer Planets, and Near Interstellar Space, M.W. Hunter II

AAS 67-200 Science Goals Effectiveness for Scientific Space Missions, D.A. Cohen

AAS 67-201 Lunar Surface Nuclear Experiment Package (Abstract)*, D.B. Ebeoglu

AAS 67-202 A Lunar and Planetary Petrology Experience (Abstract)*, A.A.Loomis

AAS 67-203 Conceptual Design of an Active Seismic Experiment for the Lunar Surface (Abstract)*, A Baker, V. Head, G. Richards

AAS 67-204 Rough Landing Planetary Payload (Abstract)*, D.N. Tompkins

AAS 67-205 Convolution Techniques (Abstract)*, A. DiNardo

AAS 67-206 Vehicular Design for a Planetary Atmosphere Measurement Probe (In microfiche only) P.A. Giragosian, E.T. Myers

AAS 67-207 Mission Reliability for Planetary Surveys (In microfiche only) A.H. Hevesh

AAS 67-208 Trajectory Simulation Methods for Preliminary Interplanetary Mission Planning (In microfiche only), S.W. Wilson, Jr. J.J. Miller

---

* Abstract in Vol. 17 Science and Technology. Complete paper in Vol. 5 AAS Microfiche Series

VOL. 10, AAS MICROFICHE SERIES, 1969
ROCKY MOUNTAIN RESOURCES FOR AEROSPACE SCIENCE AND TECHNOLOGY
(AAS Rocky Mountain Section Symposium, July 13-14, 1967, Denver, Colo.)

| | |
|---|---|
| AAS 67-209 | The Role of Government in the Development of Rocky Mountain Resources for Aerospace Science and Technology, D.G. Brotzman |
| AAS 67-210 | The Role of Industry, A.J. Kullas |
| AAS 67-211 | The NORAD Space Detection and Tracking System, T.E. Sledge, Jr. |
| AAS 67-212 | Keeping Score in Space--USAF Spacetrack, H.F. Duncan, J.J. Duncan, H.R. Headley |
| AAS 67-213 | A Survey of the Operational Problems Associated with Earth Oriented Satellite Systems, D.B. Cross |
| AAS 67-214 | An Application of Finite Difference to Orbital Motion, P.J. Short, F.J. Haugt |
| AAS 67-215 | A Digital Simulation of Attitude Dynamics of a Spin Stabilized Spacecraft, T.W. Spencer |
| AAS 67-216 | The Force XXIV Degradation and Capability Simulation Program, R. Dohrmann, P. McKowen |
| AAS 67-217 | The Dual Redundant Concept of Cumputer Operation, R. Carney, G. Lovell |
| AAS 67-218 | Effects of Subsystem Orientations on Communication Capability and Evidential Meaning, C. Halterman, G.P. Foster |
| AAS 67-219 | The Effects of Sterilization on Long Life Reliability, A.H. Kent, J.W. Carter |
| AAS 67-220 | An Analysis for Sterilization Modeling, J.P. Brannen |
| AAS 67-221 | Heat Sterilization -- Its Effect on Space Flight Hardware, M.R. Mesnard |
| AAS 67-222 | Integrated Approach to Biological Analysis and Heat Sealable Plastic Methodology Applied to the Sterile Insertion Concept, J.C. Arnett, R.D. Knight, E.R. Jones |
| AAS 67-223 | A New Approach to the Microbiological Sampling of Surfaces: The Vacuum Probe Sampler, V.L. Dugan, W.J. Whitfield, J.J. McDade, J.W. Beakley, F.W. Oswalt |
| AAS 67-224 | Electronic Packaging Design for Sterilization, R.F. Peluso |
| AAS 67-225 | Sterilizable Batteries, C.W. Brooke, Jr. |
| AAS 67-226 | A Procedure to Study the Ultrastructural and Cytochemical Alterations Produced by Weightlessness in the Protozoan, R.W. Price |
| AAS 67-227 | Lunar Mapping with Surveyor Television Pictures (Abstract), R.M. Batson |
| AAS 67-228 | Lunar Photometric Observations through the Surveyor Television System (Abstract), H.E. Holt |

AAS 67-229 Statistical Studies of Craters and Fragmental Material on the Lunar Surface with Surveyor Television Pictures (Abstract), E.C. Morris

AAS 67-230 Research of the Lunar Surface by Analysis or Backscattered Radar Signals, P. Beckmann

AAS 67-231 A Lunar Seismic Excitation Source, M.W. Burnham, D.K. Parks

AAS 67-232 Mineral Analysis of the Lunar Surface by Luminescence, J.B. Blizard

AAS 67-233 A Method of Determining Some Geological Parameters of the Surface of Planets, V.K. Smith

AAS 67-234 Role of the Geologist in Advanced Mission Analysis, C.L. Fair

AAS 67-235 The Search for Extraterrestrial Life, W.W. Bender

AAS 67-236 Heat Exchanger Design for Rocket-Borne Cryogenic Air Samplers, R.B. Jacobs

AAS 67-237 Instrumentation for Atmospheric Density and Composition Measurements Above 100 Kilometers, R.O. Woods

AAS 67-238 Results of the Uncertainty in H: of CN on Shocked $CO_2$-$N_2$-A, W.F. Rumpel, H.J. Deacon, Jr.

AAS 67-239 A Plasma Frequency Probe for High Resolution Electron Density Measurements, E.F. Pound, D.A. Burt

AAS 67-240 Antenna Breakdown in Martian Atmosphere, C.H. Brockmeyer

AAS 67-241 Basic Concepts of Synthetic Antenna Radar, G.L. LaPrade

AAS 67-242 Proposed Magnetic Conjugate Point Studies by Synchronous Satellites, J.H. Pope

AAS 67-243 Statistical Determination of Planetary Atmospheres, D.G. Stechert

AAS 67-244 Turbulent Mixing Layers in Modern Flight, R.J. Heyman, P.C. Steel, R.J. Sanderson

AAS 67-245 Nuclear Instrumentation for Nuclear Rocket Engine Testing, R.T. Perry, Jr.

AAS 67-246 Modern Control Applications to Nuclear Propulsion, R.R. Mohle

AAS 67-247 A Lunar Seismic Excitation Source, M.W. Burnham, S.J. Marshal

AAS 67-248 The Electrodynamic Acceleration of A Plasma Considering the Effects of Resistance and Mass Change, J.M. Murphy

AAS 67-249 Electric Propulsion for Satellites, W.R. Mickelsen

AAS 67-250 Electric Propulsion for Interplanetary Spacecraft, W.R. Mickelsen

AAS 67-251 Data Compression Concepts and Philosophies, W.A. Stevens, R.W. McCoy

AAS 67-252 Storage Media and Data Compaction for Space Application, K.A. Endsley

AAS 67-253 Self-Consistent Solutions of the Inhomogeneous Wave Equation in Shock Layer Plasma-Radio Frequency Field Interaction, C.A. Andrade

AAS 67-254 Coding for Deep Space Communication, R.B. Blizard, J.L. McKinney

AAS 67-255 Minimum Power - Low Bit Rate Communications from Unmanned Deep Space Probes, R.B. Knox

AAS 67-256 Transmission Medium Coherency Bandwidth Measurements, R.J. Richardson

AAS 67-257 The Magnetic Tape Recorder as a Factor in Data Analysis, L.J. Girard

AAS 67-258 Transient Spherical Waves, K.D. Granzow

AAS 67-259 Antenna Problems Related to Rocket and Spacecraft Experimentation (Abstract), H.W. Haas

AAS 67-260 Microelectronics in Spacecraft Subsystem, W.A. Fiedler

AAS 67-261 Command System for Space Applications, F. Hull, D. Leck

AAS 67-262 Pointing Reference System White Light Coronagraph Experiment, C.E. Rose

AAS 67-263 Investigation of a Deorbit Attitude Control System for a Planetary Entry Vehicle via Optimal Control Techniques, R.F. Webber, R.S. Jackson, W.C. Freested

AAS 67-264 The Optimum Design of Magnetic Torquing Coils, D. Sonnabend

AAS 67-265 Sandia De-Spun Attitude Control System, L.M. Bossart

AAS 67-266 Nutation Damping for Dual-Spin Spacecraft, T.M. Spencer

AAS 67-267 The Analysis of the Outer Loop of the Reaction Wheel Control, H.F. Zimbelman

AAS 67-268 Landing Stability Studies for Planetary Exploration Vehicles, I.J. Jaszlics, J.O. Dow

AAS 67-269 Spherical Planetary Impacter Using Crushable Materials -- Analysis and Experiment, F.A. Penning

AAS 67-270 Implementation of a Soft-Landing Test Program for Planetary Soft-Lander Probes, M.R. Mesnard

AAS 67-271 Automatic Optimum Design of Hyperstatic Trusses, E. Rich

AAS 67-272 Perforated Multilayer Insulation, R.K. McMordie

AAS 67-273 Testing of Supercritical Helium Pressurization System for Manned Spacecraft, W.E. Parente, R.E. Boyland

AAS 67-274 The Development of Titanium and Inconel Cryogenic Pressure Vessels, R.J. Balthazer, H.E. Sutton

AAS 67-275 New Concepts in Fuel Containment, W.C. McDonald

AAS 67-276 Metal - Organic Acid Corrosion and the Effect of Corrosion Products on Microorganisms, E.E. Staffeldt, O.H. Calderon

AAS 67-277 Dynamic, Thermoelectric, and Thermionic Space Power Systems, A.A. Sorensen

AAS 67-278 A Review of Rankine Cycle Power Systems for Space Applications, R.E. Barber

AAS 67-279 Turbomachinery for Space Applications, K.E. Nichols, B. Nichol

AAS 67-280 Design of a Facility for Testing Hydrogen Oxygen Fuel Cells, M.E. Slee, M.S. Imamura

AAS 67-281 Engineering Aspects of Silicon Solar Cells, M.S. Imamura, W.B. Collins

AAS 67-282 Design Consideration of a Silicon Solar Cell Panel, J.J. Leuthard, M.S. Imamura

AAS 67-283 Outdoor Solar Array Test Facility, J.H. Martin

AAS 67-284 Development of a Light Source for Laboratory Testing of Solar Cell Panels, D. Waddington

AAS 67-285 to AAS 67-300 Not assigned

VOL. 4, AAS MICROFICHE SERIES, 1968
SATURN V / APOLLO AND BEYOND
(National Meeting, June 11-14, 1967, Huntsville, Alabama)

AAS 67-301 /- 302 Not available

AAS 67-303 Management Lessons of the Past - Applications of the Future, J.L. Bowers

AAS 67-304 A Crossroad in Space?, C.J. Burton

AAS 67-305 The Saturn Instrument Unit, A.E. Cooper

AAS 67-306 /- 307 Not available

AAS 67-308 At the Threshold, H.M. Graham

AAS 67-309 Command Control and Communications for the National Space Program, H. Harris

AAS 67-310 /- 311 Not available

AAS 67-312 Planning for the Nuclear Space Age, W.E. Johnson

AAS 67-313 /- 314 Not available

AAS 67-315 F-1, J-2 and Advanced Rocket Engine Development, J.P. McNamara

AAS 67-316 Not available

AAS 67-317 Graduate Work at Huntsville, E. Rodgers

AAS 67-317A Application of Nerva, Snap 8, and the 260" Solid Systems, C.C. Ross

AAS 67-318 Spacecraft Development in the Next Decade, A.K. Thiel

AAS 67-319 /- 320 Not available

AAS 67-321 An Earth-Reference Lunar Gravimeter, G.C.Henderson R.M. Iverson

AAS 67-322 An Hypothesis on the Physiological Effects of the Absence of Electromagnetic Noise Energy During Deep Space Flight, E.R. Graf, F.E. Cole, C.D. Weathers, R.J. Sims, P. Johnson

AAS 67-323 Life Support For Lunar Exploration, G.M. Freedman

AAS 67-324 A New Approach to Estimating the Effect of the Meteoroid Hazard on Spacecraft Reliability, A.B. Deyarmond, G.E. Ingram

AAS 67-325 A Construction of Probability Envelopes of Flux-Energy Spectra, T.S. Dollman

AAS 67-326 Study of the Lunar Thermal Environment, F.B. Tatom, K.V. Ramakrishna, O.H. Vaughan

AAS 67-327 The Environment of Mercury, O.H. Vaughan, Jr.,K. Weidner

AAS 67-328 The Problem of Secondary Ejecta Near the Lunar Surface, H.A. Zook

AAS 67-329 A New Perspective on Martian Polarimetric Measurements, W.G. Egan, K.M. Foreman

AAS 67-330 Thermoluminescence of the Moon, J.B. Blizard, J.D. Plunkett

AAS 67-331 The Role of the Bell Aerospace Company in the Saturn V / Apollo Program and Beyond, R.M. Hurst

AAS 67-332 Not available

AAS 67-333 Interplanetary Meteoroid Environment and Hazard, D.J. Kessler

AAS 67-334 Near-Earth and Interplanetary Meteoroid Flux and Puncture Models, C.C. Dalton

AAS 67-335 Evidence for Recoverable Water at the Surface of the Moon, D.P. Elston

AAS 67-336 /- 339 Not available

AAS 67-340 Space is the Hole we are in, J.B. Mendaris

AAS 67-341 Configuration Selection and Perturbation Effects for a Large Orbiting Solar Array, J.A. Rebman

AAS 67-342 An Explicit Solution for the Lagrange Multipliers on the Singular Subarc of an Optimal Trajectory, R.E. Burns

AAS 67-343 The Secular Perturbations of the Orbit of a Satellite due to the Oblateness of the Central Body for the m-th Zonal Harmonics and due to the Gravitational Attraction of a Third Body, H.G.L. Krause

AAS 67-344 A Study of the Velocity Requirements for Aborting Manned Missions to the Planets Mars and Venus between 1977 and 1986, A.G. Mona

AAS 67-345 The Effect of Finite Thrust on the Synergetic Plane Change, L.G. Tramonti

AAS 67-346 Simultaneous Integration of N-Bodies by Analytic Continuation with Recursively Formed Derivatives, J.G. Hartwell

AAS 67-347 The Librational Motion of a Satellite with Large Panels, C.P. Liu

AAS 67-348 The Optimal Solution of Trajectory Problems Consisting of Several Extremal Subarcs by the Generalized Newton-Raphson Algorithms, M.S. O'Mahony, C.D. Eskridge

AAS 67-349 Minimum-Time Transfers Between Coplanar Circular Orbits under a Propellant Constraint, D.A. Davis, A.G. Bennett

AAS 67-350 Conceptual Study of a Flyby Probe for Determining Planetary Atmospheres, D.G. Stechert

AAS 67-351 The Tangential Thrust Hyperbolic Spiral: A Unique Gravitational Escape Trajectory, J.A. Jungmann

AAS 67-352 The Effect of Linearity Assumptions and Unknown Noise Sources on Kalman Filtering in Orbit Determination, W. O'Dwyer, J. Mendelsohn, A. Kaercher

AAS 67-353 Some Observations of Ballistic Anomalies and Spin Stabilization, W.G. Andrews, D.R. Reed

AAS 67-354 Astronautical Charts and Optical Devices in Space Travel, /A. Nyman

AAS 67-355 /- 360 Not available

AAS 67-361 Evolutionary Steps in S-IVB Development, T.J. Gordon

AAS 67-362 Boeing -- Apollo and Beyond, R.J. Helberg

AAS 67-363 Not available

AAS 67-364 Revisit and Reuse in Apollo Applications, R.I. Schneyer

AAS 67-365 Uprated Saturn I Program, V.J. Vehko

AAS 67-366 Not available

AAS 67-367 Martin-Marietta's Role in Space, R.S. Williams

AAS 67-368 Not available

AAS 67-369 Saturn Scientific Payloads, B.M. McCormac

AAS 67-370 Post Apollo Lunar Program Synthesis, J.O. Matzenauer

AAS 67-371 Impulsive Velocity Requirements for Aborting Mars Stopover Missions While in the Martian Sphere of Influence During the 1978-1986 Time Period, C.W. Mead, B.G. Noblitt

AAS 67-372 Manned Planetary System Comparisons Weight and Cost, V.E. Godsey, W.E. Johnson, C.P. Martin

AAS 67-373 Not available

AAS 67-374 Alternate Manned Planetary Flyby Missions, R.G. Riedesel, J.L. Woodworth

AAS 67-375 A Building Block Approach to Saturn V Uprating Using Solid Propellant Rocket Motors, J.W. Monroe, C.J. Corso

AAS 67-376 Orbital Launch Vehicles for the 1970's, L.G. Lane

AAS 67-377 A Nuclear Space Plane Concept, G.J. Schoessew, S. Smith

AAS 67-378 Decision Analysis of Space Projects: Voyager Mars, J.E. Matheson, W.J. Roths

AAS 67-379 Computerized Systems Engineering Aids for Program Management, P.T. McCuiston, Jr.

AAS 67-380 /- 381 Not available

AAS 67-382 An Integrated National Space Program Based on a Single Mission Mode, R.R. Titus

AAS 67-383 Orbital Remote Sensors Data Processing for Geoscience Investigations, G.W. Dalke, J.W. Rouse, Jr.

AAS 67-384 Mission Risk Factors - The Bio Radiological Component, G.M. Angleton

AAS 67-385 Dynamics of the Orbital Centrifuge, R.R. Reed

AAS 67-386 Interplanetary Trajectory Selection Criteria for Voyager Mars Missions, W. Stavro

AAS 67-387 Satellite Orbit Selection Criteria for Voyager Mars Missions, D.A. Tito

AAS 67-388 Synthesis of Trajectory and Capsule Design Parameters for a Venus Spacecraft-Flyby/Capsule-Entry Probe Mission, C.H. Bursey, Jr.

AAS 67-389 Not available

AAS 67-390 The Voyager Space Craft: Mission Optimization via the Selection of Spacecraft Redundancy, L.E. Hargrave, Jr., E.L. Berger

AAS 67-391 Application of Apollo Hardware Unmanned Mars Missions, B. Karen

AAS 67-392 Space Station Constellation Concept, M.F. Clark

AAS 67-393 Meteoroid Shielding Requirements of the Post-Apollo Command Module Heat Shield, O.K. Salmassy, D.M. Moodie, Jr., R.W. Billing, Jr., R. Ames

AAS 67-394 /- 399 Not available

AAS 67-400 Orbital Rescue, J.H. Bell, Jr., E.C. Zewe

AAS 67-401 Mariner Mars 1969 Inflight Guidance Equations, J.B. Shaul

AAS 67-402 /- 410 Not available

AAS 67-411 Design of a Multiple Start Mid-Course Correction Propulsion System for Unmanned Planetary Spacecraft, A.D. Cohen, E.H. Youmans

AAS 67-412 A Survey of Material and Structural Damping, E.J. Rodgers, J.D. Warrington

AAS 67-413 Saturn V Dynamic Test Program, N.A. Seaholm, W.R. Covington

AAS 67-414 POGO Stability and Sensitivity for the Saturn V Second Flight-Stage, B.W. Licht, A.C. Park

AAS 67-415 A Mechanical Impedance Prediction Technique, C.M. Pearson, J.R. Holms

AAS 67-416 Structural Analysis of Aerodynamic Deceleration Systems by the Finite Element Method, J.T. Oden, T. Sato

AAS 67-417 On a Finite Beam Resting on a Viscoelastic Foundation with Intersurface Friction, E.C. Ma

AAS 67-418 Vibration and Stability of Rockets, C.L. Huang, H.S. Walker

AAS 67-419 Mechanical Signature Analysis, Listening for the Future, T.F. Smyth, J.R. Gottshall, L.F. Sturgeon

AAS 67-420 A Numerical Analysis of Orthotropic Shells of Revolution, C.L. Yen

AAS 67-421 The Nonlinear Elastic-Plastic Rotationally Symmetric Dynamic Response of Spherical Shells, J.C. Simonis, C.E. Stoneking

AAS 67-422 Not available

AAS 67-423 Research Investigation of Bulkhead Cylindrical Junctions Exposed to Combined Load and Internal Pressure, C.A. Sciammarella, H. Wagner

AAS 67-424 Non-linear Analysis of Truss Frames, B.L. Gunnin, R.W. Furlong

AAS 67-425 Not available

AAS 67-426 Propellant Reliquefaction for a Nuclear-Propelled Manned Mars Stopover Vehicle, D.V. Merrifield, W.G. Offik, D.H. Ogden

AAS 67-427 Multiple-Purpose Solar-Powered Electric-Propulsion Spacecraft for Unmanned Interplanetary Missions, J.H. Molitor, R.N. Olson

AAS 67-428 Variation of Vacuum Thermal Contact Conductance with Time Duration Contact Pressure History, J.M. Trummel, J.C. Maerschalk, W.R. Stubstad

AAS 67-429 /- 430 Not available

AAS 67-431 Thermal Management of Reaction-Control Rocket Engine Systems, R.C. Stechman

AAS 67-432 An Application of Solar-Electric Power to a Sophisticated Planetary Surveillance System, L.H. Wood, N. Stone, W. Fleck

AAS 67-433 The Entrainment of Confined Liquids by Air Streams in Partially Filled Piping Systems, E.H. Bishop, R.E. Forbes

AAS 67-434 /- 435 Not available

AAS 67-436 Numerical Formulation of a Class of Problems in Nonlinear Viscoelasticity, J.T. Oden

AAS 67-437 Elastic and Plastic Response of Spherical Shells to Axisymetric Time-Dependent Loads, J.C. Simonis, C.E. Stoneking

AAS 67-438 Dynamic Analysis of Large Complex Space Systems, Y.Y. Yu

AAS 67-439 A Nonlinear Theory of Alkali Metals Based on Quantum Mechanics, G.M. Arkilic, J. Eftis

AAS 67-440 /-441 Not available

AAS 67-442 On the Deflections of an Orthotropic Homogeneous Plate, C.E.S. Ueng

AAS 67-443 An Analysis of Venting Gas into an Outside Supersonic Stream, S.C. Zien

AAS 67-444 Not available

AAS 67-445 Experimental Study of Liquid Motion in a Two-Phase System During Pressurization, F.P. Kirkland, R.J. Schoenhals, E.R.F. Winter

AAS 67-446 Dimensional Techniques Applied to Propellant Pressurization, Stratification, and Interface Stability, J.R. O'Loughlin

AAS 67-447 Transient Characteristics of a Gas Discharging from a Vessel, L.J. Havard, Jr., O.A. Arnas

AAS 67-448 /- 450 Not available

AAS 67-451 Use of Stellar References in Space Navigation, J.C. Martin, K.R. Dunipace, L.S. Miller

AAS 67-452 Manual Approach to Space Navigation in Planetary Orbit, R.M. Silva

AAS 67-453 Planetary Approach Guidance for a Martian Mission, L.T. Seaman, H.E. Brown

AAS 67-454 Powered Landing Guidance-and-Navigation Systems for LM, B.A. Kriegsman, D.E. Gustafson

AAS 67-455 A Technique for Stabilizing the ATM Vehicle for Extended Time Periods by Use of Momentum Exchange Devices, C.R. Hubbard

AAS 67-456 RC Transfer-Function Synthesis and Digital Computer Application, J.W. Fontenot, D.H. Townsend

AAS 67-457 Use of Simulation Techniques to Remove Test-Peculiar Nonlinear Effects from Control System Response, C.T. Hagstrom, H.S. Curtis

AAS 67-458 On an Automatic Magnetic Attitude Control Scheme for a Wheel-Barrow Spacecraft, P.R. Dahl, E. Foxman

AAS 67-459 Computer Optimization of Control Moment Gyro Design, L.P. Davis

AAS 67-460 Not available

AAS 67-461 On the Trajectory Sensitivity of Linear Optimal Control Systems, J.Y.S. Luh, M.J. Wozny

AAS 67-462 On the Problem of On-Board Fast Generation of Optimal and Homing Guidance Signals to Distant Celestial Bodies Near or Beyond the Edge of Solar System, S.S. Hu

AAS 67-463 Approximation of Lagrange Multipliers and Determination of Optimum Trans-Lunar Flight Programs, J.F. Andrus

AAS 67-464 Evaluation of a Nonlinear Slosh Model, J.R. Roberts

AAS 67-465 /- 470 Not available

AAS 67-471 Analysis and Criteria for Operation of a Heat Sink for Space Application, E.R.F. Winter, C.C. Oliver, T.G. Daehler

AAS 67-472 Radiative-Convective Coupling with Self-Absorption and Variable Absorption Coefficients in the Boundary Layer, C.C. Oliver, R.W. Crossley

AAS 67-473 Not available

AAS 67-474 An Implicit Numerical Solution for Two- and Three-Dimensional Heat Conduction in a Solid Body, C.P. Williamson

AAS 67-475 /- 476 Not available

AAS 67-477 Forces and Moments on Cylinders and Cones in a Two-Phase Plume, B.K. Adler, P.M. Schwartz, K.D. Korkan

AAS 67-478 Aerothermodynamic Considerations in Ballistic/Lifting Entry Vehicle Design, P.A. Giragosian, E.T. Myers, P.D. Sones

AAS 67-479 An Analytical Method for Estimating Heating Rates and Skin Friction on Surfaces Exposed to Separated or Wake Flows, E.T. Myers, J.D. Irwin

AAS 67-480 A Monte Carlo Technique for Determining Free Molecule and Transition Aerodynamic Characteristics, S.J. Robertson

AAS 67-481 Not available

AAS 67-482 An Analytical Study of Thermal Stratification of Propellants in Partially Filled Spherical Containers in Reduced Gravity Fields, D.O. Barnett

AAS 67-483 Two- and Three-Dimensional Boundary Layer Analysis and Highly Cooled Bodies, E.T.Myers

AAS 67-484 Not available

AAS 67-485 Aerodynamic Survey of Lifting Planetary Entry Vehicles, P.A. Girogosian

AAS 67-486 /-490 Not available

AAS 67-491 Design and Integration Study of an ATG Powered Voyager Spacecraft, A.S. Kirpich

AAS 67-492 The Effect of Coefficient Variations on the Solution of Linear Systems, L.J. Feeser, C.C. Feng

AAS 67-493 Techniques for Analyzing Unified Communication System Performance, R. Shreve

AAS 67-494 Simultaneous Solution of Nonlinear Algebraic Equations, F.B. Tatom, T. Thomas, R. Schroeder

AAS 67-495 Error-Control Coding for Inter-Planetary Spacecraft, G. Huffman, R. Pahmeier

AAS 67-496 Communications Systems for Manned Interplanetary Explorations S. Riter, R.D. Shelton

AAS 67-497 On a Solution Algorithm for Matrix Graphs, T.E. Kirchgessner, C.C. Feng, L.J. Feeser

AAS 67-498 Landing Dynamics of a Legend Planetary Exploration Vehicle, I.J. Jaszlics, J.O. Dow

AAS 67-499 Iterated Series Inversions for Computing Roots of Polynomials, M.L. Thompson

AAS 67-500 Formulation of Lunar Scientific Programs, R.H. Olds, Sr., R. Ehrlich, W. Crosmer

AAS 67-501 ASSET, An Engineer's Programming Tool for Structural Analysis, M.E. McCoy

AAS 67-502 Space Radiation Tolerant Polymeric Materials, C.L. Brock

AAS 67-503 Deflagration of Materials in Simulated Space Cabin Environments, G.T. Keys

AAS 67-504 Thermal Endurance Testing of the LEM Rendezvous Radar, F.W. Gorman

AAS 67-505 Soil-Spacecraft Interaction During Impact, D.R. Reichmuth

AAS 67-506 Crossed Beam Remote Sensing Technique Applied to Microwave Range Using Atmospheric Water-Vapor Microwave Spectra, T.S. Chang

AAS 67-507 Preliminary Design Study of a Lunar Gravity Simulator, R.B. Wysor

AAS 67-508 Measurement Techniques for the Determination of Thermal Radiation Characteristics of Lunar Material, R.C. Birkebak, C.J. Cremers

AAS 67-509 Comments on the Communication and Data Problems Associated with a Space Trip on a Conjunction Phase, J.C. Hoagland

AAS 67-510 Procedure for Compatibility Analysis and Scheduling of Experiments in Earth (Planetary) Orbit, O.T. Guffin, J.M. Duffie, Jr.

AAS 67-511 Early Orbital Space Station Systems, T.J. Cogdill, C.R. Darwin

AAS 67-512 The Air Force Manual Space Navigation Experiment on Apollo with Results and Lessons from Gemini Experiment D-9, T.R. Jorris, A.E. Barth, E.M. Vallerie III, R.M. Silva

AAS 67-513 Remote Sensing for Agriculture and Natural Resources from Space, R.B. MacDonald, D.A. Landgrebe

AAS 67-514 Thermal Problems Associated with Voyager Missions, M.J. Rosenberg, C. Lankton

AAS 67-515 The Adjection Discharge Phenomenon, J.J. Ehrlich, J.C. Bryner, W.J. Coleman

AAS 67-516 Not available

AAS 67-517 Design, Development, and Fabrication of a Portable Spectrometer-Type Monitor, E.L. Grove, W.A. Loseke, E.S. Gordon

AAS 67-518 /- 520 Not available

AAS 67-521 The Energy Wave Hypothesis, N. Engel

AAS 67-522 /- 523 Not available

AAS 67-524 A New Criterion in the Quest for Life in Our Solar System, E.R. Graf, F.E. Cole, R.W. Cole, R.J. Coleman, M.D. Fahey, R.A. Heaton, P. Johnson, F.R. McDivitt

VOL. 6 AAS MICROFICHE SERIES, 1969
ASTRONAUTICS INTERNATIONAL
(November 27-29, 1967, New York, N.Y.)

AAS 67-525 Not available

AAS 67-526 Patterns of International Cooperation in Space, A.W. Frutkin

AAS 67-527 Not available

AAS 67-528 A Single Global Communications System, K. Corrigan

AAS 67-529 Navigation Satellites, H.R. Gordinier

AAS 67-530 Application of Space Technology to Meteorology, P. Morel

AAS 67-531 Practical Uses of Space Technology - Agriculture and Forestry, A.B. Park

AAS 67-532 ESRO Programs, J.R. Bertrand

AAS 67-533A United Kingdom Space Programme, Part 1: Rockets, Boosters, Satellites, G.K.C. Pardoe

AAS 67-533B United Kingdom Space Programme, Part 2: Ground Stations and Equipment, G.K.C. Pardoe

AAS 67-534 Not Available

AAS 67-535 Space Systems Design: Unmanned Space Programs - Canadian Programs, R.S. Rettie

AAS 67-536 Not Available

AAS 67-537 U.S. Civilian Space Exploration and Application with Automated Spacecraft, L. Jaffe

AAS 67-538 Space Astronomy and Astrophysics, F.L.Whipple

AAS 67-539 Space Vehicles and Their Use in Oceanographic Research, A.G. Alexiou

AAS 67-540 /- 543 Not Available

AAS 67-544 Launch Vehicles: ELDO Programmes, W.M. Stevens

AAS 67-545 Not Available

AAS 67-546 Manned Space Flight from Mercury to Apollo, R.R.Gilruth, W.E. Stoney, Jr.

AAS 67-547 U.S. Launch Vehicle Programs, V.L. Johnson

AAS 67-548 Not Available

AAS 67-549 Apollo Applications -- A Progress Report, C.W. Mathews

AAS 67-550 /- 574 Not Assigned

VOL. 19 SCIENCE AND TECHNOLOGY SERIES, 1969
EXOBIOLOGY - THE SEARCH FOR EXTRATERRESTRIAL LIFE
(AAS/AAAS Symposium, December 30, 1967, New York, N.Y.)

AAS 67-575 Frontiers in Solar System Exobiology, C. Sagan

AAS 67-576 Automated Microbial Metabolism Laboratory, G.V. Levin

AAS 67-577 Life Detection Techniques, R.S. Young

AAS 67-578 Techniques of Interstellar Communication, E.M. Hafner

AAS 67-579 The Origin of Microbial Life on Earth and its Implications for Extra-Terrestrial Forms, H.D. Isenberg

AAS 67-580 Biochemical Bases for Life in Extra-Terrestrial Environments, W. Vishniac

AAS 67-581 Sociological Aspects of Exobiology, B.M. Wagner

AAS 67-582 Panel Discussion, A. Szutka

AAS 67-583 Panel Discussion, C. Sagan

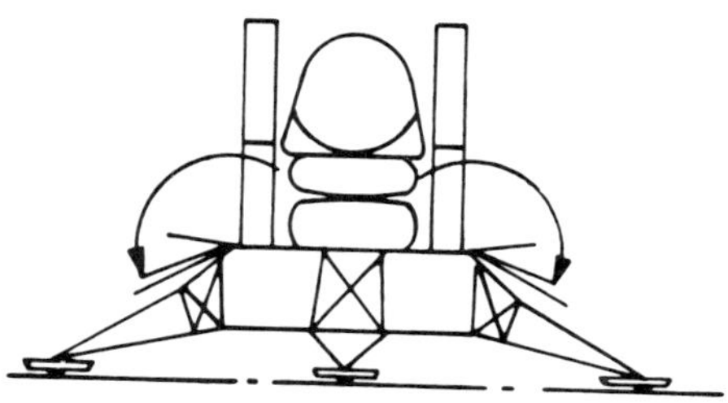

# AAS TECHNICAL PAPERS
# 1968

ORBITAL RESONANCE
(January 15-16, 1968, Redondo Beach, California)
Chairman: G.S. Gedeon, TRW Systems
No proceedings volume available

AAS 68-001 to 68-013 Not available

VOL. 18 SCIENCE AND TECHNOLOGY
TECHNOLOGY AND SOCIAL PROGRESS - SYNERGISM OR CONFLICT?, 1969
(Sixth Goddard Memorial Symposium, Mar. 12-13, 1968, Washington, D.C.)

| | |
|---|---|
| AAS 68-014 | Keynote Address, F. Pohl |
| AAS 68-015 | The Dimensions of Progress in Historical Perspective, A. Kramish |
| AAS 68-016 | Big Technology, the Technology Gap, and a Dangerous Policy Pitfall, R.R. Nelson |
| AAS 68-017 | Direct and Indirect Effects of Large Technology Programs, L.S. Silk |
| AAS 68-018 | Long-Term Impacts of Big Technology, M.L. Weidenbaum |
| AAS 68-019 | Space as an Innovative Technology, D.J. Fink |
| AAS 68-020 | Not available |
| AAS 68-021 | National Programs and the Progress of Technological Societies, T.J. Gordon, A.L. Shef |
| AAS 68-022 | The European Outlook, L.G. Napolitano |
| AAS 68-023 | Designing a Space Program, R.L. Bisplinghoff |
| AAS 68-024 | Technology of Social Progress, J.E. Karth |
| AAS 68-025 | Opening Remarks, B.A. Schriever |



VOL. 25 SCIENCE AND TECHNOLOGY SERIES
GEOLOGICAL PROBLEMS IN LUNAR AND PLANETARY RESEARCH, 1971
WATER RESOURCES OF THE INNERPLANETS
(April 10-12, 1968, Las Vegas, Nevada)
(See also paper sequences AAS 68-300 to -325, AAS 69-001 to -019, and AAS 69-670 to -685 also included in this volume.)

| | |
|---|---|
| AAS 68-026 | Not available |
| AAS 68-027 | Geological Habitat of Water, Carbon, and Rare Gases on the Moon, D.P. Elston |
| AAS 68-028 | Not available |
| AAS 68-029 | Simulation of Lunar Permafrost, J. Bruman |
| AAS 68-030 | Water on Mars, N.R. Mukherjee |

AAS 68-031 to AAS 68-033 Not available

AAS 68-034 Comments on the Origin and Surficial Aspects of Hydrothermal Alteration, P.F. Kerr

AAS 68-035 Not available

AAS 68-036 An Optimization Criterion for Multi-Spectral Photography, W.M. Adams

AAS 68-037 The Water Resources Program of the United States Department of the Interior, S.F. Singer

AAS 68-038 Remote Detection of Water-bearing Minerals, J.W. Salisbury, G.R. Hunt

AAS 68-039 Not available

AAS 68-040 A Nuclear Magnetic Resonance Spectrometer for the Measurement of the Moisture Content of Soils and Rocks, D.D. Elleman, S.L. Manatt

AAS 68-041 Rapid Measurement of Sub-Surface Moisture and Frost by Landed Planetary Probes and Surface Vehicles, J.C. Cook

AAS 68-042 Water Analysis by a Combination Neutron Experiment, W.W. Mills, W.W. Givens, R.L. Caldwell

AAS 68-043 Electromagnetic Measurement of Lunar Subsurface Water, L.I. Linlor, S.H. Ward, G.R. Jiracek

AAS 68-044 Electromagnetic Depth Sounding on Lunar Traverses, G.W. Hohmann, S.H. Ward

AAS 68-045 / -046 Not available

AAS 68-047 Water Resources of the Inner Planets, G.N. Katterfeld, V.F. Dörpholz

AAS 68-048 Not available

VOL. 24 ADVANCES IN THE ASTRONAUTICAL SCIENCES
EXPLOITATION OF SPACE FOR EXPERIMENTAL RESEARCH, 1969
FOURTEENTH ANNUAL MEETING
(May 13-15, 1968, Dedham, Mass.)

AAS 68-049 Introductory Remarks, L. Larmore

AAS 68-050 Scientific/Technical Experiment Analysis for Deriving Requirements for Astronaut Functions, T.M. Cagle

AAS 68-051 The Space Research Centrifuge Design and Evolution of Experiments, B.D. Newsom

AAS 68-052 A Proposed Experimental Method for Determining the Existence of other Planetary Systems, E.R. Graf, F.A. Ford, M.D. Fahey, R.J. Coleman

AAS 68-053 Free-Fall Experiments to Determine the Newtonian Gravitational Constant (g), D. Berman, R.L. Forward

AAS 68-054 Not Available

AAS 68-055 The Use of Nimbus Satellites for Measurements of Atmospheric Ozone over Polar Areas, J.A. Duardo, R.E. Hill, C.E. Palmer, S.V. Venkateswaran

AAS 68-056 The Application of Radar Scatterometry to Ocean Physics W.L. Floyd

AAS 68-057 Soil Temperature Studies in a Simulated Martian Substrate Environment, L.A. Wilson

AAS 68-058 A Potential Orbital Facility for Soft X-Ray Astronomy, M.C. Nilson

AAS 68-059 A Report on the Orbital Astronomy Support Facility (OASF) Study, T.J. Gordon, H.L. Wolbers

AAS 68-060 On the Observation of Stars, Galaxies, and Quasars from a Space Observatory, H.D. Greyber

AAS 68-061 Free-Fall Interstellar Gravitational Radiation Antennas (Abstract), R.L. Forward

AAS 68-062 Format Instructions for Stabilization of Manned Space Telescopes, A. Nyman

AAS 68-063 Use of Heuristic Timeline Computer Program to Derive Manned Space-Station Requirements, F.W. Burns, D.S. Edmonds, G.A. Shaw

AAS 68-064 Subsystem Reliability in a Manned Spacecraft, P. Nekrasov, R. Corprew

AAS 68-065 Mechanization and Vehicle Integration of the Space Research Centrifuge, J.E. Stumm

AAS 68-066 Martian Orbital Photographic Mission and System Analyses, R. Bashe, S. Kennedy

AAS 68-067 Partial Shunt Regulation, P. S. Nekrasov

AAS 68-068 to AAS 68-079 Not assigned

VOL. 7 AAS MICROFICHE SERIES
ASTRODYNAMICS SPECIALIST CONFERENCE, 1969
AAS/AIAA CONFERENCE
(September 3-5, 1968, Jackson, Wyoming)

AAS 68-080 Approximate Finite-Thrust Trajectory Optimization, F.T. Johnson

AAS 68-081 Properties of the Conjugate Gradient and Davidon Methods, G.E. Myers

| | |
|---|---|
| AAS 68-082 | Minimum-Time Intercept of a Maneuvering Orbital Target with Constant Thrust Acceleration, R. Rosenbaum |
| AAS 68-083 | A Singular Problem in Trajectory Optimization, H.E. Rauch |
| AAS 68-084 | Optimum Three-Impulse Transfer Between an Elliptic Orbit and a Noncoplanar Escape Asymptote, R.J. Gerbacht, P.A. Penzo |
| AAS 68-085 | A New Method for Constructing Periodic Orbits in Nonlinear Dynamical Systems, A. Bennett, J. Palmore |
| AAS 68-086 | Stability of Periodic Orbits in the Elliptic Restricted Three-Body Problem, R. Broucke |
| AAS 68-087 | Construction of Orbits Asymptotic to a Periodic Orbit, A. Deprit, J. Henrard |
| AAS 68-088 | Not available |
| AAS 68-089 | Rapid Analysis of Translunar and Lunar Abort Missions, J.C. Walker, J.E. Lancaster, F.I. Mann |
| AAS 68-090 | The Stability of the Triangular Points in the Elliptic Restricted Problem of Three-Bodies, K.T. Alfriend, R.H. Rand |
| AAS 68-091 | A Survey of Methods and Results in the Determination of Fuel-Optimal Impulsive Space Maneuvers, A.C. Robinson |
| AAS 68-092 | The Use of Green's Theorem to Find Globally Optimal Solutions to a Class of Impulsive Transfers, G.A. Hazelrigg, |
| AAS 68-093 | An Analytic Solution of the Optimal Two-Impulse 180 Transfer Between Noncoplanar Orbits and the Optimal Orientation of the Transfer Plane, F.T. Sun |
| AAS 68-094 | Minimum-Impulse Time-Free Transfer Between Neighboring Noncoplanar Elliptical Orbits with Major Axes Perpendicular to the Line of Nodes, J.V. Breakwell, W. Heine |
| AAS 68-095 | How to Open a Launch Window, F.W. Gobertz, J.R. Doll |
| AAS 68-096 | An Approximate Solution for the Short-Term Motion of a Lunar Satellite, G.H. Born, B.D. Tapley |
| AAS 68-097 | On Some Possible Simplifications and Changes in Hansen's Lunar Theory, P. Musan |
| AAS 68-098 | On the Comparison of the Newtonian and General Relativistic Orbits of a Point Mass in an Inverse Square Law Force Field, C.B. Solloway, H. Lass |
| AAS 68-099 | Trajectory Optimization Using Regularized Variables, B.D. Tapley, V. Szebehely, J.M. Lewallen |
| AAS 68-100 | A Numerical Solution of Kepler's Problem in Universal Variables, R.C. Blanchard, P.E. Zadunaisky |
| AAS 68-101 | Not available |
| AAS 68-102 | Periodic Orbits for Interplanetary Flight, W.M. Hollister |
| AAS 68-103 | Galactic Jupiter Probe Trajectory Design and Midcourse Guidance Analysis, R.T. Groves, C.R. Newman, R.E. Coady |

AAS 68-104 Optimization of a Low-Thrust Solar-Electric Planetary Orbiter Mission, C.G. Sauer, Jr.

AAS 68-105 Effect of Mission Mode and Orbit Eccentricity on Unmanned Outer Planet Capture Missions, S.M. Norman

AAS 68-106 Trajectory Requirements and Performance Comparisons of Single-Stage Electrically Propelled Space Vehicles, R.V. Ragsac

AAS 68-107 The Use of Apogee Bias to Reduce Spacecraft Velocity Correction Requirements for the Synchronous Satellite - A Statistical Analysis, J.W. Browning

AAS 68-108 Comparison of Preflight and Postflight Error Analysis of Lunar Orbiter Missions, P.B. Liebelt, J.P. Dyer

AAS 68-109 Guidance Analysis of the Multiple Outer Planet (Grand Tour) Mission, A.L. Friedlander

AAS 68-110 Lander-to-Orbiter Communication Link Geometry Characteristics for Highly Elliptical Orbits, W.J. Pragluski, R.T. Gamber

AAS 68-111 Application of State Space Transformation Theory to Orbit Determination and Prediction, S.P. Altman, J.S. Pistiner

AAS 68-112 A Simplified Approach for Determining the Information Content and Critical Error Sources of Earth-based Radio Tracking Data, D.W. Curkendall, R. McReynolds

AAS 68-113 Trajectory Analysis of a 1975 Mission to Mercury via an Impulsive Flyby of Venus, R.A. Wallace

AAS 68-114 Triple-Planet Ballistic Flybys of Mars and Venus, A.A. VanderVeen

AAS 68-115 Mission Window Definition for Jupiter Swingbys to the Outer Planets, J.L. Horsewood, J.D. Vickery

AAS 68-116 Jupiter Gravity-Assisted Trajectories, D.A. Klopp, J.C. Niehoff

AAS 68-117 Circumjovian Free and Powered Return Trajectories, A.B. Sergeyevski

AAS 68-118 Gravity-Turn Descent Trajectories to the Surface of Mars, J.A. Jungmann

AAS 68-119 Payload Maximization for a Mars Soft Lander, G.R. Knowles

AAS 68-120 Use of Aerobraking for Mars Capture, W.E. Johnson

AAS 68-121 Considerations of Applying Continuous Thrust Through Synergetic Plan Changing, F.S. Nyland

AAS 68-122 Approach Navigation and Entry Corridors for Planetary Atmosphere Braking, E.M. Repic, A.M. McDermott

AAS 68-123 Stationary Resonant Pitching Motions of a Controlled Satellite, J.A. Morrison

AAS 68-124 Nonlinear Attitude Motion Near Resonance, D.L. Hitzl

AAS 68-125 The Stability of the Force-Free Motions of Dual-Spin Spacecraft, R. Pringle, Jr.

AAS 68-126 Solar Pressure and Thermal Effects on the Radio Astronomy Explorer Satellite, J.L. Farrell, J.K. Newton

AAS 68-127 Not available

AAS 68-128 Doppler Tracking of Near-Synchronous Satellites, B.B. Holland

AAS 68-129 Autonomous Orbit Determination at Synchronous Altitude, W.L. Brogan, J.L. Lemay

AAS 68-130 Improvement of Astronomical Constants and Ephemerides from Pioneer Radio Tracking Data, J.D. Anderson

AAS 68-131 Consistency of Lunar Orbiter Residuals with Trajectory and Local Gravity Effects, P.M. Muller

AAS 68-132 The Lunar Gravitational Field as Determined from the Tracking Data of the Lunar Orbiter Series of Spacecraft, J.P. Gapcynski, W.T. Blackshear, H.R. Compton

AAS 68-133 Laser Tracking and Interplanetary Probe Orbit Determination, A.G. Lubowe

AAS 68-134 Mariner Mars 1969 Flight Path Design and Mission Analysis, C.E. Kohlhase,Jr.

AAS 68-135 Mission Analysis Study for the 1972 Venus Launch Opportúnity, D.P. Fields, C.K. Wilkinson,Jr.

AAS 68-136 Orbit Synthesis, H.K. Karrenberg, E. Levin, R.D. Luders

AAS 68-137 The Selection of a Satellite System for Determining the Geopotential, C.B. Winn

AAS 68-138 Secondary Mission Optimization, J.D. Mason, D.B. Smith, W.D. Dickerson

AAS 68-139 A New Criterion for Mission Selection - Abort Capability, R.R. Titus

AAS 68-140 Discussion of Calculations and Computer Techniques Leading to the Reobservation of the Tenth Satellite of Jupiter, M.L. Charnow

AAS 68-141 A Polynomial Force Model Representation for Determining Near-Equatorial, Near-Synchronous Satellite Orbits, N.S. Hall, R.H. Gersten

AAS 68-142 An Algorithm for Integrating Lifetime Orbits in Multi-revolution Steps, D. Boggs

AAS 68-143 Orbit Determination Using Analytic Partial Derivatives of Perturbed Motion, J.L. Arsenault, P.E. Koskela, K.C. Ford

AAS 68-144 Orbit Determination for Powered Flight Space Vehicles in Deep Space Missions, J.F. Jordan

AAS 68-145 The Role of Correlations in Navigation Satellite Error Analysis, D.A. Conrad, S.Y. Itoga

AAS 68-146 Automated Mission Analysis Using a Parametric Sensitivity Executive Program, R. Gold, S. Ross

AAS 68-147 Optimization of "Ephemeridal" Parameters for Minimum Propellant Requirements on Multiplanet Roundtrip Swingby-Stopover Missions, C.W. Mead, A.C. Young

AAS 68-148 Automated Mission Analyses Trajectory Determination Procedures, R.S. Patrick, D.D. Van Winkel

AAS 68-149 Vest Pocket Mission Analysis, D. Sonnabend

AAS 68-150 An Empirical Simulation of the Translunar Injection Burn for Apollo, D.T. Martin, R.F. Sievers

AAS 68-151 A Simulation of Automatic Station Keeping for Synchronous Satellites, M.C. Crocker II

AAS 68-152 to AAS 68-176 Not assigned

VOL. 25 ADVANCES IN THE ASTRONAUTICAL SCIENCES
ADVANCED SPACE EXPERIMENTS, 1969
(September 16-18, 1968, Ann Arbor, Michigan)

AAS 68-177 Planetary Orbital, Entry, and Fly-by Experiments -- Introduction, R.F. Fellows

AAS 68-178 Realism About Space, E.M.Cortright

AAS 68-179 Not available

AAS 68-180 Our Future in Space, G. VanderJagt

AAS 68-181 Scientific Space Research - The Next 10 Years, J.W.Findlay

AAS 68-182 Bistatic Radar Astronomy, V.R. Eshleman

AAS 68-183 Planetary Atmosphere Mass Spectrometry, N.W. Spencer

AAS 68-184 Planetary Ultraviolet Experiments, C.A. Barth

AAS 68-185 S-Band Occultation Experiments, A.J. Kliore, B. Seidel

AAS 68-186 Planetary Magnetic Field Experiments, E.J. Smith

AAS 68-187 Direct Measurements of Planetary Atmospheres by Entry Probes, A. Seiff

AAS 68-188 Planetary Surface Experiments -- Introduction, R.V. Meghreblian

AAS 68-189 Biological Studies of Mars: Theoretical Considerations and a Practical Experiment, F. Fanale, C. Stembridge, N.H. Horowitz

AAS 68-190 Planetary Atmospheres: Compositional and Other Changes Associated with Presence of Life, J.E. Lovelock, C. Giffin

AAS 68-191 Relevance of Future Space Missions to the Origin of the Solar System, J. Witting

AAS 68-192 A Planetary Study and Its Implementation, D.G. Rea

AAS 68-193 Not available

AAS 68-194 Lunar Orbital Mission Plans, R. Allenby

AAS 68-195 Not available

AAS 68-196 Orbital Infrared Experiments, J.W. Salisbury, G.R. Hunt

AAS 68-197 An Integrated Experiment for Compositional Analysis from an Orbiter, I. Adler, J. Trombka

AAS 68-198 Exploration of Gravitational Anomalies of Planets from Orbiting Vehicles, P.H. Savet

AAS 68-199 Determination of Electrical Parameters of the Moon, S.H. Ward

AAS 68-200 Lunar Surface Experiments -- Introduction, A.H. Chidester

AAS 68-201 Surveyor Results (Abstract), A.R. Hibbs

AAS 68-202 Lunar Geological Field Investigations, G.A. Swann

AAS 68-203 Lunar Geophysics, G. Simmons

AAS 68-204 Post-Apollo, Passive Seismic Experiment, G.V. Latham

AAS 68-205 The Velocity Structure of the Moon R.L. Kovach

AAS 68-206 Magnetometer Experiment, P. Dyal, C.P. Sonett

AAS 68-207 The Lunar Heat-Flow Experiment, M.G. Langseth

AAS 68-208 The Future of Lunar Exploration, N. Hess

AAS 68-209 Particles and Fields Experiments -- Introduction, G.H. Ludwig

AAS 68-210 The Interplanetary Magnetic Field Solar Origin and Terrestrial Effects (Abstract), J.M. Wilcox

AAS 68-211 Low Frequency Waves in the Magnetosphere (Abstract), R.A. Helliwell

AAS 68-212 The Structure of the Magnetosphere, L.J. Cahill, Jr.

AAS 68-213 Auroral Phenomena (Abstract), B.J. O'Brien

AAS 68-214 Low-Energy Charged Particles in the Earth's Magnetosphere (Abstract), L.A. Frank

AAS 68-215 Astronomy in Space -- Introduction, O.C. Mohler

AAS 68-216 X-Ray Sources, W.A. Hiltner

AAS 68-217 Space Observations of the Solar Corona, R.M. MacQueen

AAS 68-218 Optical Space Astronomy, W.G. Tifft

AAS 68-219 The Results and Interpretation of Some of the Harvard OSO-IV Observations (Abstract), L. Goldberg, R.W. Noyes, W.H. Parkinson, E.M. Reeves, G.L. Withbroe

AAS 68-220 Interpretation of Venus Lyman - Emission as Observed from Mariner V, M.B. McElroy, L. Wallace

VOL. 20 SCIENCE AND TECHNOLOGY SERIES
BIOENGINEERING AND CABIN ECOLOGY, 1969
AAS/AAAS SYMPOSIUM
(December 30, 1968, Dallas, Texas)

AAS 68-221 Spaceship Earth -- a Global View of Ecology, S.F. Singer

AAS 68-222 The Present Status and Future Needs of Biomedical Engineering Training, J.E. Jacobs

AAS 68-223 Industry's Utilization of Human Factors and Bioengineering, J.A. Kraft

AAS 68-224 Bioastronautics in the Soviet Union (includes Bibliography), C.M. Bertone

AAS 68-225 Effects of Isolation on Man's Performance, (Includes Bibliography), S.B. Sells, J.R. Rawls

AAS 68-226 Space and Undersea Comparison, Prelude to Tektite I, E.B. Thompson, E.S. Burcher

AAS 68-227 Life-Support System Tests in a Manned Space Cabin Simulator, W.F. Ardt, Jr., J.K. Jackson

VOL. 8 AAS MICROFICHE SERIES
SPACE PROJECTIONS, 1969
SPACE PROJECTIONS FROM THE ROCKY MOUNTAIN REGION
(AAS Rocky Mountain Section, July 15-16, 1968, Denver, Colorado)

AAS 68-228 Upper Atmosphere of Venus, C.A. Barth

AAS 68-229 Atmosphere of Venus, Analysis of Recent Observations, A. Barger

AAS 68-230 Search for Life in Planetary Evolution, R. Young

AAS 68-231 Ice Caps on Venus, W.F. Libby

AAS 68-232 Radiophysics of Jupiter, J. Warwick

AAS 68-233 Planetary Atmosphere Determination Error Analysis Using Kalman Filter, A.C. Serold

AAS 68-234 Satellite System for Determining the Geopotential, C.B. Winn

AAS 68-235 Lineament Trend Analysis of Red Sea Gemini Photography, J. Bridwell

AAS 68-236 Measurement of Air Density, C.D. Westlund, O.C. Haycock

AAS 68-237 Studies of Earth Airglow, D.J. Baker, A.M. Despain

AAS 68-238 Anticipated Antenna Problems on Mars and Venus Probes C.H. Brockmeyer

AAS 68-239 Electrical Measurements of an Erectable Parabola for Space Missions, D. Holst

AAS 68-240 Retro-Directive Antenna Array System, G.E. Johnson

AAS 68-241 Multichannel Data Compression Simulation, W.A. Stevens

AAS 68-242 Proper Digital Filters, C.A. Halijak

AAS 68-243 Investigation of Synchronous-Wave R-F to D-C Conversion, D.C. Watson, R.W. Crow, C.C. Johnson

AAS 68-244 Adaptive Multiplexing in Spacecraft Data Management, N.R. Sheppard

AAS 68-245 The Mechanics and Applications of the Planetary Swingby Technique for Optimization of Interplanetary Trajectories, G.A. Flandro

AAS 68-246 Stability of Highly Elliptic Venus Orbits Under the Influence of Solar Gravitational Perturbations, D.G. Roos

AAS 68-247 Two-Dimensional Representation of Planetary Probe Targeting Considerations, S.G. Homic

AAS 68-248 Launch Period, Targeting and Error Analysis for Mars Direct and Orbital Missions, D.W. Marquet

AAS 68-249 Payload Maximization and Mission Mode Comparison for Hard and Soft Martian Landers, G.R. Knowles, D.A. Howard

AAS 68-250 Mission Influence on the Aerothermodynamic Environment for a Venus Entry Vehicle, H.C. Norman, P.M. Hart

AAS 68-251 VELA Satellite Particle Observations at 18 Earth Radii, M.D. Montgomery

AAS 68-252 Testing X-Ray Telescope Performance, R.S. Wriston

AAS 68-253 Geomagnetic and Interplanetary Field Variability, Mariners II, IV and V Compared, D. Jones

AAS 68-254 The Use of Positive Feedback to Control Spring Mass Systems, R.C. Culver

AAS 68-255 Spacecraft Spin Axis Motion With Active Internal Devices, D.E. Maas, J.M. Wilson

AAS 68-256 Guidance and Control Design for a Six-Engine Martian Soft Lander, R.N. Ingoldby, P. Stafford

AAS 68-257 Applications of the Laser Gyro, K. Phelps

AAS 68-258 Reaction Jet Damping for an Aerodynamically Stable Entry Vehicle, R.P. Seibel

AAS 68-259 A Microcellular Computer Embodying a Discrete Kalman Filter, D.A. Rudberg, L. Cannon

AAS 68-260 Control of Powered Split Structures, M.L. Moe

AAS 68-261 U.S. Space Program Projections, K.P. Timmons, C.A. Braunwarth

AAS 68-262 Buoyant Venus Stations for the Early 70's, S.R. Sadin, R.E. Frank

AAS 68-263 Mars Entry and Landing Capsule, E.K. Casani, J. Gerpheide

AAS 68-264 Free Vibration Analysis by the Modal Supstitution Method, R.L. Bajan, C.C. Feng

AAS 68-265 Elasticity Solution by a Modified Galerkin Method, D.O. Blackketter

AAS 68-266 Meteoroid Penetration Detector, S. Russak

AAS 68-267 Ablative Heat Shields for Planetary Entry Vehicles, E.L. Strauss, H.E. Sparhawk

AAS 68-268 Radiative Heating of Venusian Atmosphere, H.J.Deacon,Jr. W.F.Rumpel

AAS 68-269 Triangulated Pantograph Beam Applications, M.L. Clevett, Jr.

AAS 68-270 Navigation Satellite Program, T. Wyatt

AAS 68-271 Transferability of Aerospace Management Technology, J.G. Milliken, J.S. Gilmore

AAS 68-272 Hypersonic Propulsion Facility at Ogden, Utah, R.O. Boe

AAS 68-273 Space Simulation Tests of Planetary Spacecraft, B. Hartel

AAS 68-274 Residual Gas Analyzer for Monitoring Sublimation from Spacecraft Materials, R.W. Jackman

AAS 68-275 Acquisition of Data for the Apollo Crew Motion Experiment, W.J. Maes, C.H. Murrish

AAS 68-276 A Trailing Camera Technique for Studying Projectile Penetration into Earth Materials, J.L. Colp, W.B. Pepper

AAS 68-277 White Sands Missile Range, A Rocky Mountain Regional Space and Missile Test Facility, A.L. Vick

AAS 68-278 Pulse Motor for Spacecraft Attitude Control, D. Suddeth, D.Blackwood

AAS 68-279 Space Applications of Beryllium Propellants, C.R. Voris

AAS 68-280 Testing a Sterilizable Liquid Propulsion System, J.B. Keough

AAS 68-281 Optical Measurements Through a Rocket Plume, J. Pernicka

AAS 68-282 Space Mechanism Lubrication, B.J. Perrin, R.W. Mayer

AAS 68-283 Penetration of Terrestrial Materials by Instrumented Projectiles, J.L. Colp

AAS 68-284 The Magnetotropometer, R.P. Woolley

AAS 68-285 Design Evolution of the Wolf Trap Life Detector, D.E. Buckendahl

AAS 68-286 Some Biological Problems in the Search for Extraterrestrial Life, H.P. Klein

AAS 68-287 Non-Existence of a "Biota-Cloud" Recontamination Hazard for a Planetary Lander, J.F. Vandrey

AAS 68-288 Determination of Terminal Sterilization Cycle for Mars Capsule, A.R. Hoffman

AAS 68-289 New Fast Techniques for Bio Assay, M. Rothenberg

AAS 68-290 Biological Monitoring of Full-Scale Simulated Mars Lander, A.A. Rothstein, H. Kaufman

VOLUME 11 AAS MICROFICHE SERIES 1979
SPACE AS A NATIONAL RESOURCE
(October 24, 1979, Los Angeles, California)

AAS 68-291 Space as a National Resource -- Introduction, D.T. Perkins

AAS 68-292 Potential of Unmanned Space Probes, J.P. Burke

AAS 68-293 Contributions of Man in Space, A.M. Mayo

AAS 68-294 Not available

AAS 68-295 Contributions of Space Technology to Civilian Needs, G.W. Hoover

AAS 68-296 The Space Program as a National Priority, A. Bell

AAS 68-297 The Impacts of the Space Budget on the Economy, B. Augenstein

VOL. 25 SCIENCE AND TECHNOLOGY SERIES
GEOLOGICAL PROBLEMS IN LUNAR AND PLANETARY RESEARCH, 1971
THE INTERRELATIONSHIP OF LUNAR PROBE DATA II
(January 25-26, 1968, Huntington Beach, California)
(See also paper sequences AAS 68-026 to -048, AAS 69-001 to -019, and AAS 69-670 to -685, also included in this volume.)

AAS 68-298 to -300 Not available

AAS 68-301 A Tectonic Analysis of the Moon, B.B. Brock

AAS 68-302 Comments on the Figure of the Moon, S.R. Runcon

AAS 68-303 Not available

AAS 68-304 Crater-Age Classification and Relative Age Determination of Lunar Surfaces, or "Son of Interplanetary Correlation of Geologic Time", H.A. Pohr, T.W. Offield

AAS 68-305 Not available

AAS 68-306 Statistical Analysis of Lunar Crater Chains, G. Fielder

AAS 68-307 to -309 Not available

AAS 68-310 Lunar Sinuous Rills, W.S. Cameron

AAS 68-311 to -314 Not available

AAS 68-315 Albedo and Photometry of Surveyor Footprints, J.D. Halajian

AAS 68-316 to -319 Not available

AAS 68-320 Lunar Terrain Roughness with Respect to Roving Vehicles, O.H. Vaughan

AAS 68-321 Not available

AAS 68-322 Applications of Luminescence Techniques to the Study of the Lunar Surface, R.T. Greer

AAS 68-323 Not available

AAS 68-324 The Worth of Lunar Probe Data in Planning Lunar Colonization, H.A. Shiffer

AAS 68-325 Not available

VOL. 23-1, AAS MICROFICHE SERIES
FIRST INTERNATIONAL SPACE RESCUE SYMPOSIUM, 1975
(19th Congress of the International Astronautical Federation, October 14, 1968, New York City, New York, USA)

*IAF 68-01 Survey of Space Rescue Capabilities, R. Fleisig, G.W. Heath

IAF 68-02 An Approach to Effective International Cooperation in Space Rescue and Safety, A.M. Mayo

IAF 68-03 Low Earth Orbit Escape and Rescue Concepts, J.W. Wild, E. Perchonok

IAF 68-04 A Unified Approach to Lunar Base and Earth Orbital Rescue, K. Speiser, R. Fleisig

IAF 68-05 The Main Physiological Problems Raised by the Rescue of Astronauts, R. Grandpierre

IAF 68-06 Possible Trends of Rescue of the Incapacitated Astronaut, J. Dvorak

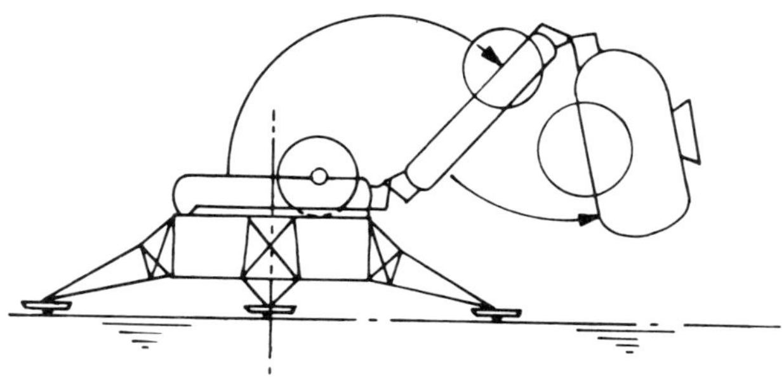

* The designation IAF has been used to identify papers presented at certain International Astronautical Federation/International Academy of Astronautics Conferences but assigned to the American Astronautical Society for publication.

# AAS TECHNICAL PAPERS
# 1969

VOL. 25, SCIENCE AND TECHNOLOGY SERIES
GEOLOGICAL PROBLEMS IN LUNAR AND PLANETARY RESEARCH, 1971
LUNAR TRANSIENTS
(February 17-18, 1969, Huntington Beach, California)
(See also paper sequences AAS 68-026 to -048, AAS 68-300 to -325, and AAS 69-670 to -685, also included in this volume.)

AAS 69-001 Tidal Mechanisms and Event Morphologies, B.M. Middlehurst, W.B. Chapman

AAS 69-002 Relationships of Tectonic Processes of the Earth and Moon, N.A. Kozyrev

AAS 69-003 to -006 Not available

AAS 69-007 Observations of Lunar Transient Phenomena in the Aristarchus Region, W.S. Cameron

AAS 69-008 / -009 Not available

AAS 69-010 Evidence of Repeated Activity of the Lunar Crust, S. Miyamoto

AAS 69-011 T.L.P. and Luminosity during Terrestrial Earthquakes, C.M. Botley

AAS 69-012 to -014 Not available 1969

AAS 69-015 Flares on the Moon, J. Classen

AAS 69-016 to -019 Not available

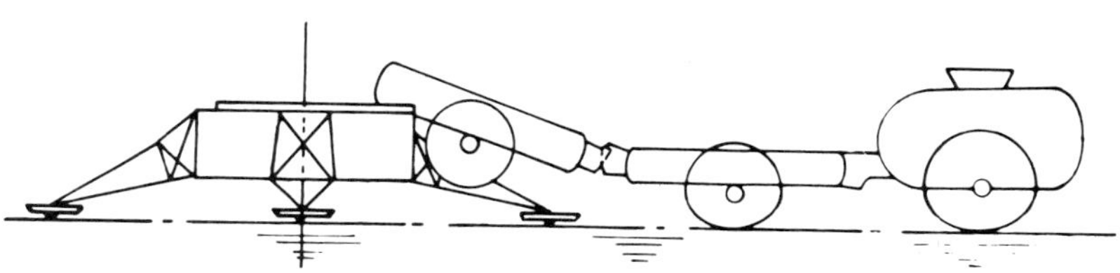

VOL. 21, SCIENCE AND TECHNOLOGY SERIES
REDUCING THE COSTS OF SPACE TRANSPORTATION, 1969
(Seventh Goddard Memorial Symposium, March 4-5, 1969, Washington, D.C.)

| | |
|---|---|
| AAS 69-020 | Keynote Address, C.W. Mathews |
| AAS 69-021 | Economic Factors of Space Transportation, D.S. LeVine, R.H. Lea |
| AAS 69-022 | Low-Cost Booster Propulsion - Technology and Management, D.M. Ross |
| AAS 69-023 | Aircraft Design and Development Experience Related to Reusable Launch Vehicles, F.J. Dore |
| AAS 69-024 | An Approach to Low-Cost Space Transportation, R.R. Kearton |
| AAS 69-025 | Transportation to Earth Orbit: 1975-1985, L. Roberts |
| AAS 69-026 | Progress Report on a Launch Vehicle Cost Breakthrough, H.J. McClellan |
| AAS 69-027 | Reusable Space Transportation Systems: 1970/1980 - An Evolutionary View, F. Raymes, J.I. Dodds |
| AAS 69-028 | Comparison of Six Designs of Low-Cost Launch Vehicles, P. Dergarabedian, J.B. Kendrick |
| AAS 69-029 | Toward Efficient Shuttle Service to Earth Orbit, E.S. Love |
| AAS 69-030 | Low-Cost Orbital Transportation: How Big a Step!, C.D. Marks, R.G. Quest |
| AAS 69-031 | Concluding Luncheon Address, W. vonBraun |
| AAS 69-032 | Not assigned |

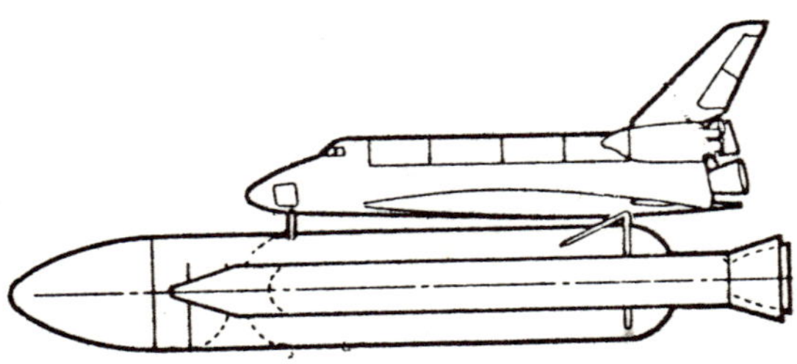

VOL. 22, SCIENCE AND TECHNOLOGY SERIES
PLANNING CHALLENGES OF THE 70'S IN THE PUBLIC DOMAIN, 1970

VOL. 26, ADVANCES IN THE ASTRONAUTICAL SCIENCES
PLANNING CHALLENGES OF THE 70'S IN SPACE, 1970

VOLS. 13 THROUGH 19, AAS MICROFICHE SERIES

- VOL. 13 SUPPL. TO VOL. 22 S&T, 1971
- VOL. 14 SUPPL. TO VOL. 26 Advances, 1971
- VOL. 15 OR - MANAGEMENT APPLICATIONS, 1971
- VOL. 16 OR - URBAN PROBLEMS, 1971
- VOL. 17 OR - MATHEMATICAL & COMPUTER METHODS, & MILITARY APPLICATIONS, 1971
- VOL. 18 SPACECRAFT & SPACE TECHNOLOGY, 1971
- VOL. 19 SPACE EXPLORATION, SPACE SCIENCE, & APPLICATIONS, 1971

AAS 15TH/ORS 35TH ANNUAL MEETINGS
(June 17-20, 1969, Denver, Colorado)

AAS 69-033 Specific Aspects of Venus Experimental Studies, U.K. Khodarev, L.N. Lupichov (v26 Adv)*

AAS 69-034 Game Approach to the Guidance Problem of Space Vehicle Descent in Atmospheres, G.M. Zuev, V.M. Ivanin, Y.N. Ivanov (v14 Mic)*

AAS 69-035 Keynote Address: The Application of Knowledge, W.M. Schirra, Jr. (v26 Adv)

AAS 69-036 Future Manned and Automated Earth Orbital Science Missions, H.J. Smith (v26 Adv)

AAS 69-037 The Soft Data Problem: A Case Study of Education, M.H. Halbert (v22 S&T)*

AAS 69-038 Problems of Managing Criminal Justice, R.D. Trujillo (v13 Mic)

AAS 69-039 Development of a Central Information System within the Context of a Model Cities Program: Opportunities and Obstacles, F.G. Caro (v13 Mic)

AAS 69-040 Frontiers in the Planning and Management of Mental Institutions, P.R. Binner (v22 S&T)

---

* The papers appear in Vol. 22, Science and Technology, Vol. 26 Advances in the Astronautical Sciences, or in a set of the AAS Microfiche, as indicated. The following abbreviations are used:

  Adv = Advances in the Astronautical Sciences
  Mic = AAS Microfiche Series
  S&T = Science and Technology Series

AAS 69-041 An Analytical Model of a Welfare Case Load, H.A. Morse (v22 S&T)

AAS 69-042 Liability for Space Activities, V.P. Nanda (v19 Mic)

AAS 69-043 to -046 Not available

AAS 69-047 A General Description of an Environmental Information Systems Program, S.H. Cohen (v17 Mic)

AAS 69-048 Bayesian Risk Analysis by Decision Networks, A. Vazsonyi (v17 Mic)

AAS 69-049 / -050 Not available

AAS 69-051 No-Loss Cryogenic Storage on the Lunar Surface, J.H. Bell, Jr. (v19 Mic)

AAS 69-052 to -057 Not available

AAS 69-058 Results of Earth Resources Surveys of the Earth and Plans for Future Satellites, W.A. Fischer (v26 Adv)

AAS 69-059 Applications of Earth Orbital Remote Sensing to Forest and Range Problems, R.N. Colwell (v26 Adv)

AAS 69-060 Earth Resource Information Systems, J.R. Shay (v14 Mic)

AAS 69-061 Aerospace Applications in Agriculture and Forestry, A.B. Park (v26 Adv)

AAS 69-062 Synoptic Oceanography: Remote Sensing Studies of the Ocean, J.W. Sherman (v26 Adv)

AAS 69-063 On-Line Evaluation of Planned Population Programs, E.H. Palmer (v13 Mic)

AAS 69-064 The Use of Multi-State Input-Output Analysis in the Analysis of Population Flow Patterns, R.C. Koza (v22 S&T)

AAS 69-065 A Mathematical Theory for the Exploitation of Biological Population, S.H. Mann (v13 Mic)

AAS 69-066 The Pioneer Spacecraft Program, G.A. Reiff (v19 Mic)

AAS 69-067 / -068 Not available

AAS 69-069 Solar Activity and Planetary Positions, J.B. Blizard (v19 Mic)

AAS 69-070 to -078 Not available

AAS 69-079 On Non-Parametric Sensitivity Analysis in Linear Programming, R.N. Braswell, F.M. Allen (v17 Mic)

AAS 69-080 to -082 Not available

AAS 69-083 Planning Future Space Programs. Are We on the Right Track?, H.S. Becker (v26 Adv)

AAS 69-084 What It Will Take to Sell a Space Station, R. Casey (v26 Adv)

AAS 69-085 Not available

AAS 69-086 Computerized Forest Management, M. Lazara (v13 Mic)

AAS 69-087 Some Myths about World Hunger, and Why They Must be Dispelled, R.C. Liebenow (v13 Mic)

AAS 69-088 Microbes as a New Source of Food for the Expanding World Population, D.M. Updegraff (v22 S&T)

AAS 69-089 Maximum Productivity from Industry-Owned Forest Lands, J.L. Clutter, J.E. Bethune (v13 Mic)

AAS 69-090 Not available

AAS 69-091 The Atmosphere of Mars and the Mariner Mars '69 UV Experiments, C.W. Hord (v19 Mic)

AAS 69-092 to -094 Not available

AAS 69-095 An Application of Linear Integer Programming Using Two Simultaneous Objective Functions and Non-Linear Constraints, E.K. Kutchma (v17 Mic)

AAS 69-096 A Network Isolation Algorithm, M. Bellmore (v17 Mic)

AAS 69-097 A Simplices Decomposition Algorithm for Integer Linear Programming, H. Eto (v17 Mic)

AAS 69-098 to -102 Not available

AAS 69-103 Probabilistic Analysis of a Two-Task Multiprogramming Environment, F.D. Knight, R.S. Wood, J.E. Suich (v17 Mic)

AAS 69-104 Not available

AAS 69-105 Technological Forecasting - What Effect Does it Have on Aerospace Planning, P.H. Neukirch (v15 Mic)

AAS 69-106 to -109 Not available

AAS 69-110 Current and Future Range Considerations, D.M. Jones (v26 Adv)

AAS 69-111 / -112 Not available

AAS 69-113 Planning Space Activities in the 1970's, K.P. Heiss (v14 Mic)

AAS 69-114 Basic Problems in the Development of National Water Data System, E.L. Hendricks (v13 Mic)

AAS 69-115 Reclamations Water Challenges for the 70's, F. Dominy (v22 S&T)

AAS 69-116 Optimal Estuarine Water Quality Management, S.K. Mukherjee (v22 S&T)

AAS 69-117 An Algorithm for Irrigation Project Planning, P. Rogers, D.V. Smith (v22 S&T)

AAS 69-118 Not available

AAS 69-119 Water Resources Development for Denver Metropolitan Area, R.E. Wiedemann (v13 Mic)

AAS 69-120 Some Systems Possibilities for Global Atmospheric Research Program, S. Ruttenberg, M. Tepper (v19 Mic)

AAS 69-121 / -122 Not available

AAS 69-123 Weather Modifications in the 1970's, A.M. Kahan (v19 Mic)

AAS 69-124 to -136 Not available

AAS 69-137 Portfolio Analysis in New Product Screening and Development, P.F. Hase (v15 Mic)

AAS 69-138 to -140 Not available

AAS 69-141 Habitability of Manned Spacecraft, C.C. Johnson (v14 Mic)

AAS 69-142 Prominence of Space Suits and Life-Support Systems in Space, L.E. Bell, (v14 Mic)

AAS 69-143 Recent Advances in Life-Support System Concepts, J. Shapira (v26 Adv)

AAS 69-144 to -147 Not available

AAS 69-148 Analysis and Comparison of Various Lower Bounds on Schedule Times for the Solution of the Flow Shop Problem, S. Ashour, M.N. Quraishi (v17 Mic)

AAS 69-149 / -150 Not available

AAS 69-151 The Application of Flowgraphs to Elastomechanical Systems, T. Enger, C.C. Feng (v18 Mic)

AAS 69-152 Large Space-Erectable Structures, D.H. Vaughan (v18 Mic)

AAS 69-153 Reusable Lifting Center - Body Launch Vehicle Concept for Pressure-Fed Engines, H.F. Wuenscher (v18 Mic)

AAS 69-154 to -158 Not available

AAS 69-159 Modeling the Community Health Service System, A.H. Packer, (v16 Mic)

AAS 69-160 The Maternal and Infant Care Simulation Model, F. Kennedy (v16 Mic)

AAS 69-161 to -163 Not available

AAS 69-164 Automation of Launch Vehicle Adaptive Costs in Support of Long-Range Space Planning, R.E. Drowns (v15 Mic)

AAS 69-165 / -166 Not available

AAS 69-166A Fiscal Forecasting for Space Vehicles, J.D. Graham (v15 Mic)

AAS 69-167 / -168 Not available

AAS 69-169 An Experiment i the Application of Systems Analysis to Innovational Program Planning in Education, M. Hershkowitz, S.I. Finkel (v17 Mic)

AAS 69-170 to -172 Not available

AAS 69-173 Thermal Design for the OSO-F University of Minnesota Experiment, R.R. Anderson (v18 Mic)

AAS 69-174 Thermal Control of Four Apollo Telescope Mount Instruments, D.E. Regenbrecht (v18 Mic)

AAS 69-175 to -179 Not available

AAS 69-180 Motivation for Future Manned Planetary Missions, S.F. Singer (v26 Adv)

AAS 69-181 Technology for Manned Planetary Missions, J.L. Sloop (v26 Adv)

AAS 69-182 Not available

AAS 69-183 Wildlife Resources Management and the Human Population, H.R. Woodward (v13 Mic)

AAS 69-184 Recreation in the Year 2,000, M.D. Arnold (v13 Mic)

AAS 69-185 to -188 Not available

AAS 69-189A Growing Need for the Specialist with a Universal Mind, J.W. Clark (v17 Mic)

AAS 69-189 / -190 Not available

AAS 69-191 The Selection and Utilization of Instructional Resources: A Systems Approach, B.H. Banathy (v17 Mic)

AAS 69-192 Not available

AAS 69-193 Systems Analysis in Hospital Extended Care Planning, M.E. Cohen (v16 Mic)

AAS 69-194 to -196 Not available

AAS 69-197 Branch and Bound Methods for Optimizing, C.M. Shetty (v17 Mic)

AAS 69-198 On Modeling the Resource Allocation and Control Problem, R.N. Braswell (v17 Mic)

AAS 69-199 to -203 Not available

AAS 69-204 A New Approach to Logistics Management, C.J. Hall (v17 Mic)

AAS 69-205 Not available

AAS 69-206 Logistics Requirements for the Apollo Applications Program, R.F. Thompson (v17 Mic)

AAS 69-207 Lunar Surface Exploration, O.L. Tiffany (v26 Adv)

AAS 69-208 Extraterrestrial Construction, B. Hall (v26 Adv)

AAS 69-209 The Challenge of Lunar Exploration, D.U. Wise (v14 Mic)

AAS 69-210 Lunar Orbital Exploration, C.L. Kober (v14 Mic)

AAS 69-211 Management of Fuel Resources, R.J. Cameron (v13 Mic)

AAS 69-212 Management on Non-Fuel Resources, A. Kaufman (v13 Mic)

AAS 69-213 Planning to Meet the Needs for Mineral Resources, C. Mottley (v22 S&T)

AAS 69-214 Planning for Construction Minerals in the Urban Environment, J.R. Dunn, W.A. Wallace (v22 S&T)

AAS 69-215 Not available

AAS 69-216 Minimization of Training Cost and Quantity of Multi-skilled Personnel when Requirements are Uncertain, K.W. Haynam (v17 Mic)

AAS 69-217 On the Adjustment of Indirect Fires by Use of Quasilinearization, W.D. O'Neil, J.R. Holdsworth (v17 Mic)

AAS 69-218 Cost Effective Army Material Testing, G. Benz (v17 Mic)

AAS 69-219 to -222 Not available

AAS 69-223 Agricultural Resources Information System - The Users Point of View, H.G. Lorsch (v19 Mic)

AAS 69-224 Not available

AAS 69-225 Linear Programming in Occupational Medicine, J. Modi (v16 Mic)

AAS 69-226 A Heuristic Distribution and Allocation Model for Whole Blood, C.C. Pegels, A. Jelmert (v16 Mic)

AAS 69-227 Cost-Effectiveness Analysis of Artificial Kidney Systems, J. Modi (v16 Mic)

AAS 69-228 to -232 Not available

AAS 69-233 The Need for Automatic Test Facilities for Large-Scale Arrays, E. Sarkisian (v17 Mic)

AAS 69-234 / -235 Not available

AAS 69-236 An Algebraic Theory of the Analysis and Testing of Digital Networks, A. Brown, D.D. Johnson, H. Young (v17 Mic)

AAS 69-237 Combinatorial Analysis for Check-Out Techniques, E.C. Neu (v17 Mic)

AAS 69-238 Long Life Reliability Problems and Solutions for Valves under Spatial Environments, J.C. DuBuisson (v18 Mic)

AAS 69-239 Machine Reading of X-Ray Film, J. Neri, J.C. Pernicka (v18 Mic)

AAS 69-240 Parts Selection for Long-Life Spacecraft, G.E. Neuner (v18 Mic)

AAS 69-241 Economical Planetary Transfers in the Real Planetary Cases, J.P. Gravier, C. Marchal, R.D. Culp (v18 Mic)

AAS 69-242 Optimum Transfer between Hyperbolic Asymptotes about a Planet of Finite Size, J.M. Walton, C. Marchal, R.D. Culp (v18 Mic)

AAS 69-243 Decision Policies for Midcourse Corrections to the Trajectory of a Space Vehicle, J.T. Wagner (v18 Mic)

AAS 69-244 Non-Linear Longitudinal Dynamics of an Orbital Lifting Vehicle, N.X. Vinh, A. Dobrzelecki (v18 Mic)

AAS 69-245 Not available

AAS 69-246 Lunar Sinuous Rilles, H. Masursky (Abstract Only) (v19 Mic)

AAS 69-247 The Effect of Use of Lunar Resources on Lunar Logistics and Earth-Lunar Transportation, E.A.Steinhoff (v19 Mic)

AAS 69-248 Not available

AAS 69-249 Systems Analysis and Urban Problems, M. Kurtz (v13 Mic)

AAS 69-250 Not available

AAS 69-251 One Answer from the Aerospace Industry on Low-Cost Housing, D.H. Christensen (v13 Mic)

AAS 69-252 Climate as a Factor in Community Environmental Management, J.F. Clarke, D. Baulch (v22 S&T)

AAS 69-253 Dynamic Land Inventory System, J. Moss (v22 S&T)

AAS 69-254 A Radiometer for Measuring Solar Flux Variations, R.C. Linton (v19 Mic)

AAS 69-255 Deployment of a Tethered Orbiting Interferometer, C.J. Swet, J.M. Whisnant (v19 Mic)

AAS 69-256 / -257 Not available

AAS 69-258 Investigation of an Early Flyby Sequence for Imaging of the Planet Mercury, L.A. Manning (v19 Mic)

AAS 69-259 / -260 Not available

AAS 69-261 Conus Air Defense Engagement Simulation (CADENS), O.V. Fedoroff, H.C. Lorck, R.A. Patchett (v17 Mic)

AAS 69-262 to -265 Not available

AAS 69-266 A Systems Analysis Model of Urbanization and Change: An Experiment in Interdisciplinary Education, C.F. Steinitz, P. Rogers (v16 Mic)

AAS 69-267 to -269 Not available

AAS 69-270 Utilization of an Empiric-Programming Model for Investigations of Urban Development Planning Strategies, J.W. Dickey, J.T. Harrison (v16 Mic)

AAS 69-271 Optimal Economic Design of a Queueing System, I.W. Kabak (v17 Mic)

AAS 69-272 / -273 Not available

AAS 69-274 On a Class of Queueing Problems and Discrete Transforms, S.S. Rao, N.J. Jaiswal (v17 Mic)

AAS 69-275 Automatic Test - An Overview and Classification, W.A. Heffner (v17 Mic)

AAS 69-276 to -281 Not available

AAS 69-282 Swept-Frequency Microwave Measurements in Simulated Reentry Environments, R.P. Davidson, R.J. Churchill, J.P. Rybak (v19 Mic)

AAS 69-283 / -284 Not available

AAS 69-285 Brains and Personality, J.H. Engel (v22 S&T)

AAS 69-286 Challenge of Planetary Exploration, P.G. Thome (v26 Adv)

AAS 69-287 Not available

AAS 69-288 The 1973 Mariner-Mercury Mission, P.K.Eckman (v26 Adv)

AAS 69-289 Not available

AAS 69-290 The Pioneer Mission to Jupiter, C.F. Hall, H.F. Matthews (v26 Adv)

AAS 69-291 Reconnaissance Missions to the Outer Planets, J.E. Long (v26 Adv)

AAS 69-292 Optimum Public Policies for Introducing Transportation Innovations, E.B. Cohen (v22 S&T)

AAS 69-293 Optimal Container Inventory and Routing, W.A. Horn (v13 Mic)

AAS 69-294 Design of Inter-City Ground Transportation Systems, E.S. Diamant, J.R. Vadeboncoeur (v22 S&T)

AAS 69-295 Railroad Network Simulation Models, J.N. Cetinich (v13 Mic)

AAS 69-296 to -299 Not available

AAS 69-300 Low-Cost Solid Propellant Boosters, R. Wasel (v18 Mic)

AAS 69-301 to -303 Not available

AAS 69-304 Solar Cell-Electrochemical Power Supplies-10W to 50 KW, A.H. Smith (v18 Mic)

AAS 69-305 Applications and Development of Space Nuclear Power Systems, J.D. LaFleur, Jr. (v18 Mic)

AAS 69-306 Not available

AAS 69-307 PPBS in City, State, and County, S.J. Mushkin (v15 Mic)

AAS 69-308 to -313 Not available

AAS 69-314 Safety Considerations for a High-Density Automated Vehicle System, M. Lenard (v16 Mic)

AAS 69-315 The Analysis of a T-Type Traffic Intersection Using GERT, G.E. Whitehouse, C.A. Roth (v16 Mic)

AAS 69-316 to -319 Not available

AAS 69-320 Missions to the Comets, D.L. Roberts, A.L.Friedlander (v14 Mic)

AAS 69-321 Design of an Advanced Meteoroid Probe of the Asteroid Belt, S.L. Russak (v14 Mic)

AAS 69-322 Early Exploration of the Asteroids Region by Solar-Powered Electrically Propelled Spacecraft, J.R. Wrobel, D.J. Kerrisk (v26 Adv)

AAS 69-323 Supersonic Transport in the Future, J.C. Maxwell (v22 S&T)

AAS 69-324 State Department of Transportation, T. Bryant (v13 Mic)

AAS 69-325 VSTOL Aircraft, C. von Kaan (v13 Mic)

AAS 69-326 Jumbo Jets, J. Steiner (v22 S&T)

AAS 69-327 Economic Impact on Domestic Air Traffic: An Econometric Approach, B.R. Lippke, J.R. Stewart (v22 S&T)

AAS 69-328 / -329 Not available

AAS 69-330 Nuclear Propulsion for Exploration of the Solar System, J.P. Layton (v18 Mic)

AAS 69-331 Strategic Superiority, Deterrence, and Arms Control, J.I. Coffey (v17 Mic)

AAS 69-332 Not available

AAS 69-333 The Promise of UHF Satellite for Mobile, Broadcast, and Austere Services, J.L. Hult (v18 Mic)

AAS 69-334 / -335 Not available

AAS 69-336 Deployment Latch-up Dynamics of an Erectable Truss Antenna, M. Downing, H.A. Mitchell (v18 Mic)

AAS 69-337 to -339 Not available

AAS 69-340 How to Solve Ill-Structured Problems, F. Seliz (v17 Mic)

AAS 69-341 to -346 Not available

AAS 69-347 A Comparison of Economic Lot Size and Minimum-Cost Production Methods, D.S. Plymale (v17 Mic)

AAS 69-348 to -351 Not available

AAS 69-352 Experimental Concepts for Life-Detection Methods Applicable to Mars, V. Oyama (v19 Mic)

AAS 69-353 Not available

AAS 69-354 The Coming Information Explosion, H.J. Hilton (v13 Mic)

AAS 69-355 A Matrix Approach to Management Information System Design, C.S. Hwa (v22 S&T)

AAS 69-356 Ultrafiche and Its Application to Colleges and Universities, J.P. Kottenstette (v13 Mic)

AAS 69-357 Coming Information Explosion: Information Storage Models, F.F. Leimkuhler (v22 S&T)

AAS 69-358 Not available

AAS 69-359 NASA Space Program Possibilities and Upper-Stage Implications, B. Maggin (v18 Mic)

AAS 69-360 to -364 Not available

AAS 69-365 Solving Classes of Infinite Games by Linear Programming, T.W. Ruefli (v17 Mic)

AAS 69-366 A systems Approach to Data Management, J.C. Helland (v19 Mic)

AAS 69-367 Not available

AAS 69-368 Automatic Maintenance Aspects of Data Processing in Unmanned Spacecraft, A. Avizienis, D.A. Rennels (v19 Mic)

AAS 69-369 Science Data Management Automation Systems for Spacecraft, R.H. Clausen (v19 Mic)

AAS 69-370 to -373 Not available

AAS 69-374 A Control System Excitation-Response Model for Import-Proportion in a National Economy, P.S. Satsangi, S.S. Sengupta, J.B. Ellis (v15 Mic)

AAS 69-375 The Application of Taylor's Theorem in Obtaining the Probability Distribution of Long-Range Profit, D.E. Smith (v15 Mic)

AAS 69-376 to -381 Not available

AAS 69-382 The Simulation of Intercity Travel Demands, P.M. Pearson (v16 Mic)

AAS 69-383 / -384 Not available

AAS 69-385 The Application of OR to Stop Selection and Sequencing for the Aeromedical Evacuation Problem, T. E. Kowalsky, J. Griffin, G.M. Cusack (v16 Mic)

AAS 69-386 Accessible Regions Beyond the Solar System, M.W. Hunter (v26 Adv)

AAS 69-387 Evolution of Interstellar Operations, K.A.Ehricke(v14 Mic)

AAS 69-388 Interstellar Travel, a Round-Trip Propulsion System with Relativistic Velocity Capabilities, P. Norem (v14 Mic)

AAS 69-389 The Investment in Natural and Human Resources, J.M. King (v13 Mic)

AAS 69-390 Improving the Analysis of Water Resources, I.A. Watson (v13 Mic)

AAS 69-391 to -394 Not available

AAS 69-395 Use of Multiple Outlets for Reducing the Effect of Stratification on Propellant Supply Temperature Variations, R.J. Schoenhals, E.R.F. Winter, R.N. Nelson (v18 Mic)

AAS 69-396 Why People Play Games - Report of a Survey, C.J. Thomas, G.M. McNichols (v17 Mic)

AAS 69-397 Experimental Uses of the Political Military Exercise, J.D. Steinbruner (v17 Mic)

AAS 69-398 to -400 Not available

AAS 69-401 Flight Evaluation Radar-Inertial Systems, R.J. Madigan (v18 Mic)

AAS 69-402 Low-Cost Inertial Navigation, W.J. Whitesell (v18 Mic)

AAS 69-403 Guidance Scheme for Low Thrust Parabolic Insertion, R.A. Schultz, S.S. Hu (v18 Mic)

AAS 69-404 Simplified Performance Analysis of Space and Missile Guidance, J.T. Wagner, D.F. McAllister (v18 Mic)

AAS 69-405 Navigation for Surface and Aerospace Vehicles by a Very Low Frequency Radio Network, H.W. Turney (v18 Mic)

AAS 69-406 Contribution of Navigation Analysis of Onboard Systems to Mission Planning in the 70's, J.L. Knoedler (v18 Mic)

AAS 69-407 Optimal Attitude Control System for Spinning Aerospace Vehicles, V.S. Sohoni, D.R. Guild (v18 Mic)

AAS 69-408 Not available

AAS 69-409 Decade Forecasting Methodology Based upon Information Progression, J.C. Hoagland, G.E. Smith (v15 Mic)

AAS 69-410 to -419 Not available

AAS 69-420 The Recent Advances in Navigation by Satellite, E. Ehrick (v26 Adv)

AAS 69-421 Meteorological Satellites - Beyond the First Nine Years, R.L. Haley (v26 Adv)

AAS 69-422 Advanced Operational Meteorological Satellites, C.A.Spohn (v14 Mic)

AAS 69-423 A Mean Variance Approach to Optimal Policy Choice under Uncertainty, C. Corrado (v13 Mic)

AAS 69-424 Decentralized Decision-Making for National Development, I. Adelman, F.L. Golladay (v13 Mic)

AAS 69-425 Macro Economic Decision Model for National Planning, J.Y. Kuark (v13 Mic)

AAS 69-426 to -432 Not available

AAS 69-433 Systems Requirements Methodology for Long-Duration Apollo Logistics Spacecraft, H.S. Alexander (v18 Mic)

AAS 69-434 / -435 Not available

AAS 69-436 Maneuvering Unit Development Programs, H.J. McMann (v18 Mic)

AAS 69-437 Not available

AAS 69-438 State of the Art of Space Tools - EVA/IVA, C.C. Kubokawa (v18 Mic)

AAS 69-439 to -443 Not available

AAS 69-444 Some New Models for Inventory and Production Planning Operations, I.I. Karayalcin (v15 Mic)

AAS 69-445 A Quantitative Approach to the Span of Control Problem, I.I. Karayalcin (v15 Mic)

AAS 69-446 Evaluation of Organization Criteria, I.I. Karayalcin (v15 Mic)

AAS 69-447 to -453 Not available

AAS 69-454 Earth Benefits and System Restraints, J.S. Smith (v26 Adv)

AAS 69-455 Not available

AAS 69-456 Group-Practice and Efficient Health Services Delivery, E.L. Dolinsky (v13 Mic)

AAS 69-457 Health Data Systems in the Offing: A Statistical Approach, O.K. Sagen (v13 Mic)

AAS 69-458 Imperatives for Health Systems Research, M.R. Greenlick (v13 Mic)

AAS 69-459 CORE City Health Care - A Resource Management Challenge, D.L. Cowen (v22 S&T)

AAS 69-460 to -463 Not available

AAS 69-464 Perception of Laser Light, C.F. Asiala, Jr. (v19 Mic)

AAS 69-465 Contractor Basic Accounting Records as a Data Source for Cost Analysis, J. String (v19 Mic)

AAS 69-466 / -467 Not available

AAS 69-468 Econometric Analysis of Earth Orbital Transportation Systems, C. Rutland (v19 Mic)

AAS 69-469 Problems of Abort from Manned Spacecraft, A.L. Greensite (v18 Mic)

AAS 69-470 EVA/IVA Flexible Lightweight Umbilicals, D.A. Myers, H.J. Richter (v18 Mic)

AAS 69-471 Not available

AAS 69-472 EVA Stabilization Shoes, A.L. Greensite, E.H. Wrench (v18 Mic)

AAS 69-473 Gravity Independent Liquid Collection and Fluid Phase Separation System, W.E. James, D.J. Holecek (v18 Mic)

AAS 69-474 to -476 Not available

AAS 69-477 Modeling Air Antisubmarine Tactics with Data, F.G. Stahl (v17 Mic)

AAS 69-478 Field Tests of Operations Research Models, A.S. Bolthouse, A.M. Johnson (v17 Mic)

AAS 69-479 to -482 Not available

AAS 69-483 Potential Contributions of Orbital Earth Resources Data to Urban Planning, C.T.N. Paludan (v26 Adv)

AAS 69-484 From Space Technology to Profits, M.F. Rudy (v14 Mic)

AAS 69-485 Space Photography - Window to Discovery, W.V. Trollinger (v14 Mic)

AAS 69-486 Manufacturing in Space: Low and Zero Gravity Manufacturing in Orbit, H.F. Wuenscher (v26 Adv)

AAS 69-487 Not available

AAS 69-488 Management of Power and Energy Resources, E. Ewald (v15 Mic)

AAS 69-489 Management of Well Information for the Petroleum Industry, J.M. Forgotson, Jr. (v22 S&T)

AAS 69-490 Management of Waste Heat Discharges of Power Facilities, N.S. Lynn, L.W. Ross (v22 S&T)

AAS 69-491 The Impact of Apollo Applications on Space Station Engineering, W.D. Green (v19 Mic)

AAS 69-492 Manned Exploration of Mars and Venus Using Chemical Propulsion, R.R. Titus (v19 Mic)

AAS 69-493 Not available

AAS 69-494 Technology Flow for Future Space Activities, C.A. Braunwarth (v19 Mic)

AAS 69-495 to -498 Not available

AAS 69-499 Educating for the RFP, P. Levine (v18 Mic)

AAS 69-500 Lunar Landing Decisions in American Decision Making, J. Logsdon (v18 Mic)

AAS 69-501 An Historical Note in the Genesis of Manned Planetary Flight, R.B. Merrifield (v18 Mic)

AAS 69-502 / -503 Not available

AAS 69-504 Macro-Analysis of the American Education System, A.M. Mood (v16 Mic)

AAS 69-505 Expansion of University Facilities to Accommodate Increasing Enrollments, R. Sanderson (v16 Mic)

AAS 69-506 to -510 Not available

AAS 69-511 Structuring and Dimensioning Urban Activity Models, E.F. Sullivan (v16 Mic)

AAS 69-512 Cost Effectiveness of Health and Welfare Programs, the Case of Child Health Care, T.W. Hu (v16 Mic)

AAS 69-513 Not available

AAS 69-514 Electromagnetic Experiments at the Lunar Surface, J.B. Arnett, G.C. Henderson, A.J. Hoffman (v19 Mic)

AAS 69-515 Not available

AAS 69-516 EVA Maneuvering Unit Comparison, C.E. Whitsett, Jr., A.E. Wudell, W.H. Tobey (v14 Mic)

| | |
|---|---|
| AAS 69-517 | Orbital EVA, K. Sheffield (v14 Mic) |
| AAS 69-518 | Safety for Large Space Stations, J.W. Wild (v26 Adv) |
| AAS 69-519 | Not available |
| AAS 69-520 | Looking Ahead: Unmanned Spacecraft Systems Safety, L.C. Montgomery (v14 Mic) |
| AAS 69-521 | System Safety for Astronaut Maneuvering Units, G.B. Mumma (v14 Mic) |
| AAS 69-522 | System Safety - A Management System for Safety Used in Manned Space Flight, P.H. Bolger (v26 Adv) |
| AAS 69-523 | The Role of Nitrogen Oxides from Automobiles in Smog, R.D. Cadle (v22 S&T) |
| AAS 69-524 | Follow the Leader - To Pollution Control, K.J. Schiager (v22 S&T) |
| AAS 69-525 | The Restoration of Natural Waterbodies, H.V. Anthony G.P. Fulton (v13 Mic) |
| AAS 69-526 | Denver's Pollution Problems, W.J. Walker (v13 Mic) |
| AAS 69-527 | Industry's Role in Air Pollution Abatement Programs, L.T. Jensen (v13 Mic) |
| AAS 69-528 | An Evaluation of the Use of Systems Analysis to Solve a Waste Management Problem, I. Price, D.H. Stimson (v22 S&T) |
| AAS 69-529 to -533 | Not available |
| AAS 69-534 | Optimum Operations of a Tandem Rolling Mill, C.F. Wang (v15 Mic) |
| AAS 69-535 | Attrition Rates for Probabilistic Combat Models, S.R. Kimbleton (v17 Mic) |
| AAS 69-536 | Evaluation of V/STOL Aircraft in a Multi-Mission Scenario Using Statistical Distributions for Mission Frequencies, D.R. Limaye (not available) |
| AAS 69-537 | Graphic Comparison of Cost Effectiveness in a Multi-Mission Scenario, D.R. Limaye (not available) |
| AAS 69-538 | Assessment of Weapons System Effectiveness with Combined Monte Carlo-Expected Value Type Simulations, J.C. Stamper, G.L. Gallagher (not available) |
| AAS 69-539 | Possibility of Producing a Dense Thermonuclear Plasma by an Intense Field Emission Discharge, F. Winterberg (not available) |
| AAS 69-540 | Not available |
| AAS 69-541 | Application of the Learning Curve to Maintainability Tests and Demonstrations, R.L. Street (v18 Mic) |
| AAS 69-542 | Not available |
| AAS 69-543 | Two Simulation Models for Planning and Evaluating Aircraft Avionics Maintenance Systems, A.P. Feddersen (v18 Mic) |

AAS 69-544 A Stochastic Model for the Analysis of Epidemics, M.E. Thomas (v16 Mic)

AAS 69-545 / -546 Not available

AAS 69-547 An Assessment of the Contribution of Operations Research to Educational Planning, H. Weitz (v16 Mic)

AAS 69-548 Not available

AAS 69-549 Application of Baysian Decision Theory to Large Fire Control Systems, J.B. Davis (v17 Mic)

AAS 69-550 Aerospace Technology and Urban "Disorders", C.J. Harris (v17 Mic)

AAS 69-551 Subjective Measures of Merit Involving Multiple Criteria of Evaluation, L.H. Smith (v17 Mic)

AAS 69-552 Not available

AAS 69-553 A Methodology to Compare Policies for Exploring the Solar System, R.G. Chamberlain, L. Kingsland, Jr. (v19 Mic)

AAS 69-554 to -560 Not available

AAS 69-561 The World Game, R.B. Fuller (v22 S&T)

AAS 69-562 Planning for Strategic Deterrence in the 70's, R.C. Seamans, Jr. (v26 Adv)

VOL. 23 SCIENCE AND TECHNOLOGY SERIES
SPACE TECHNOLOGY AND EARTH PROBLEMS, 1970

VOL. 12 AAS MICROFICHE SERIES - SUPPL. TO VOL. 23 S&T, 1970
(Las Cruces, New Mexico, Oct. 23-25, 1969)

AAS 69-563 Horizontal Elevator New Concept in Rapid Transit, E.O. Haltom

AAS 69-564 Optimum Design of Vehicular Drive Systems, L.P. Hajdu, J. Peschon, P. Henault

AAS 69-565 Aerospace Research Support of High-Speed Tracked Air-Cushion Vehicle Program, D.G. Stephens

AAS 69-566 Not available

AAS 69-567 Electron Density Measurements in the Upper Atmosphere*, J. Smith

AAS 69-568 The Thermistor Measurement of Temperature in the 30-65 KM Atmospheric Region*, H.N. Ballard, B. Rofe

AAS 69-569 A Cryogenic Sampler for the Determination of Stratosphere Composition*, M. Izquierdo, H. Ballard, A. Haire, M. Thiel

AAS 69-570 A Parachute-Borne Beta Ray Densitometer*, J.A. Whitacre, H.N. Ballard, M. Izquierdo, B. Sellers

---

* In abstract in Volume 23 Science & Technology and in full in Volume 12 AAS Microfiche Series.

AAS 69-571 Faraday Rotation System for Measuring Electron Densities*, R. Rubio

AAS 69-572 The Analysis of Earth Resources Using Space Photography*, R.N. Colwell

AAS 69-573 Land Use and Natural Resources Inventories*, D.J. Belcher

AAS 69-574 Bonanza - A Systems Approach to Earth Resources*, W.W. Godwin, L.T. Grose

AAS 69-575 Analysis of Data from Remote Sensing Systems*, R.F. Pascucci, G.W. North

AAS 69-576 Land Use Mapping of the Southwestern United States from Satellite Imagery*, N.J.W. Thrower, L.W. Senger

AAS 69-577 Simulating the Consequence of Cloud Cover on Earth-viewing Space Missions*, S.C. Brown

AAS 69-578 Precision Pointing of a High-Gain Antenna from a Planetary Lander*, R.C. Brown

AAS 69-579 Space Photographs of the Earth in the Study of Geotectonics*, E.Y. Kedar

AAS 69-580 Methods of Orbital Experiments Synthesis, V. Gradecak

AAS 69-581 Functional Requirements for the Spaceborne Computer System of a Mid-70's Space Station, D.O. Baechler, P.S. Schaenman

AAS 69-582 A Multiband Photographic System*, B. Chang

AAS 69-583 Potential Benefits from Remote Sensing of Agricultural, Forest, and Range Resources, E.E. Hardy

AAS 69-584 Space Systems for Earth Resources Information*, D.M. Waltz

AAS 69-585 The Potential Contribution of Remote Sensing to World Food Production, S. Baker

AAS 69-586 Systems Planning: A Spaceborne Methodology for Earth-Bound Problems, E.A. Gardner

AAS 69-587 Earth Resources Satellites: Capabilities Versus Requirements, A.J. Stratton

AAS 69-588 Remote Sensing - Sense or Nonsense?*, R.C. Wilson

AAS 69-589 City Planning Using an Urban Performance Model*, A. Brown

AAS 69-590 Space Technology Applications to Highway Safety*, M.A. Wilson, J.S. Youtcheff

AAS 69-591 Application of Aerospace Guidance and Control Concepts to High-Speed Train Transportation Systems, S.A. Steele

AAS 69-592 Truck Vibrations - An Old Problem with a Modern Solution Via Computer*, H.S. Walker, G.R. Potts

AAS 69-593 Lightweight Expandable Bridge, M.L. Clevett

AAS 69-594 Not available

AAS 69-595 Man's Progress into the Sea, W.M. Nicholson

AAS 69-596 Potential Use of Earth Satellites for Solving Problems in Oceanography and Hydrology, E.P. McClain

AAS 69-597 A Systems Approach to Water Basin Management Planning*, R.F. Jewett

AAS 69-598 Application of Large-Scale Computers to Abatement of Water Pollution*, J.M. Jeglic, G.D. Pence, Jr.

AAS 69-599 Ocean-Wave Pneumatic Attenuation Systems*, S.N. Sherk

AAS 69-600 to -602 Not available

AAS 69-603 Multisensor Correlation Technique Applied to Hydrology, J.M. Kennedy

AAS 69-604 A Tradeoff Study in Space Geodesy Altimetry*, E. Weiss

AAS 69-605 Precision and Accuracy of Test Geometry*, F. Hanson

AAS 69-606 Planning for Integrated Urban Communications, J.O. Campbell

AAS 69-607 Radar Signature Analysis Using an Equivalence Class Technique*, G.W. Gruver

AAS 69-608 An Estimation Theory Approach to Bit Synchronization*, A.L. McBride

AAS 69-609 Reducing Apollo Backup Voice-Telemetry Data Interference Using a Phase Feedback Detection Process, R.J. Panneton, W.B. Warren

AAS 69-610 Linear Phase Detection Using Frequency-Modulation Demodulation Techniques, G.D. Arndt, S.W. Novosad

AAS 69-611 Unlimited Water and Power with Solar Energy Energizing Electronic, Steam, and Turbines in Systems*, J.F. Crane

AAS 69-612 Application of an Atmospheric Model to the Ocean Environment*, G.S. Ruggles, L.N. Walker

AAS 69-613 Underwater Range Tracking at Barking Sands Tactical Underwater Range, Hawaii*, G.A. Nussear, P. Seelinger

AAS 69-614 Tidal Basin Resonance for Power, M.H. Cobble

AAS 69-615 Not available

AAS 69-616 Product Assurance in Oceanics-- A Natural Outgrowth of the Aerospace Industry*, E.R. Britton

AAS 69-617 A Study of the Prediction Accuracies of Several Atmosphere Models Used on a Medium Altitude Satellite*, J.A. Ward, Jr.

AAS 69-618 Application of Space Technology to Atmospheric Pollution Measurement, R.S. Wiltshire, W.H. Clohessy

AAS 69-619 Experiments in Dynamical Weather Prediction, K. Miyakoda

AAS 69-620 Potential for Application of Earth-Oriented Satellites to Air Pollution, A.P. Altshuller, R.A. McCormick

AAS 69-621 The Role of Extraterrestrial Surveillance in Cloud and Weather Modification, C.L. Hosler

AAS 69-622 The Cold Earth, W.L. Webb

AAS 69-623 A Comparison between the Helicone and Conical Horn Antennas, K.R. Carver

AAS 69-624 Long-Range Solar-Flare Prediction, J.B. Blizard

AAS 69-625 Sea Robin, An Ocean Buoy/Satellite Communications Experiment*, C.E. Mortlock, Jr.

AAS 69-626 Communication Antennas for Ocean-Based Data Acquisition System*, J.R. Dearden

AAS 69-627 Not available

AAS 69-628 Demand Forecasting for Information-Transfer Satellites*, R.W. Hough

AAS 69-629 Not available

AAS 69-630 The Impact of Satellite Monitoring on Natural Resource Public Policy Issues, I. Clark

AAS 69-631 Applications of Aerospace Sciences and Technological Advances to the Delivery of Rural and Remote Area Health Care, A.H. Schwichtenberg

AAS 69-632 Not available

AAS 69-633 Keynote Address: The Systems Challenge in Transportation, J.E. Beggs

AAS 69-634 Space Technology and the Weather, P. Dergarabedian

AAS 69-635 A Progress Report on the Lifting-Body Flight-Test Program, M.O. Thompson

AAS 69-636 to -649 Not assigned

WHITHER LUNAR AND PLANETARY EXPLORATION IN THE 1970'S?
(AAS/AAAS Meeting, December 26-27, 1969, Boston, Mass.)

No proceedings volume available

AAS 69-650 to -669 Not available

VOL. 25 SCIENCE AND TECHNOLOGY SERIES
GEOLOGICAL PROBLEMS IN LUNAR AND PLANETARY RESEARCH, 1971
SUPPLEMENTARY PAPERS, 1969
(See also paper sequences AAS 68-026 to -048, AAS 68-300 to -325, AAS 69-001 to -019.)

AAS 69-670 Thermography of Salt Pools, Bad Water, Death Valley, a Martian Analog?, J. Green

AAS 69-671 Application of Optical Fluorescence for the Characterization of Lunar Materials, R.T. Greer

AAS 69-672 Use of the Lunar Probes to Recognize Shocked Materials on the Moon's Surface, N.M. Short

AAS 69-673 Details of the Apollo 12 Landing Site from Surveyor III Photography, J. Green

AAS 69-674 The Structural Evolution of the Moon's Surface, G.P. Tamrazyan

AAS 69-675 Rejuvenation Processes on the Lunar Surface, L.B. Ronca, R. R. Green

AAS 69-676 Subsurface Waters of the Moon and Mars, G.N. Katterfeld, P.M. Frolov

AAS 69-677 Some Lunar and Martian Analogs of Terrestrial Volcanic Structures, J. Green

AAS 69-678 The Exploration of the Planets from Space, J. Green, L. Larmore

AAS 69-679 The Effects of Lunar Probe Data on Lunar Drilling Equipment Design, D.S. Crouch

AAS 69-680 Crystallization Temperatures of Rocks from Tranquillity Base Compared with Experimental Crystallization of Dry Silicate Melts, L. Kopecký

AAS 69-681 Remarks on Planetological Terminology, K. Beneš

AAS 69-682 Comments on "Apollo 11 Observations of a Remarkable Glazing Phenomenon on the Lunar Surface by T. Gold", J. Green

AAS 69-683 to -685 Not issued

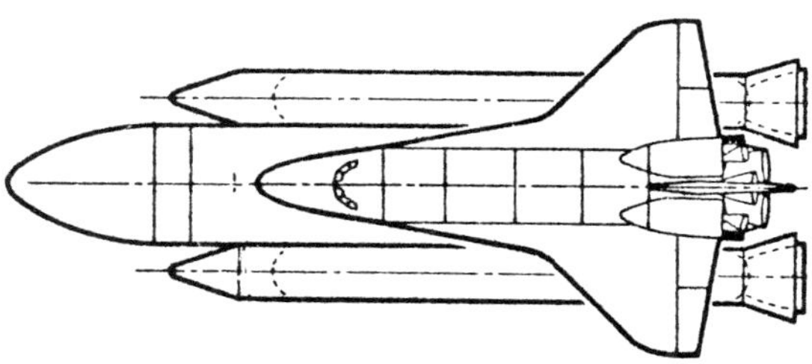

VOL. 23-2, AAS MICROFICHE SERIES
SECOND INTERNATIONAL SPACE RESCUE SYMPOSIUM, 1975
(20th Congress of the International Astronautical Federation, October 5-12, 1969, Mar del Plata, Argentina)

IAF 69-01 Survey of Space Rescue Capabilities, R. Fleisig, G.W. Heath

IAF 69-02 Control of Hazards of Manned Space Flight, P.H. Bolger

IAF 69-03 Potentials for International Contribution of Astronaut Emergency Recovery, A.R. Nagy, Jr., E. Perchonok, S.T. Chu

IAF 69-04 Space Rescue: Requirements for Recovery of Astronauts Returning to Earth Under Contingency Conditions, G.W. Heath

IAF 69-05 Potential Causes and Consequences of Space-Flight Emergencies, S.H. Dole

IAF 69-06 Space Escape Systems - A Potential Solution, F.X. Kane

IAF 69-07 Rescue from Earth Orbit, R.H. Francis

IAF 69-08 Rescue at Sea, C. Dorian

IAF 69-09 Earth-Based Rescue Feasibility, E. Perchonok, E.J. Rattin

IAF 69-10 Low-Earth-Orbit Emergency - Escape Vehicle, R.H. Bradley, W.K. Carter

IAF 69-11 Possibility of Diagnostics of the State of Incapacitated Astronauts by "Off-Body" Sensors, J. Cmiral, J. Dvorak, J. Sulc

IAF 69-12 Spacecraft Damage Containment and Control Procedures, E.L. McCabe, J.W. Wild

IAF 69-13 Flammability Characteristics of Simulated Space Cabin Wall Penetrated by a Micrometeoroid, E.S. Mills, G.V. Colombo

IAF 69-14 Decompression and Recompression in Space Rescue, H.J. von Beckh

IAF 69-15 Protective Effect on an Air-Bag Cushion on Crash Injury, S. Yokobori, R. Yurugi, I. Matsumoto

IAF 69-16 Space Rescue and Radioprotection, S. Baranski, P. Czerski

# AAS TECHNICAL PAPERS
# 1970

AEROSPACE SYSTEMS DEVELOPMENTS: IMPLICATIONS FOR THE LAW
(Eighth Goddard Memorial Symposium, March 1970, Washington, D.C.)
No proceedings issued

AAS 70-001 to -019 Not available

VOL. 27 ADVANCES IN THE ASTRONAUTICAL SCIENCES, 1970
SPACE STATIONS

VOL. 28 ADVANCES IN THE ASTRONAUTICAL SCIENCES, 1970
SPACE SHUTTLES AND INTERPLANETARY MISSIONS

SPACE STATION & MANNED PLANETARY EXPLORATION
SIXTEENTH ANNUAL MEETING
(June 8-10, 1970, Anaheim, California)

AAS 70-020 An Overview of NASA's Space Station Program, D.R. Lord, R.L. Lohman, R.F. Lovelett (v27 Adv)

AAS 70-021 Space Station Subsystems and Technology, C.L. Gould (v27 Adv)

AAS 70-022 The Case for an Early Manned Lunar Orbit Station, H.E. Bauer, D.E. Lockwood (v27 Adv)

AAS 70-023 Manned Planetary Spacecraft Commonality with Space Station, A.L. Jones, K.A. Ehricke (v27 Adv)

AAS 70-024 /-025 Not available

AAS 70-026 The Effect of Transportation Systems on Manned Planetary Precursor Missions, M.W. Hunter II (v28 Adv)

AAS 70-027 Missions to Mercury (1973-1985), D.L. Roberts, D.A. Klopp, W.C. Wells (v28 Adv)

AAS 70-028 The Outer Planets: Challenge for the Seventies, E.M. Repic, J.R. Eyman, E.J. Dazzo (v28 Adv) 1970

AAS 70-029 A Theory on the Nature and Origin of Comets with Implications for Space Mission Planning, D.H. Robey (v28 Adv)

AAS 70-030 Unmanned Exploration of Mars, E.B. Pritchard, E.F. Harrison (v28 Adv)

AAS 70-031 Operational Concepts for a 10-Year Space Station, R.E. Holmen, F.C. Runge (v27 Adv)

AAS 70-032 Space Station Habitability: Its Form Relationship to Man, F. Toerge, C.A. O'Donnell (v27 Adv)

AAS 70-033 Ten-Year Space Station Experiment Program, R.E. Smith, D.L. Erickson, W.P. Pratt (v27 Adv)

AAS 70-034 Experiment Module Concepts, W.W. Withee (v27 Adv)

AAS 70-035 Astronaut Capabilities for Scientific Observations, T.L. Page (v27 Adv)

AAS 70-036 Commercial Use of Space Station, C.L. Kober (v27 Adv)

AAS 70-037 Perspective and Systems Engineering Manned Planetary Flight, K.A. Ehricke (v28 Adv)

AAS 70-038 The Long-Term Motion of Artifical Planetary Satellites, C. Uphcff (v28 Adv)

AAS 70-039 An Earth Departure Technique for Use in Manned Interplanetary Missions Using the Nerva Engine, S.F. McAdoo, Jr., J. Funk (v28 Adv)

AAS 70-040 Manned Planetary Missions with Reusable Nuclear Shuttles, R.G. Riedesel, R.S. Cowls (v28 Adv)

AAS 70-041 Not available

AAS 70-042 A Space Shuttle Design Consideration, C.B. Peterson (v28 Adv)

AAS 70-043 Some Operational Aspects of the Space Shuttle, S.T. Beddingfield (v28 Adv)

AAS 70-044 Space Shuttle Main Propulsion, F.M. Stewart, R.L. Wetherington (v28 Adv)

AAS 70-045 The Characteristics of the Space Shuttle, O.C. Jean, R.D. Scott (v28 Adv)

AAS 70-046 Progress in Technology for Space Shuttles, R.W. Rainey (v28 Adv)

AAS 70-047 Atmospheric Maneuvers for Space Shuttles, P.F. Holloway, E.B. Pritchard (v28 Adv)

AAS 70-048 Space Shuttle Mission Trajectory Optimization, V.S. Sohoni (v28 Adv)

AAS 70-049 Not available

AAS 70-050 Space Shuttle-Space Station Integration Issues, E.T. Williams (v28 Adv)

AAS 70-051 Apollo Safety Experience, C.M. McGuire (v27 Adv)

AAS 70-052 Not available

AAS 70-053 Tekite Program Safety Planning, J.B. Tenney, Jr. (v27 Adv)

AAS 70-054 Radiation Model of Man for Analysis of Future Space Missions, P.G. Kase, L.M. Ryan (v27 Adv)

AAS 70-055 Space Shuttle System Safety Program, J.W. Griswold (v27 Adv)

AAS 70-056 Space Station Safety, R. Aller, M. Shaw (v27 Adv)

AAS 70-057 Experiment Information Management, R.M. Adams, J.S. Quilty, J.F. Spitzer (v27 Adv)

AAS 70-058 Earth-Moon Transportation, K.A. Ehricke (v27 Adv)

AAS 70-059 Minis, Maxis, and Mustard -- Considerations in the Sizing of International Space Transport Systems, R.F. Creasey (v27 Adv)

AAS 70-060 U.S./European Cooperation for Manned Space Projects - Will It Come?, J. Lambrecht (v27 Adv)

AAS 70-061 Not available

AAS 70-062 European Participation in the Post-Apollo-Program, H. Tolle (v27 Adv)

AAS 70-063 Not available

AAS 70-064 The Challenge of Space Research, D.H. Menzel (v27 Adv)

AAS 70-065 The Value of a Strong Safety Program, N.J. Ryker (v27 Adv)

AAS 70-066 to -069 Not available

VOL. 26, SCIENCE AND TECHNOLOGY SERIES, 1971
TECHNOLOGY UTILIZATION IDEAS FOR THE 70'S AND BEYOND
(October 30, 1970, AAS/AIAA Meeting, Winrock, Arkansas)

AAS 70-070 Pollution Control Using Advanced Energy Generation Systems, R.R. Barthelemy

AAS 70-071 Organophosphorus Insecticide Decontaminant, B.C. Wolverton, R. Richter, S.M. Lefstad

AAS 70-072 Oil-Water Separation and Oil Harvesting System, D.J. Graham, R.L. Johnson, P.G. Bhuta

AAS 70-073 Portable Mass Spectrometer/Computer for Air Pollution Analysis and Monitoring, J.C. Pernicka, G.F. McGowan

AAS 70-074 Not available

AAS 70-075 Rover - An Application of Aerospace Technology to Ground Transportation, W.A. DeTally, J.A. Benson

AAS 70-076 Application of Aerospace Data Acquisition Technology to Atmospheric and Meteorological Investigations, M.R. Smith, R.A. Johnson

AAS 70-077 Aerospace-Developed Entropy Correlation Applied in Evaluating Combustor-Incinerator Design, C.J. Harris

AAS 70-078 An Isotope X-Ray Device, D.R. Russell

AAS 70-079 Measurement and Analysis of Vibration in Transportation Systems, S.A. Clevenson, J.J. Catherines

AAS 70-080 General Electric Phenol Analysis Equipment, R.J. Starkey, Jr., J.J. Shull, E.D. Orr

AAS 70-081 Rankine Cycle Power Systems for Automotive Application, J.A. Peoples

AAS 70-082 Utilization of Homogeneous Chemiluminescent Gas Reactions in Air Pollution Monitoring and Control, A. Fontijn, A.J. Sabadell, R.J. Ronco

AAS 70-083 Nonpolluting Central Power Stations, F.L. Robson, A.J. Giramonti

AAS 70-084 On Properties of Mature Severe Vortical Storms, P. Dergarabedian, F.E. Fendell

AAS 70-085 A Gradiometer System for Gravity Anomaly Surveying, M.B. Trageser

AAS 70-086 Engineering for Ecological Systems with Specific Reference to Fisheries, E.B. Cohen, E.R. Glaser, E. Kaplan

AAS 70-087 Manual and Automated-Computerized System in Side-Looking Radar Applications for Land Use Survey, E.Y. Kedar

AAS 70-088 Power from Space - Technology Transfer for Human Survival, P.E. Glaser

AAS 70-089 Applications of Systems Engineering for Solutions to Social and Economic Problems, R.R. Smith, G. Priftis

AAS 70-090 Molecular-Beam Sampling of Gases in Engine Cylinders, W.S. Young, Y.G. Wang, W.E. Rodgers, E.L. Knuth

AAS 70-091 Utilization of Aerospace Technology in the Stability of Motor Vehicles, J.D. Iversen

AAS 70-092 A Strategy for Remote Sensor Use in Mining, Hydrology, and Pollution Control, E.I. Curtis

AAS 70-093 to -099 Not issued

SURFACE FEATURES OF MARS
(November 17, 1970, Huntington Beach, California)
(Proceedings to be issued as a volume in the Science & Technology Series)

AAS 70-100 to -119 Not available

AAS/AAAS MEETING ON LUNAR SAMPLE ANALYSIS
(December 26-31, 1971, Chicago)
(No Proceedings issued)

AAS 70-120 to -139 Not available

VOL. 33 SCIENCE AND TECHNOLOGY SERIES, 1974
ORBITAL INTERNATIONAL LABORATORY, Third and Fourth OIL Symposia
(International Astronautical Federation, October 5-6, 1970, Constance, Germany)

AAS 70-140 Third Orbital International Laboratory Symposium - Opening Statement, A.H. Abdel-Ghani(printed in book as A.H.Abel-Ghani)

AAS 70-141 Development of Large Earth Orbital Space Station, K.A. Ehricke

AAS 70-142 The Large Earth Orbital Space Station - an International Program, R.B. Demoret, G.W. Morgenthaler

AAS 70-143 Scientific and Technical Experiments Carried Out on the Soyuz 9 Spacecraft, V.I. Sevastyanov

AAS 70-144 Spectrophotometry of the Earth from Manned Spacecraft, K. Ya. Kondratyev, et al.

AAS 70-145 The Sociological Needs of Large Developing Countries, H.G.S. Murthy, K.S. Karnik

AAS 70-146 Management of Large Multinational Space Operations, E.A. Lauschner

AAS 70-147 Results of the Visual Observations of Night, Twilight, and Day Horizons of the Earth from Soyuz 9 Spacecraft, K. Ya. Kondratyev, et al.

AAS 70-148 Not available

AAS 70-149 Legal Aspects of Orbiting International Laboratories, V.S. Vereshchetin

AAS 70-150 The Orbital Laboratories in the Framework of the Space Treaty, P. Magno

AAS 70-151 Psychological Aspects of Operating Large Cooperative Space Ventures Involving Multinational Crews, O.G. Gazenko

AAS 70-152 Psychological Problems of Recruiting International Crews of Space Stations, M.A. Novikov, Yu. A. Senkevich

AAS 70-153 The Experience of ESRO in Managing International Space Projects, J.A. Vandenkerckhove

AAS 70-154 Television Broadcast Satellite Direct-to-Home, S.K. Sarkar

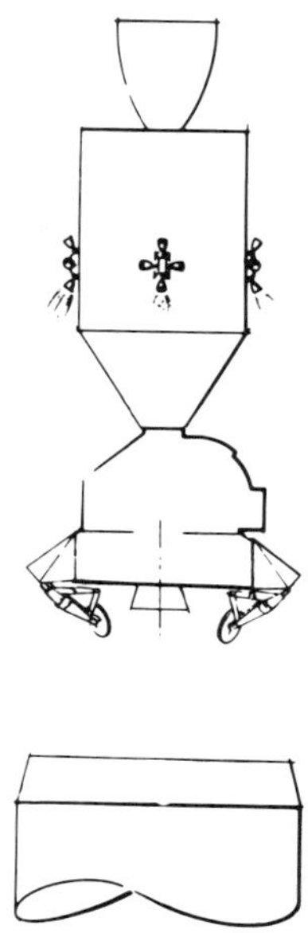

VOL. 23-3, AAS MICROFICHE SERIES
THIRD INTERNATIONAL SPACE RESCUE SYMPOSIUM, 1975
(21st Congress of the International Astronautical Federation, October 4-9, 1970, Constance, Germany)

IAF 70-01 Survey of Space Escape/Rescue/Survivability Capabilities, R. Fleisig, G.W. Heath, P.H. Bolger

IAF 70-02 Evolution of Emergency Systems for Manned Space Flight, P.H. Bolger

IAF 70-03 Space Rescue Operations, J.W. Wild, H. Schaefer

IAF 70-04 The Standardization of Space Vehicle Components, a Conditio Sine Qua Non for Successful International Space Rescue Operations, H.J.von Beckh

IAF 70-05 Communication and Rescue Time Constraints for Emergency Astronaut Return, A.R. Nagy, Jr., S.T. Chu

IAF 70-06 New Materials for Manned Spacecraft, Aircraft, and Other Applications, M.I. Radnofsky

IAF 70-07 A Lunar Orbiting Vehicle for Emergency Rescue, W.H. Jones

IAF 70-08 Lunar Mission Escape and Rescue Concepts, E. Perchonok

IAF 70-09 Application of Thermal Guidance Systems and Techniques to Space Rescue Preparations, J.D. Taylor

IAF 70-10 Space Emergency Re-entry Vehicle, C.J. Huang, S.L. Dickerson

IAF 70-11 Rescue and the Mini-Shuttle, R.H. Francis

IAF 70-12 Relative Importance of Different Biological Parameters for the Rescue of Astronauts, R. Grandpierre

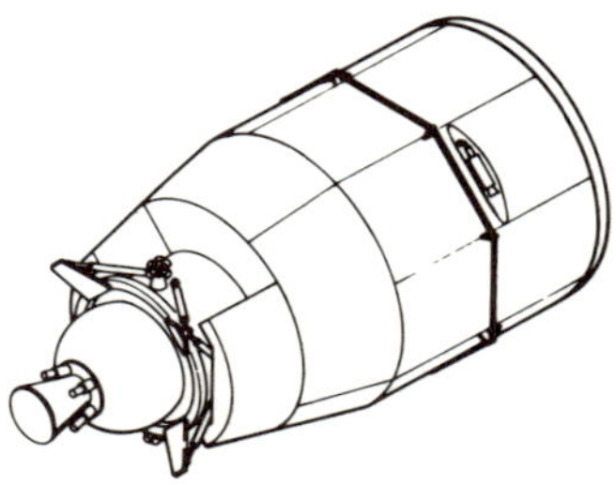

# AAS TECHNICAL PAPERS
# 1971

VOL. 27, SCIENCE AND TECHNOLOGY SERIES
INTERNATIONAL COOPERATION IN SPACE OPERATIONS AND EXPLORATION, 1971
(Ninth Goddard Memorial Symposium, March 10-11, 1971, Washington, D.C.)

AAS 71-001 International Cooperation in Space, H. Bondi

AAS 71-002 NASA's Plans for the 70's, W. von Braun

AAS 71-003 European Views on the NASA Programme for the 70's and 80's, J.P. Causse

AAS 71-004 European Industrial Views on NASA's Plans for the 70's, Y. Demerliac

AAS 71-005 U.S. Industrial Views on NASA's Plans for the 70's, W.F. Burke

AAS 71-006 Establishing the Permanent Intelsat, A. Washburn

AAS 71-007 Critical Systems and Technologies for the Future, C.S. Draper

AAS 71-008 Advanced Air Traffic Control Systems, J.A. Dinkespiler

AAS 71-009 Earth Resources Survey Systems - the Activities of Developing Countries, H.J. Ricciardi

AAS 71-010 Benefits from Space Applications in the 70's, W. von Braun

AAS 71-011 Lessons Learned from a Decade of International Space Cooperation, R.J.H. Barnes

AAS 71-012 Why Cooperate in Space Activities - the U.S. Diplomat's View, H. Pollack

AAS 71-013 Why Cooperate in Space Activities - a European Viewpoint, R.C. di Carrobio

AAS 71-014 An Assessment of US-USSR Cooperation in Space, M.L. Harvey



VARIABLE GEOMETRY AND EXPANDABLE STRUCTURES
(AAS/AIAA Meeting, April 21-23, 1971, Anaheim, California)
No proceedings volume.

CAPABILITIES FOR THE DESIGN OF MANNED AND UNMANNED SPACE MISSIONS
(AAS/JPL Meeting, May 26-27, 1971, Pasadena, California)
No proceedings volume.

AAS 71-015 to -099 Not available

VOL. 29 I & II, ADVANCES IN THE ASTRONAUTICAL SCIENCES
THE OUTER SOLAR SYSTEM, 1971 *
(AAS 17th Annual Meeting, June 28-30, 1971, Seattle, Washington)

AAS 71-100 Solar System Exploration--A Strategy for the 1970's, R.S. Kraemer

* See note next page.

| | |
|---|---|
| AAS 71-101 | The Pioneer F and G Missions, F.D. Kochendorfer, C.F. Hall |
| AAS 71-102 | Grand Tour Surveys of the Outer Solar System, J.W. Keller |
| AAS 71-103 | Jupiter Orbiters and Probes: A Preliminary Assessment of Requirements, J.E. Long |
| AAS 71-104 | An Assessment of Comet and Asteroid Missions, J. Niehoff |
| AAS 71-105 | Our Present State of Ignorance of the Outer Planets and Their Satellites, D.G. Rea |
| AAS 71-106 | The Masses of the Principal Planets, W.J. Klepczynski, P.K. Seidelmann, R.L. Duncombe |
| AAS 71-107 | Galilean Satellite Eclipse Observations: The Scientific Potential and Results of Recent Observations, T.F. Greene, R.W. Shorthill, L.G. Despain |
| AAS 71-108 | Radio Observations of Jupiter, J.K. Alexander |
| AAS 71-109 | Radio Emission from the Major Planets--The Thermal Component, S. Gulkis |
| AAS 71-110 | Grand Tour Meteoroid Environment Analysis, J.W. Haffner |
| AAS 71-111 | Pioneer Jupiter Precursor Probe to the Outer Planets, T.M. Lough |
| AAS 71-112 | State-of-the-Art Approach to Outer Planet Exploration, C.H. Guttman, J.E. Martin, Jr. |
| AAS 71-113 | Comet Rendezvous with Outer Planet S/C Via Jovian Powered Swingby, D.H. Robey |
| AAS 71-114 | 1977 JSP and 1979 JUN Combined Orbiter/Flyby Missions for Outer Planet Exploration, J. Walter, J. Archer, J. Crouch, L. McWhorter |
| AAS 71-115 | Optical and Electronic Imaging Systems for Solar System Exploration, P.A.M. Hawley |
| AAS 71-116 | Guidance and Navigation for Comet Rendezvous Missions, D. Fraser, H.L. Malchow, L. Sackett |
| AAS 71-117 | Navigation Analysis of a Jupiter--Uranus--Neptune 1979 Mission, T. Hendricks, R. Boain |
| AAS 71-118 | An Investigation of the Effects of the Major Earth-Based Navigation Error Sources for an Outer Planet Mission, D.W. Curkendall, J.P. McDanell |
| AAS 71-119 | Challenges in Developing an Approach Guidance Instrument for the Outer Planets, R.H. Stanton |
| AAS 71-120 | Approach System Guidance Sensor, G.E. Townsend, J.A. Hill, P.R. Rupert |
| AAS 71-121 | Terminal Guidance from a Spinning Spacecraft, R.A. Park, K.R. Jenkin, D.M. Layton |

*Papers AAS 71-100 through -135, and -167 through -169 appear in Part I; papers 71-136 through -166 appear in Part II.

AAS 71-122 Planetary Quarantine Considerations for Outer Planet Missions, W. Stavro, C. Gonzalez

AAS 71-123 Magnetic Field Measurements on the Grand Tour Missions to the Outer Planets: Objectives, P.J. Coleman, Jr.

AAS 71-124 The Plasma Experiment for the Outer Planet Missions, J.H. Wolfe

AAS 71-125 A Plasma Wave Instrument for the Outer Planets Grand Tour Missions, F.L. Scarf, A.M.A. Frandsen, D.A. Gurnett, R.A. Helliwell, R.E. Holzer, P.J. Kellogg, E.J. Smith, E. Ungstrup

AAS 71-126 Energetic Particles, S.M. Krimigis

AAS 71-127 Combined Zodiacal Experiment, R.K. Soberman

AAS 71-128 Radio Science Experiments on the Outer Planet Grand Tours Missions, V.R. Eshelman

AAS 71-129 Not Available

AAS 71-130 An X-Ray Experiment for the Outer Planet Grand Tour Mission, K.A. Anderson

AAS 71-131 The Infrared Investigations on the Outer Planets Grand Tour, C.B. Farmer

AAS 71-132 / -134 Not Available

AAS 71-135 Photopolarimetry (Abstract)

AAS 71-136 Post Encounter Mission Options for Pioneer Jupiter Flybys, W.J. Dixon

AAS 71-137 The Outer Planets - What Should the Missions Be?, P.A. Penzo

AAS 71-138 Mission Optimization for Satellite Imagery on Outer Planet Missions, A.B. Sergeyevsky

AAS 71-139 Effects on Multiple Outer Planet Missions of Using a Saturn-Jupiter Instead of a Jupiter-Saturn Flyby Sequence, J.W. Young, M.E. Hannah

AAS 71-140 Jupiter Orbiter Missions, J.C. Niehoff, W.C. Wells

AAS 71-141 Mission Analysis Performance Parameters for 1978-1980 Jovian Probe Oriented Missions, D.P. Fields, M.M. Moge, C.K. Wilkinson, Jr.

AAS 71-142 Preliminary Analysis of an Atmosphere-Entry Probe Mission to Jupiter, B.L. Swenson

AAS 71-143 Design of a Probe for Deep Penetration into Jupiter's Atmosphere, J. Pettus, D. Cross, J. Mellin, P. Reznicek

AAS 71-144 Aerodynamic Heating for Probe Vehicles Entering the Outer Planets, W.A. Page

AAS 71-145 Heat Protection for Atmospheric Entry into Saturn, Uranus, and Neptune, M.E. Tauber

AAS 71-146 Ablative Heat Shields for Jupiter Entry Probes, R.M. Wakefield, J.H. Lundell

AAS 71-147 Reflecting and Ablating Heat Shields for Radiative Environments, P.R. Nachtsheim, D.L. Peterson, J.T. Howe

AAS 71-148 Scientific Questions and Measurement Approaches for a Probe Entering the Atmosphere of Jupiter, A. Seiff

AAS 71-149 Configuring the Grand Tour Spacecraft, J.R. Eyman, E.G. Clegg

AAS 71-150 Integration of Radioisotope Thermoelectric Generators for the Pioneer F/G Program, G.J. Nothwang, A.C. Wilbur

AAS 71-151 Systems Impact of a Conical Antenna for Planetary Communications, T. Bright, R. Evey, R. Sheppard, R. Munson

AAS 71-152 Centralized Data Handling for Grand Tour Missions, R.A. Easton

AAS 71-153 Integration of a Jupiter Entry Probe on Outer Planet Spacecraft, J.R. Mellin

AAS 71-154 Propulsion Subsystem Design for Jupiter Orbiter Missions, R.L. Phen, J.P. Don

AAS 71-155 Design Considerations for Improvement in Spacecraft Reliability for Long Duration Space Missions, C.F. White, P.M. Hooten

AAS 71-156 Reliability Improvement Techniques for Long Life Missions, C.L. Stumpf

AAS 71-157 Skewed Versus Orthogonal Reaction Wheels for Outer Planet Exploration, J.A. Cummings, A.K. Goodfellow, R.F. Vaughan

AAS 71-158 Long Duration Mission Reliability Via Multiprocessing, M.L. Johnson

AAS 71-159 Tops Contribution to Long Life Design, P.O. Chelson, C.C. Wertz

AAS 71-160 Extended Lifetime Design in Radioisotope Thermoelectric Generators, H. Jaffe, P.A. O'Riordan

AAS 71-161 Parts Qualification and Acceptance for Long-Life Outer Planet Mission Spacecraft, J.J. Hendrick

AAS 71-162 Use of the Boeing Computer Simulator for Logic Design Confirmation and Failure Diagnostic Programs, A.W. VanAusdal

AAS 71-163 Scientific Goals of Missions Beyond the Solar System, L.G. Despain, J.P. Hennes, J.L. Archer

AAS 71-164 The Ultraplanetary Probe, K.A. Ehricke

AAS 71-165 The First Step Beyond the Solar System, A.J. Dessler, R.A. Park

AAS 71-166 The Scientific Exploration of Near Stellar Systems, J.L. Archer, A.J. O'Donnell

AAS 71-167 Pioneer-Class Mission Capabilities, H.F. Matthews, R.W. Jackson

AAS 71-168 Planetary Radio Astronomy on the Outer Planets Grand Tours, J.W. Warwick

AAS 71-169 Technology and a Time of Crisis - III, E.M. Cortright

AAS 71-170 to -299 Not assigned

VOL. 20, AAS MICROFICHE SERIES
AAS/AIAA 1971 ASTRODYNAMICS SPECIALIST CONFERENCE, 1972
(August 17-19, 1971, Ft. Lauderdale, Florida)

AAS 71-300 Advances in the Analysis of Shuttle Operations Between Earth and Lunar Orbits, E.D. Webb, P.J. Van Zytveld

AAS 71-301 Mission Analysis of Shuttle Operations Between Stations Orbiting the Earth and the Moon, R.D. Sugar, R.A. Winneberger

AAS 71-302 Alternate Operational Techniques Which Increase the Useful Payload of the Earth-to-Orbit Shuttle, S.F. McAdoo, Jr., J. Funk

AAS 71-303 Accommodation of Astronomy and Earth Observations by Space Shuttle Orbital Sortie Missions, D.E. Davin

AAS 71-304 Extended Launch Windows for Ground-Based Rescue Missions, W.A. Fey

AAS 71-305 A Trajectory Aiming Plane to Assist in Satellite Orbit Selection for Planetary Reconnaissance Missions, J.A. Borras

AAS 71-306 Optimal Analytic Multiburn Trajectories, D.J. Jezewski

AAS 71-307 Equivalence of the Minimum Norm and Gradient Projection Constrained Optimization Techniques, J.L. Junkins

AAS 71-308 Applications of a Mathematical Programming Technique to Finite Thrust Rocket Trajectory Optimization, A.G. Childs, E.S. Armstrong, A.T. Markos

AAS 71-309 Branched Trajectory Optimization by the Method of Steepest Descent, W.J. Thurneck, Jr.

AAS 71-310 Comparison of Minimum-Propellant Guidance Laws for Impulsive and Bounded-Thrust Spacecrafts, A.L. Kornhauser, P.M. Lion

AAS 71-311 A Mission Analysis Software System (MASS), W.L. Sachs, G.A. Hazelrigg, Jr., R.G. Brusch

AAS 71-312 Investigation of the Masses of the Outer Planets, P.K. Seidelmann, R.L. Duncombe, W.J. Klepczynski

AAS 71-313 A Nonlinear Stability Problem in the Three Dimensional Restricted Three Body Problem, K.T. Alfriend

| | |
|---|---|
| AAS 71-314 | Not Available |
| AAS 71-315 | Regularization of Grand Tour Trajectories, P.E. Nacozy |
| AAS 71-316 | Not Available |
| AAS 71-317 | An Extended Version of the Herrick-Gibbs Preliminary Orbit Determination Method, R.E. Coons |
| AAS 71-318 | Planetary Quarantine Constraints on Orbital Parameter Selection, R.J. Reichert |
| AAS 71-319 | Flight Path and Mission Strategies to Satisfy Outer Planet Quarantine Constraints, C. Gonzalez, W. Stavro |
| AAS 71-320 | Integrating the Relay Link Design and Trajectory Design for the Viking '75 Mission, D.W. Marquet, J.W. Berry, O.W. Smith |
| AAS 71-321 | Relay-Link Communications Geometry for a Jupiter Atmospheric Probe, R.B. Fischer |
| AAS 71-322 | An Analysis of the Bi-Injection Mode for Outer Planet Missions, P.R. Odom |
| AAS 71-323 | Characteristics of Mars Orbit Insertion Statistical $\Delta V$ Requirements for 1977 Landing Missions, A.L. Satin, M.M. Grogan |
| AAS 71-324 | A Characteristic Surface Representation of the Effects of Off-Optimum Propulsion Parameters on Low-Thrust Performance, K.L. Atkins |
| AAS 71-325 | Optimal Boost Trajectories for the Shuttle Vehicle, H.C. Sullivan, B.R. Uzzell |
| AAS 71-326 | A General Branched Trajectory Optimization Algorithm with Applications to Space Shuttle Vehicle Mission Design, H.L. Rozendaal |
| AAS 71-327 | Not Available |
| AAS 71-328 | Near Optimal Shuttle Trajectories Using Accelerated Gradient Methods, I.L. Johnson, J.L. Kamm |
| AAS 71-329 | Optimal Shuttle Trajectory - Vehicle Design Using Parameter Optimization, J.L. Kamm, I.L. Johnson |
| AAS 71-330 | Space Tug Performance Optimization, J.D. Mason |
| AAS 71-331 | Effects of Physical Librations of the Moon on the Orbital Elements of a Lunar Satellite, A.J. Ferrari, W.G. Heffron |
| AAS 71-332 | Periodic Solutions to Hamiltonian Systems with Infinitely Deep Potential Wells, W.B. Gordon |
| AAS 71-333 | On Some Reference Differential Equations with Applications to Celestial Mechanics, N.X. Vinh, E.V. Laitone |
| AAS 71-334 | A Novel Method for Averaging Equations of Motion, F. Morrison |

AAS 71-335 A Comprehensive and Trouble-Free Reformation of the Transfer Time Equation, R.F. Burt

AAS 71-336 Out-of-Plane Motion about Libration Points: Nonlinearity and Eccentricity Effects, T.A. Heppenheimer

AAS 71-337 Selenodetic Data Determined from Lunar Orbiter Satellites, W.R. Wells

AAS 71-338 Ellipsoidal Gravity Coefficients of the Geopotential, H.G. Walter

AAS 71-339 Earth Dynamic Filtering of Geodetic Data, P. Mennemeyer

AAS 71-340 Errors of Quadrature Connected with the Simple Layer Model of the Geopotential, K.R. Koch

AAS 71-341 Refined Geodetic Results Based on Doppler Satellite Observations, R.J. Anderle

AAS 71-342 A Fourier Series Approach to Gravity Anomaly Data Representation, M.H. Wallenstein

AAS 71-343 Information Content of the Mariner Mars 1971 Radio Tracking Data, J. Lorell

AAS 71-344 A Unified State Model of Orbital Trajectory and Attitude Dynamics, S.P. Altman

AAS 71-345 The Stability of Motion of Force-Free Spinning Satellites with Flexible Appendages, L. Meirovitch, R.A. Calico

AAS 71-346 Application of Canonical Transformation Theory to a Nonconservative Astrodynamical System, J.E. Cochran

AAS 71-347 Stability of a Dual-Spin Spacecraft with Energy Dissipation in a Flexible Momentum Wheel, P.M. Bainum, J.V.Fedor, P.G. Fuechsel

AAS 71-348 Viking Thrust Vector Control Dynamics Using Hybrid Coordinates to Model Vehicle Flexibility and Propellant Slosh, G.E. Fleischer, L.F. McGlinchey

AAS 71-349 Not Available

AAS 71-350 Analysis of an Experiment to Determine the Gravitational Constant in an Orbiting Space Laboratory, J.P. Vinti

AAS 71-351 Satellite Break-ups, J.R. Gabbard, H.B. Wackernagel

AAS 71-352 The Solar Radiation Pressure Force and Torques Model, R.M. Georgevic

AAS 71-353 Swathing Patterns of Earth-Sensing Satellites and their Control by Orbit Selection and Modification, J.C. King

AAS 71-354 Not Available

AAS 71-355 Automatic Multi-Satellite Ephemeris Maintenance, J.A. Ward, Jr.

AAS 71-356 Not Available

AAS 71-357 MESS - Misalignment Estimation Software System for In-Flight Alignment and Calibration of Spacecraft Attitude Sensors, R. desJardins

AAS 71-358 Grand Tour Mission Set Selection: Evaluation and Optimization, K.W. Ledbetter

AAS 71-359 Satellite Encounter Opportunities of the Grand Tour Missions, P.A. Penzo

AAS 71-360 Asteroid Encounters on Grand Tour and Other Missions Which Tranverse the Region Beyond Mars, D.F. Bender

AAS 71-361 Characteristics of Trajectories that Fly By Jupiter and Saturn and Return to Earth, V.R. Bond, K.W. Anson

AAS 71-362 Fuel Requirements for a Second Encounter with Mercury Subsequent to a Venus/Mercury Mission, D.V. Byrnes, G.C. Snyder

AAS 71-363 Halley's Comet Flythrough and Rendezvous Missions Via Solar Electric Propulsion, A.L. Friedlander

AAS 71-364 Gravitational Field Measurements of Planetary Bodies with Gravity Gradient Instrumentation, R.L. Forward

AAS 71-365 Optimum Successive Orbital Rendezvous, J.E. White

AAS 71-366 Application of Singular Perturbation Method to Optimum Lift Controlled Maneuvering Entry, Y.Y. Shi, L. Pottsepp

AAS 71-367 Optimal Low-Thrust Takeoff from an Orbit about an Oblate Planet, R.A. Jacobson, W.F. Powers

AAS 71-368 Transfers between Coplanar Coaxial Ellipses Using the Combination of High-and Low-Thrust Propulsion, C.J. Vahlberg, Jr., T.N. Edelbaum

AAS 71-369 Analysis of a Round-Trip Mission to Eros Using Chemical Propulsion, J.D. Adams, Jr., R.E. Drowns, G.A. Hazelrigg, Jr.

AAS 71-370 Optimal Round Trip Space Tug Trajectories for Earth Escape Missions, V.J. Weyers, F. Teren

AAS 71-371 Lunar Orbit Determination in the Presence of Unmodeled Accelerations, D.S. Ingram, B.D. Tapley

AAS 71-372 Tracking Phobos and Deimos Aboard an Orbiting Spacecraft, T.C. Duxbury, G.H. Born

AAS 71-373 Cartography and Orbit Determination from a Mars Orbiting Spacecraft, S.N. Mohan, G.A. Ransford

AAS 71-374 An Analysis of Approach Navigation Accuracy and Guidance Requirements for the Grand Tour Mission, D.W. Jones, B.D. Tapley

AAS 71-375 Orbit Prediction in the Presence of Gravitational Anomalies, S.R. Croopnick

AAS 71-376 Long-Term Orbital Trajectory Determination by Superposition of Gravity and Drag Perturbations, R.S. Pimm

AAS 71-377 Lifetime of a Lunar Orbit Space Station (LOSS), P.T. Guttman

AAS 71-378 Graphic Mission Analysis Techniques for Outer Planet Missions, R.P. Nagorski

AAS 71-379 Utility of Computer Graphics in Support of Grand Tour Mission Imaging Analyses, E.J. Dazzo

AAS 71-380 Automated Flight Path Design for Planetary Flyby Missions Using Advanced Interactive Graphics Hardware/Software, A.K. Mukhopadhyay

AAS 71-381 Calculation of Trajectories Utilizing a Slowly-Varying Jacobi Function, P.E. Russell, B. Culpepper

AAS 71-382 A Fast Method of Spacecraft Trajectory Computation in the n-Body Inverse Square Force Field, T.C. Fang

AAS 71-383 Early Solar System Escape Missions - An Epilogue to the Grand Tours, A.B. Sergeyevsky

AAS 71-384 Classical Least Squares and Sequential Estimation Techniques as Applied to the Analysis of the Mariner VI and VII Tracking Data, P.B. Esposito, C.L. Thornton, J.D. Anderson, D.O. Muhleman

AAS 71-385 An Analysis of Sequential Filtering for Estimating the State of a Planetary Orbiter, R.K. Russell, J.H. Hokhikian, D.W. Curkendall

AAS 71-386 On Erroneous Prediction of the Performance of Sequential Processors with Linear Methods, L.S. Diamant, A.C. Bond, D.M. Detchmendy, P.T. Pixley

AAS 71-387 / -388 Not Available

AAS 71-389 Slewing about Non-Orthogonal Axes, P.B. Davenport

AAS 71-390 Numerical Integration of Orbital Equations of Motion, L.J. Gulick

AAS 71-391 Orbit Determination Strategy and Results for the Mariner Mars 1971 Mission, J.W. Zielenbach, J.F. Jordan

AAS 71-392 Lunar Impact Targeting Technique Improvements and Simplifications Implemented for Apollo 14, J. Clingan, J. Crouch, L. McWhorter, J. Wyatt, J. Haussler, W. McFadden

AAS 71-393 Simplified Interplanetary Guidance Using Onboard Optical Measurements, H.A. Hamer

AAS 71-394 The Contribution of Onboard Optical Instrumentation to Deep Space Navigation, H.A. Garcia

AAS 71-395 Spectral Factorization in Time-Varying Systems Having Periodic Coefficients and Its Application to Navigation of Solar Electric Interplanetary Missions, T. Nishimura

AAS 71-396 Attitude Determination and Sensor Alignment Via Weighted Least Squares Affine Transformations, P.B. Davenport

AAS 71-397 Some Problems of Optical Orientation Determination While in Solar Inertial Attitude, L.J. Miller

AAS 71-398 Not Available

AAS 71-399 Applications of Quasi-VLBI Tracking Data Types to the Zero Declination and Process Noise Problems, V.J. Ondrasik, K.H. Rourke

VOL. 33 SCIENCE AND TECHNOLOGY SERIES
ORBITAL INTERNATIONAL LABORATORY, Third and Fourth OIL Symposia, 1974
(International Astronautical Federation, September 24-25, 1971, Brussels, Belgium)

AAS 71-400 Earth Environment and Resources Management, K.A. Ehricke

AAS 71-401 Planning for a Global Remote-Sensing Information System, J.M. Harrison, L.W. Morley, A.F. Gregory

AAS 71-402 Survey of Earth Resources Technology and Scientific Objectives, N. Takagi

AAS 71-403 Planning for an Operational Earth Exploration Satellite Service, S.K. Sarkar

AAS 71-404/-405 Not available

AAS 71-406 Some Aspects of Eventual International Cooperation for the Establishment of an Orbital International Laboratory (in French), S. Sanz Aranguez

AAS 71-407 Consequences of Space Technology and the Problems of the Developing Countries, H.K. Afshar

AAS 71-408 Not available

AAS 71-409 What Can an Orbital Space Station Do for your Nation? P. Hartl

AAS 71-410 Earth Resources Research and its Utilization by Developing Countries - Panel, C.S. Draper, Chairman

VOL. 24-1, AAS MICROFICHE SERIES
FOURTH INTERNATIONAL SPACE RESCUE SYMPOSIUM, 1975
(22nd Congress of the International Astronautical Federation, September 21, 1971, Brussels, Belgium)

IAF 71-01 Survey of Space Flight Safety Systems, R. Fleisig, P.H. Bolger

IAF 71-02 Long Range Planning for Development of Space Flight Emergency Systems, P.H. Bolger, C.W. Childs

IAF 71-03 Introduction to Recovery Session, G.W. Heath

IAF 71-04 Survey of Recovery Capabilities, G.W. Heath

IAF 71-05 The Role of the International Civil Aviation Organization in Organizing Search and Rescue Services, Air Navigation Bureau of the International Civil Aviation Organization (ICAO)

IAF 71-06 Potential Response of Maritime Services to Craft and Persons in Distress at Sea, Z.N. Sdougos

IAF 71-07 Availability of World-Wide Weather Data with Respect to Possibilities for Determining Recovery Sites, E. Dobryshman

IAF 71-08 The History and Development of World-Wide Survival Equipment, M.I. Radnofsky

IAF 71-09 Medical and Technical Aspects of Rescue and Survival of Astronauts in High Mountain and Mountainous Remote Areas, F. Bühler

IAF 71-10 Some Aspects of Survival and Rescue of Astronauts in Polar Regions, O. Wilson

IAF 71-11 Possibilities and Dangers During Very Long Working Periods in Space Rescue, J. Cmiral, J. Dvorak, J. Sulc

IAF 71-12 The Earth Orbit Shuttle as a Space Rescue Vehicle, M.G. Hinton, Jr., E. Perchonok

IAF 71-13 Space Rescue Requirements, R. Rose

# AAS TECHNICAL PAPERS
# 1972

VOLUME 28, SCIENCE AND TECHNOLOGY SERIES
ASTRONOMY FROM A SPACE PLATFORM, 1972
(December 27-28, 1971, AAS/AAAS/ORSA/American Astronomical Society Meeting, Philadelphia, Pa.)

AAS 72-001 Planetary Astronomy Objectives, C. Sagan

AAS 72-002 Solar Astronomy Objectives, F.Q. Orrall

AAS 72-003 Skylab and Its Solar Astronomy Experiments, W.C. Schneider

AAS 72-004 Solar Instrumentation, G. Newkirk, Jr.

AAS 72-005 Data Processing for Astronomy Missions, G.H. Ludwig

AAS 72-006 Stellar Astronomy Objectives, G.W. Preston

AAS 72-007 Extra-Galactic Astronomy Objectives, L.R. Doherty

AAS 72-008 Future Orbital Observatory Modules for Stellar and Galactic Astronomy, C.L. Kober

AAS 72-009 Very Long Baseline Interferometry: The Impact on Astronomy and Geophysics, I.I. Shapiro

AAS 72-010 Future Infrared Space Astronomical Instruments, W.F. Hoffmann

AAS 72-011 Future Unmanned Astronomy Spacecraft, F.P. Simmons

AAS 72-012 Unusual Objects and High Energy Astronomy, J.P. Ostriker

AAS 72-013 High Resolution Imagery with the Large Space Telescope, R.E. Danielson

AAS 72-014 The High Energy Astronomical Observatory, L.E. Peterson

AAS 72-015 X-Ray Astronomy -- Results and Instruments, H. Gursky 1972

AAS 72-016 High Energy Particle Astronomy, A. Buffington, R.A. Muller, L.H. Smith, G.F. Smoot

AAS 72-017 Synthetic Aperture Optics, R.H. Miller

AAS 72-018 Electronic Imaging Devices for Astronomy from a Space Platform, G.R. Carruthers

AAS 72-019 Automation in Astronomy, E.J. Wampler

AAS 72-020 Active Optics for Space Astronomy, H.F. Wischnia

AAS 72-021 Technology of Space Astronomical Instruments, M. Bottema

AAS 72-022 Selection of Astronomy Experiments for Space, F.L. Whipple

AAS 72-023 Not Available

AAS 72-024 Manned Versus Unmanned Space-Based Astronomy, K.G. Henize

AAS 72-025 Strategies for Space-Based Astronomy, J.E. Naugle

AAS 72-026 The Impact of the Proposed Space Astronomy Program on Our Society, W. Sullivan

AAS 72-027 to -029 Not Assigned

VOLUME 29, SCIENCE AND TECHNOLOGY SERIES
SPACE TECHNOLOGY TRANSFER TO COMMUNITY AND INDUSTRY, 1972
(AAS Eighteenth Annual Meeting and Tenth Goddard Memorial Symposium, March 13-14, 1972, Washington, D.C.)

AAS 72-030 Earth Benefits from Space and Space Technology, W. von Braun

AAS 72-031 NASA's Technology Utilization Program, J.T. Hamilton

AAS 72-032 American Astronautical Society User-Development Program, P.B. Richards

AAS 72-033 Some Practical Considerations in Technology Transfer, W.H. Pickering

AAS 72-034 Research Applied to National Needs, A.J. Eggers, Jr.

AAS 72-035 The Nixon Administration's Technology Program, E.E. David, Jr.

AAS 72-036 Not Available

AAS 72-037 Community and Industrial Challenges to Aerospace Technology in the Area of Industrial Instrumentation and Control, T.J. Williams

AAS 72-038 Aerospace Technology and the Petroleum Industry, W.E. Brown

AAS 72-039 Food and Related Bionomic Challenges to the Aerospace Industry, P.A. LaChance

AAS 72-040 The Challenge of the Textile Industry, F.X. Werber

AAS 72-041 Biomedical Research and Development, L. Laster

AAS 72-042 Experiences and Perceptions of Aerospace Technology Transfer to Health Care Delivery, C.W. Garrett, L.C. Von Renner

AAS 72-043 Application of Aerospace Technology to Electronic Medicine, J.R. Gilmer

AAS 72-044 Spacecraft Microbial Control Technology Applied to Medicine, R.H. Green

AAS 72-045 New Energy Technology, R.E. Balzhiser

AAS 72-046 DOT Philosophy of New Technology for Future Transportation Decisions, I.P. Halpern

AAS 72-047 Families of Transportation Technology Where Space Technology Can Apply, H.H. Richardson

AAS 72-048 Not Available

AAS 72-049 Automobile Design and Safety - Technology Tradeoffs and Transfer Processes, R. Brenner

AAS 72-050 to -099 Not Assigned

VOLUME 32, SCIENCE AND TECHNOLOGY SERIES
HEALTH CARE SYSTEMS, 1974
(User-Developer Conference on Health Care Systems, November 21-22, 1972, Dallas, Texas)

AAS 72-100 Conference Introduction and Overview, E.B. Konecci

AAS 72-101 AAS User-Developer Conference Concept, P.B. Richards

AAS 72-102 National Science Policy, R. Bisplinghoff

AAS 72-103 Health Care Systems: Introduction, E.B. Konecci

AAS 72-104 Keynote Address: National Health Care System, V.E. Wilson

AAS 72-105 The Manned Spacecraft Center and Medical Technology, R.S. Johnston, S.L. Poole

AAS 72-106 Practicing Physician's Point of View, H.E. Cromer

AAS 72-107 Design of Health Care Systems for the User, M.O. Breitmeyer

AAS 72-108 Rural Medicine, O.E. Bradsher

AAS 72-109 AMHT Developments, H.R. Oldfield

AAS 72-110 Systems Management for Improved Health Care, J. Smith

AAS 72-111 Pathology and Biomedical Testing, C. Thuss

AAS 72-112 Role of Material in Developing Artificial Organs: the Heart, R. Mayor, R. Pierson

AAS 72-113 Computerized Patient Management for Community Clinics, P.R. Tobias

AAS 72-114 Communication Systems for Emergency Medical Services, F.B. Vogt

AAS 72-115 Emergency Health Care: Helicopter Potential, P. Domanovsky

AAS 72-116 Technology for the Handicapped, H. Kafafian

AAS 72-117 The Electronic Medicine Concept, R.M. Lebovitz

AAS 72-118 PANEL: New Technology in the Treatment of Alcoholism, J.R. Gilmer, Moderator

AAS 72-119 Health Care Manpower, B. Locker

AAS 72-120 to -150 Not Assigned

VOLUME 30, SCIENCE AND TECHNOLOGY SERIES
SPACE SHUTTLE PAYLOADS, 1973
(AAS/AAAS/ORSA Meeting, December 27-28, 1972, Washington, D.C.)

AAS 72-151 The Space Shuttle System Description, Operations, and Payload Capability, J.C. Heberlig

AAS 72-152 Experimental Payloads: Inception to Integration, M. Bader, A.M. Worden

AAS 72-153 Astronomy from the Space Shuttle, H.J. Smith

AAS 72-154 Particle and Field Payloads, J.G. Roederer

AAS 72-155 Plasma Physics and Environmental Perturbation Laboratory, J.L. Vogl

AAS 72-156 Atmospheric Studies Payloads, R.D. Hudson

AAS 72-157 Earth Orbital Systems and Biomedical Research, R.T. Jordan

AAS 72-158 Use of Space Shuttle for Earth Resource Mapping, Inventory, and Evaluation, W.D. Carter

AAS 72-159 Use of the Space Shuttle for Communications and Navigation Experiments, R.B. Marsten

AAS 72-160 The Use of the Space Shuttle to Support Large Space Power Generation Systems, P.E. Glaser

AAS 72-161 Use of Shuttle for Test of Experimental Hardware in the Space Environment, R.L. Hammel

AAS 72-162 Use of Shuttle for Manufacturing and Materials Process Experiments in Low-G, L.R. McCreight

AAS 72-163 The MATHEMATICA Economic Analysis of the Space Shuttle System, K.P. Heiss

AAS 72-164 Low Cost Payload Studies, W.F. Moore

AAS 72-165 Teleoperators and EVA for Shuttle Missions, T.B. Malone, S. Deutsch

AAS 72-166 Handling, Control, and Servicing of Shuttle Payloads, L.O. Schulte

AAS 72-167 Research and Applications Modules (RAM), J.M. Macdonald

AAS 72-168 Planetary Exploration in the Shuttle System Era, W.J. Bursnall, G.R. Hook

AAS 72-169 Use of Shuttle in Establishing Large Space Installations, K.A. Ehricke

AAS 72-170 The Apollo/Soyuz Test Project, R.A. Petrone

AAS 72-171 International Cooperation on Space Shuttle Payloads, C.L. Kober

AAS 72-172 The European Shuttle Payload Activity, R.W. Johnson

AAS 72-173 PANEL: Space Shuttle Contributions to Technological, Economic, Political, and Defense Goals, G.W. Morgenthaler, Chairman

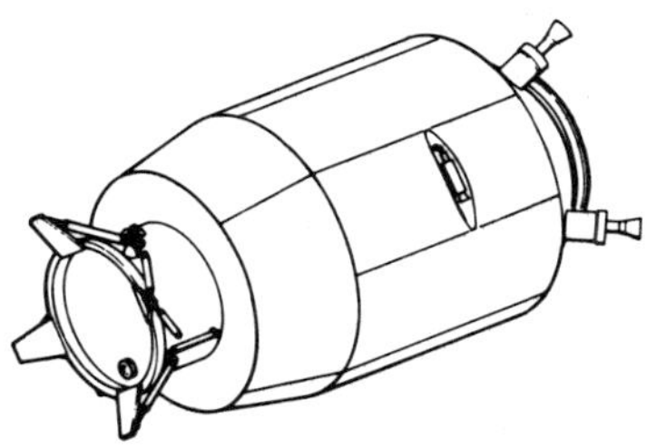

VOL. 24-2, AAS MICROFICHE SERIES
FIFTH INTERNATIONAL SPACE RESCUE SYMPOSIUM, 1975
(23rd Congress of the International Astronautical Federation, October 10, 1972, Vienna, Austria)

| | |
|---|---|
| IAF 72-00 | Human Stress Tolerances in Relation to Time and Intensity, P.A. Campbell |
| IAF 72-01 | Survey of Space Flight Safety Systems, R. Fleisig, P. Bolger |
| IAF 72-02 | Survey of Recovery Capabilities, G.W.Heath |
| IAF 72-03 | Space Rescue Operations in the Early 1980's, J.W. Wild, H. Schaefer |
| IAF 72-04 | A Global Alarm Network -- Warning and Location System for use in Space and Aviation Operations, P. Bolger |
| IAF 72-05 | Extending the Utility of the Earth Orbit Shuttle as a Space Rescue Vehicle, E. Perchonok |
| IAF 72-06 | The Space Transportation System (STS); an International Rescue System, F.X. Kane, R.J. Demshki |
| IAF 72-07 | Skylab Rescue Capability, K.S. Kleinknecht, L.G. Williams |
| IAF 72-08 | Theoretical Possibility of Technical Improvement of the De-spinning and De-tumbling Ability of Astronauts, J. Cmiral, J. Dvorak, I. Krekule |
| IAF 72-09 | De-spinning and De-tumbling Satellites in Rescue Operations, M.H. Kaplan |
| IAF 72-10 | Safety Design of Space Station Against Collision Hazards with Artificial Orbiting Bodies, M. Nagatomo, M. Matsuo K. Uesugi |

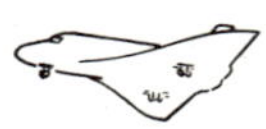

# AAS TECHNICAL PAPERS
# 1973

VOLUME 31 SCIENCE AND TECHNOLOGY
THE SECOND FIFTEEN YEARS IN SPACE, 1973
(Eleventh Goddard Memorial Symposium, March 8-9, 1973, Washington, D.C.)

AAS73-001 The Industrial Viewpoint, R. Anderson

AAS73-002 The Scientific Viewpoint, R.L. Bisplinghoff

AAS73-003 Solar Power for our Nation, T.J. Kelly, J. Mockovciak, Jr.

AAS73-004 Future National Space Efforts, F.E. Moss

AAS73-005 The Future Look in Rocket Engines, D.J. Sanchini, F.M. Kirby

AAS73-006 Computer Control for Manned and Automated Space Vehicles, A.L. Hopkins, Jr.

AAS73-007 Data Communications Technology, P.F. Glaser

AAS73-008, -009 Not Available

AAS73-010 The Viking Mission, J.S. Martin, Jr., G.A. Soffen

AAS73-011 to -013 Not Available

AAS73-014 Astronaut of 1988, D.K. Slayton

AAS73-015 European Industrial Cooperation in the Space, J. Henrici

AAS73-016 Review of European Space Projects after 1980, M. Levy

AAS73-017 Not Available

AAS73-018 An International Outlook of the Second Fifteen Years in Space, A.H. Abdel-Ghani

AAS73-019 to -049 Not Assigned

1973

VOLUME 34, SCIENCE AND TECHNOLOGY
MANAGEMENT AND DESIGN OF LONG-LIFE SYSTEMS, 1974
(AAS Symposium, April 24-26, 1973, Denver, Colorado)

AAS 73-050 The Long Life of Pioneer Interplanetary Spacecraft, W.J. Dixon

AAS 73-051 Long Life Technology Work at Rockwell International, D.K. Huzel

AAS 73-052 Highlights of the Long-Life Assurance Study (Abstract), H.L. Williams, J.B. Fox, R.W. Burrows

AAS 73-053 General Accelerated Methodology for Long-Life Mechanical Components (Abstract), D.B. Hamilton

AAS 73-054 Dependable Long-Lived Household Appliances, G.W. Jones

AAS 73-055 Design and Development Philosophy to Achieve Reliability and Long Life in Large Turbine Generators, C.W. Elston

AAS 73-056 Integrated Management Approach to Long Life in Space, E. Schaeffer

AAS 73-057 Artificial Heart Reliability Factors, E.N. Huffman, K.G. Hagen

AAS 73-058 Architectural Concepts and Redundancy Techniques in Fault - Tolerant Computers, D.A. Rennels

AAS 73-059 Design of Mechanical Components for Long-Term Performance, R.J. Salvinski

AAS 73-060, -061 Not Available

AAS 73-062 Test Programs for Long-Life Hardware: Report of the Test Working Session, D.S. Peck, R.W. Burrows

AAS 73-063 Systems Design for Long-Life Systems, R.F. Miles, Jr.

AAS 73-064 Guidelines for Long-Life Mechanical Equipment, C.H.T. Pan, R. Gates, J.K. Hoffman, D.K. Huzel

AAS 73-065 Guidelines for Long-Life Electronic Equipment: Summary, E.R. Dreher

AAS 73-066 to -099 Not Assigned

VOLUME 30, ADVANCES IN THE ASTRONAUTICAL SCIENCES
INTERNATIONAL CONGRESS OF SPACE BENEFITS, 1974
(AAS Nineteenth Annual Meeting, June 19-20, 1973, Dallas, Texas)

AAS73-100 International Cooperation in Space, F.E. Moss

AAS73-101 Opportunities in Space, H. Ziegler

AAS73-102 to -105 Not Available

AAS73-106 Educational Television in India, U.R. Rao

AAS73-107, -108 Not Available

AAS73-109 Global Aspects of Educational TV by Satellite, F. Fiario

AAS73-110 Not Available

AAS73-111 Radio Frequency Sensors, R.K. Moore

AAS73-112, -113 Not Available

AAS73-114 Sensor Development, an Overview of Recent Canadian Experience, J. MacDowall, P.A. Lapp

AAS73-115 The Use of Remote Sensing in the U.S.S.R. for the Study of Natural Resources, V.B. Komarov

AAS73-116 Applications of Observations from ERTS 1 to Earth Resources Survey and Management, W. Nordberg

AAS73-117 Not Available

AAS73-118 Canadian Interests and Activities in Space Communications and Navigation, J.H. Chapman

AAS73-119 U.K. Interests in Space Communications and Navigation, G.K.C. Pardoe

AAS73-120 The Aerosat Programme, J.P. Contzen, J.A. Vandenkerckhove

AAS73-121 Small Earth Terminal Applications for Satellite Communications, A.M. Werth, T. Sekimoto

AAS73-122 Environmental Monitoring via the ERTS-1 Data Collection System, J.E. Painter, J.E. Seitner

AAS73-123 Not Available

AAS73-124 Infrared Remote Sounding of the Atmosphere from Space, D.Q. Wark

AAS73-125 Atmospheric Sounding from an Earth Satellite, R.M. Scott

AAS73-126 Techniques for Deriving Winds from Cloud Movement, L.F. Hubert

AAS73-127 Applications of Space Data to Forecasting, W.D. Bonner

AAS73-128 Not Available

AAS73-129 Management Looks at the Canadian Program of Remote Sensing, Phase I, 1971-1975, L.W. Morley, D.J. Clough

AAS73-130 Contributions of the EROS Program to the Department of the Interior's Resources and Management Responsibilities, W.A. Fischer

AAS73-131, -132 Not Available

AAS73-133 Some Fruits from the Seeds of Space Science, C. Raphael

AAS73-134 Not Available

AAS73-135 An Apollo Compatible Cloud Physics Experiment, L.A. Eaton, A.B. Hollinden, P.R. Satterblom

AAS73-136 Study of Laser Remote Sensing Techniques from Space Platforms, J.J. Grossman, H.G. Gross, H.A. Hyatt

AAS73-137 Not Available

AAS73-138 Environmental Data -- From Sensors to Users, T.S. Austin

AAS73-139 to -143 Not Available

AAS73-144 Buoy Tracking of Ocean Currents, R.L. Molinari

AAS73-145 On a Dedicated Oceanographic Satellite, J.R. Apel

AAS73-146 Infrared Sensing of Ocean Surface Temperature, G.A. Maul

AAS73-147 Not Available

AAS73-148 Satellite Communications in the 1980's - a European View, T.F. Howell

AAS73-149 Not Available

AAS73-150 Satellite Utilization for Educational Communications in the U.S., R.P. Morgan, J.P. Singh

AAS73-151 Design Concepts for an Earth Resources Data Management System, T.M. Ragland

AAS73-152 Remote Sensing Data Management from a User's Viewpoint, W.H. Stevenson, T.M. Vanselous

AAS73-153 Analysis Research for Earth Resource Information Systems: Where Do We Stand?, D.A. Landgrebe

AAS73-154 Not Available

AAS73-155 Some Developments in Semi-Direct Broadcast Satellites and Community Receiving Systems, L.G. Ludwig

AAS73-156 to -159 Not Available

AAS73-160 The Space Shuttle Program and Its Technology, R.F. Thompson

AAS73-161 Application of Space Technology to Ground Transportation, M. Klein

AAS73-162 Space Technology Applied to Advanced Sea Transportation, M. Pitkin

AAS73-163 Global Rocket Transports: Changing Perspectives, R. Salkeld

AAS73-164 to -169 Not Available

AAS73-170 On the Choice and Optimization of Parameters for the Orbital Station "Salyut" Photography System, Yu. K. Khodarev, B.N. Rodionov, Yu. M. Chesnokov, B.S. Dunaev, Yu. I. Fivensky, Yu. L. Biryukov

AAS73-171 to -179 Not Available

AAS73-180 How to Make It Happen Telesat Canada, R.F. Chinnick

AAS73-181 to -199 Not Assigned

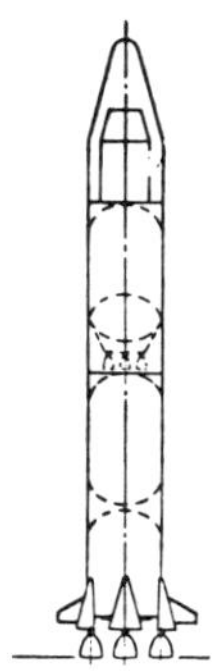

VOL. 21 AAS MICROFICHE SERIES
AAS/AIAA ASTRODYNAMICS SPECIALIST SYMPOSIUM 1973
(AAS/AIAA Astrodynamics Specialist Symposium, July 16-18, 1973, Vail, Colorado)

AAS 73-200 Not Assigned

AAS 73-201 An Orbital Radar Mapper of Venus in the 1980's - Mission Design and Analysis, S.K.Asnin

AAS 73-202 Multiple-Probe Targeting for a Spinning Spacecraft on a Planetary Encounter Trajectory, L.R.Anderson

AAS 73-203 Not Available

AAS 73-204 Mission Design for a 1980 Encke Slow Flyby Using Solar Electric Propulsion, D.F.Bender, K.L.Atkins, C.G.Sauer,Jr.

AAS 73-205 Damping Spacecraft Nutation by Means of a Despun Antenna, I.P.Leliakov, P.M.Barba

AAS 73-206 On the Rotational Motion of a Freely Rotating Body Induced by Mass Redistribution, J.E.Cochran,Jr., N.O.Speakman

AAS 73-207, -208 Not Available

AAS 73-209 Gravity Gradient Stabilization of Satellites in Highly Eccentric Orbits, R.E.Lohfeld, D.K.Anand, J.M.Whisnant

AAS 73-210 Analytical Solutions for Euler Parameters, H.S.Morton,Jr., J.L.Junkins, J.N.Blanton

AAS 73-211 Tracking and Orbit Determination of Near-Earth Orbiting Satellites Using Earth Synchronous Relay Satellites, W.C.Bryant,Jr., C.C.Goad

AAS 73-212, -213 Not Available

AAS 73-214 A Highly Efficient Error Analysis Program for the Evaluation of Spacecraft Tests of General Relativity with Application to Solar Probes, J.D.Anderson

AAS 73-215 Refinement of the Earth's Gravitational Constant Using Drag-Free Satellite Ranging Techniques, J.V.Breakwell, M.Hough

AAS 73-216 The Long-Term Motion of Artificial Jovian Satellites, C.Uphoff

AAS 73-217 Error Analysis of Multi-Conic Techniques, L.A.D'Amario, T.N.Edelbaum

AAS 73-218 Modified Multirevolution Integration Methods for Satellite Orbit Computation, O.E.Graf, D.G.Bettis

AAS 73-219 Applications of the Kustaanheimo-Stiefel Transformation of the Perturbed Two-Body Problem, V.R.Bond

AAS 73-220 The Lyapunov Stabilization of Satellite Equations of Motion Using Integrals, P.E.Nacozy

AAS 73-221 Notions of Analytic vs Numerical Stability as Applied to the Numerical Calculation of Orbits, C.E.Velez

AAS 73-222, -223 Not Available

AAS 73-224 Application of Aerospace Techniques to Water Quality Modeling and Control, R.P.Canale, W.F.Powers

AAS 73-225 Not Available

AAS 73-226 Application of Estimation Theory to the Classification of Vectorcardiograms, M.E.Womble, J.S.Halliday, S.K.Mitter, F.G.Yanowitz, M.C.Lancaster

AAS 73-227 Not Available

AAS 73-228 Effects of Mission Constraints on Viewing a Spinning Elongated Asteroid During Flyby, P.R.Odom, L.C.Allen

AAS 73-229 Maneuver Strategy for a Possible Pioneer 11 Extended Mission, R.B.Frauenholz

AAS 73-230 Computation and Analysis of Satellite Opportunities for the Mariner Jupiter/Saturn 1977 Mission, J.G.Beerer, P.A.Penzo R.A.Wallace

AAS 73-231 The Jupiter Orbiter Satellite Tour Mission, J.C.Beckman, D.B.Smith

AAS 73-232 Librational Dynamics of Satellites with Thermally Flexed Appendages, V.J.Modi, K.Kumar

AAS 73-233 Liapunov Stability Analysis of Hybrid Dynamical Systems in the Neighborhood of Nontrivial Equilibrium, L.Meirovitch

AAS 73-234 Not Available

AAS 73-235 Simulation of Flexible Appendage Interactions with Mariner Venus/Mercury Attitude Control and Science Platform Pointing, G.E.Fleischer

AAS 73-236 A Suboptimal Stochastic Controller for an N-Body Spacecraft, V.Larson

AAS 73-237 The Stability of a Dual-Spin Satellite, K.T.Alfriend, C.H.Hubert

AAS 73-238 Optimal Multiple-Impulse Direct Ascent Fixed-Time Rendezvous, L.R.Gross, J.E.Prussing

AAS 73-239 Station-Keeping for a Translunar Communication Station, J.V.Breakwell, A.A.Kamel, M.J.Ratner

AAS 73-240 Function Space Quasi-Newton Algorithms for Optimal Control Problems with Bounded Controls and Singular Arcs, E.R.Edge, W.F.Powers

AAS 73-241 Contribution to the Trajectory Optimization Problem of a Low-Thrust Space Vehicle, K.J.Schwenzfeger

AAS 73-242 A Contrained Discrete Optimal Guidance Strategy for Low-Thrust Spaceflight, R.A.Jacobson

AAS 73-243 Mariner Venus/Mercury 1973 Navigation Strategy, E.L.McKinley, J.B.Jones, M.H.Bantell,Jr.

AAS 73-244 Analytical Methods of Orbit Determination, D.W.Curkendall, V.J.Ondrasik

AAS 73-245 Navigation and Dispersion Analysis for Probe Missions to Jupiter, Saturn, and Uranus, T.C.Hendricks, E.D.Vogt, D.V.Byrnes

AAS 73-246 Advanced Pioneer Guidance and Navigation Requirements for Outer Planet Missions, C.K.Paul, R.K.Russell, J.Ellis

AAS 73-247 Mariner/Jupiter/Saturn Navigation in the Presence of Massive Planetary Satellites, C.E.Hildebrand, C.E.Kohlase, J.F.Dixon, G.A.Ransford

AAS 73-248 Mariner 9, an Instrument of Dynamical Science, J.F.Jordan, J.Lorell

AAS 73-249 Mars Gravity Derived from the Long-Period Motion of Mariner 9, A.J.Ferrari, E.J.Christensen

AAS 73-250 Mars Physical Parameters as Determined from Mariner 9 Observations of the Natural Satellites, G.H.Born

AAS 73-251 Topographical Coordinates from Mariner 9 Observations of the Martian Surface, S.N.Mohan

AAS 73-252 Mariner 9 Extended Mission Sequence Design Optimization, R.J.Reichert, W.A.Webb

AAS 73-253 The Solar Radiation Pressure on the Mariner 9 Mars Orbiter, R.M.Georgevic

AAS 73-254 A Method of Numerical Integration for Trajectories with Variational Equations, W.H. Goodyear

VOL. 24-3/-4 AAS MICROFICHE SERIES
SIXTH INTERNATIONAL SPACE RESCUE SYMPOSIUM 1973
(International Academy of Astronautics, Baku, USSR)

IAF 73-01 Survey of Recovery Capabilities, G.W.Heath (v24-3/-4 Micro)*

IAF 73-02 Reliability of Voice Communications between Space Crew Members of Different Nationalities, V.A.Popov, A.V.Nikonov, N.V.Krylova (v24-3 Micro)

IAF 73-03 Reliability of Life Support Systems as Related to General Space Flight Safety Requirements, B.A.Adamovich, G.G.Ter-Minasyan (v24-3 Micro)

IAF 73-04 Radiation Protection Reliability and Space Flight Safety, E.E.Kovalev (v24-3 Micro)

IAF 73-05 Medical Aspects of Aircrew Rescue after Emergency Landing (Ditching) in Uninhabited Area, V.G.Volovitch (v24-3 Micro)

IAF 73-06 Spacecraft Crew Safety Terms and Definitions, G.T.Beregovoi, G.P.Shibanov, V.I.Yaropolov, I.I.Baranetski (v24-3/-4 Micro)

IAF 73-07 Safety Considerations Involved in Testing, Checkout, and Launch Operations at Cape Kennedy, P.H.Bolger, W.Boyes (v24-4 Micro)

IAF 73-08 Survey of Space Flight Safety Systems, R.Fleisig (v24-4 Micro)

IAF 73-09 Techniques for Detumbling a Disabled Space Base, M.H.Kaplan (v24-4 Micro)

IAF 73-10 The Simulation of Human Reactions under Near-Vacuum Conditions: Reactions to Deep Anoxia, L.Cottl, J.Dvorak (v24-4 Micro)

IAF 73-11 Psychological Problems in a Multinational Crew during an International Space Mission, M.A.Novikov, Yu.A.Senkevich (v24-4 Micro)

IAF 73-12 An Optimized Space Rescue System, F.X.Kane (v24-4 Micro)

IAF 73-13 Engineering Principles to Assure Compatible Docking between Future Spacecraft of USA and USSR, C.C.Johnson (v24-4 Micro)

---

* Micro=AAS Microfiche Series

# AAS TECHNICAL PAPERS
# 1974

VOLUME 35, SCIENCE AND TECHNOLOGY,
ENERGY DELTA, SUPPLY VS. DEMAND, 1975; reprinted 1976
(AAS/AAAS Symposium, February 25-27, 1974, San Francisco, California)

AAS 74-001 Energy Supply and Demand Challenges and Some Possible Solutions, M.R. Thomasson

AAS 74-002 Energy Research & Development Alternatives for Future Supply, C. Starr

AAS 74-003 Time Factors in Slowing Down the Rate of Growth of Demand for Primary Energy in the United States, L. Lees, M.P. Lo

AAS 74-004 Energy Usage in the Home - Consumption and Conservation, R.B. Rosenberg

AAS 74-005 Evolution and Effects of Energy Conservation in the Transportation Sector, R.L. Strombotne, B. Yang

AAS 74-006 HUD Takes Total Energy Concept One Step Beyond (Abstract), G.S. Leighton

AAS 74-007 Industrial Energy Conservation Overview, M. Meyerson

AAS 74-008 Energy Conservation in Primary Metals Processing, A.S. Russell

AAS 74-009 Not Available

AAS 74-010 Energy Conservation in the Cement Industry, H.M. Garrett, J.A. Murray

AAS 74-011 Energy and the Structure of Adaptation (Abstract), R.A. Rappaport

AAS 74-012 Not Available

AAS 74-013 Energy and Politics, R. Engler

AAS 74-014 Energy Decline - Railroad Revival?, F. Cottrell

AAS 74-015 Outlook for Petroleum-Based Energy Sources, L.G. Stewart

AAS 74-016 Future Availability of Natural Gas, F.D. Hart, L.W. Fish

AAS 74-017 The Outlook for Shale Oil, A.E. Lewis

AAS 74-018 Not Available

AAS 74-019 Coal Gasification: A Review of Status and Technology, J.G. Conner

AAS 74-020 Environmental Impact of Large Scale Development of Coal and Oil Shale, J.P. Brennan

AAS 74-021 Development of the Nuclear-Electric Energy Economy, P.N. Ross

AAS 74-022 The Prospects for Large Scale Breeder Reactors (Abstract), B. Wolfe

AAS 74-023 The Outlook for Fusion Energy Sources - Remaining Technological Hurdles, R.F. Post

AAS 74-024 Solar Heating and Cooling of Buildings, R.D. Bourke, E.S. Davis

AAS 74-025 Roles for Solar Thermal Conversion Systems in our Energy Economy, D.F. Spencer, A.B. Greenberg

AAS 74-026 The Solar Community - Energy for Residential Heating, Cooling, and Electric Power, W.H. McCulloch, D.O. Lee, W.P. Schimmel, Jr.

AAS 74-027 Solar/Hydroelectric Combined Power Systems, F.A. Blake

AAS 74-028 Current Worldwide Utilization and Ultimate Potential of Geothermal Energy Systems, L.J.P. Muffler

AAS 74-029 New Technology Challenges in Exploration, Exploitation, and Environmental Impact of Geothermal Systems, G.V. Keller

AAS 74-030 Salt Domes, Pit Craters, and Dry Steam Fields - Heat Pipe Applications, J. Green

AAS 74-031 Ocean Thermal Power and Windpower Systems - Natural Solar Energy Conversion for Near-Term Impact on World Energy Markets, W.E. Heronemus, J.G. McGowan

AAS 74-032 Hydrogen - A Carrier of Energy, D.P. Gregory

AAS 74-033 Prospects of Photosynthetic Energy Production, B. Kok

AAS 74-034 Demand for Energy as a Function of Price, H.S. Houthakker, M. Kennedy

AAS 74-035 Energy Supply as a Function of Price, M.F. Searl

AAS 74-036 The Energy Game and The Role of Uncertainty, A.H. Packer

AAS 74-037 Panel Discussion: Recommended Strategies for Minimizing the Energy "Crunch": 1975 - 1990, and Beyond 1990, G.W. Morgenthaler

AAS 74-038 to -049 Not Assigned

VOLUME 38, SCIENCE AND TECHNOLOGY,
SKYLAB SCIENCE EXPERIMENTS, 1975
(AAS/AAAS Symposium, February 25-27, 1974, San Francisco, California)

AAS 74-050 Skylab Deep Space Experiments, R.A. Parker

AAS 74-051 Skylab Solar Astronomy Experiments, R.L. Schweickart

AAS 74-052 Skylab Earth Atmosphere Experiments, O.K. Garriott

AAS 74-053 Crystal Growth and Segregation Behavior Under Zero Gravity Conditions: InSb, H.C. Gatos, A.F. Witt, C.J. Herman, M.C. Lavine, M. Lichtensteiger

AAS 74-054 Directional Solidification of InSb-GaSb Alloys, W.R. Wilcox, J.F. Yee, M. Lin, K. Sarma, S. Sen

AAS 74-055 Crystal Growth of IV-VI Compounds by Vapor Transport in Space: A First Report, H. Wiedemeier, F.C. Klaessig,

AAS 74-056 The Influence of Man in Space - Observations, T.E. Hanes

AAS 74-057 The Response of Single Human Cells to Zero Gravity, P. O'B. Montgomery, Jr., J.E. Cook, R.C. Reynolds, J.S. Paul, L. Hayflick, W.W. Schulz, D. Stock, S. Kinzey, T. Rogers, D. Campbell, J. Murrell

AAS 74-058 Vestibular Side Effects in the Orbital Stage of Skylab II and III Missions, A. Graybiel, E.F. Miller II, and J.L. Homick

AAS 74-059 Skylab Medical Contributions and Significance to Manned Space Flight, C.A. Berry

AAS 74-060 Land Use Studies with Skylab Data, W.G. Rohde, D.S. Simonett

AAS 74-061 Skylab Data and Water Resource Management, F.C. Polcyn, T.W. Wagner

AAS 74-062 Agricultural, Forest, and Range Inventory Using Skylab Data, R.N. Colwell

AAS 74-063 Panel Discussion: Where Do We Go From Here?, G.W. Morgenthaler

AAS 74-064 to -074 Not Assigned

VOLUME 36, SCIENCE AND TECHNOLOGY
SKYLAB AND PIONEER REPORT, 1975
(Twelfth Goddard Memorial Symposium, March 8, 1974, Washington, D.C.)

AAS 74-075 The Skylab Program - an Overview, J.H. Disher

AAS 74-076 Apollo Telescope Mount Experiments Technology, W.C. Keathley

AAS 74-077 Manufacturing in Space and Materials Sciences on Skylab, E. Stuhlinger

AAS 74-078 The Medical Story, R.S. Johnston, L.F. Dietlein, E.L. Michel

AAS 74-079 Earth Resources Experiments and Results, J.G. Zarcaro

AAS 74-080 Not Available

AAS 74-081 Pioneer Missions to Jupiter, C.F. Hall

AAS 74-082 to -099 Not Assigned

VOLUME 31, ADVANCES IN THE ASTRONAUTICAL SCIENCES
THE SKYLAB RESULTS, 1975
VOLUME 22, AAS MICROFICHE SERIES
(AAS 20th Annual Meeting, August 20-22, 1974, Los Angeles, California)

AAS 74-100 to -105 Not Available

AAS 74-106 Skylab Vibroacoustic Test Program (Abstract & Microfiche)*, R.E. Jewell

AAS 74-107 Cluster Systems Integration, C.C. Hagood, K.P. Timmons

AAS 74-108 Cluster Man/System Design Requirements and Verification, H.H. Watters

AAS 74-109 MSFC Skylab System Safety Program, E.M. McNail

AAS 74-110 Skylab Contamination Control, C.M. Davis

AAS 74-111 MSFC's Utilization of Existing Design Technology in the Skylab Program, R.G. Eudy

AAS 74-112 Not Assigned

---

* Those items marked "Abstract & Microfiche" appear in Vol. 31, Advances in the Astronautical Sciences in abstract form only but are published in Volume 22, AAS Microfiche Series - a supplement to Vol. 31.

AAS 74-113 Management and Control of Skylab Missions, J.W. Roach, E.F. Kranz

AAS 74-114 Skylab Flight Plan Development, W.M. Anderson, T.W. Holloway

AAS 74-115 Not Available

AAS 74-116 MSFC Engineering Support to Skylab Operations, H.F. Kurtz, Jr., W.G. Clarke, M. Naumcheff

AAS 74-117 Experiment Planning during the Skylab Missions, J.R. Siever, R.A. Parker

AAS 74-118 Not Assigned

AAS 74-119 Flight Crew Observations of Spacecraft Systems, R.L. Crippen, R.H. Truly

AAS 74-120 Skylab Extravehicular Activity, D.C. Schultz, R.R. Kain, R.S. Millican

AAS 74-121 Skylab EVA System Development, R.T. Heckman

AAS 74-122 Repair of Major System Elements on Skylab, R.E. Pace, Jr.

AAS 74-123 Skylab Mission Task and Work Performance (Abstract), J.F. Kubis, E.J. McLaughlin

AAS 74-124 Visual Observations from Space, W.B. Lenoir, J.L. Kaltenbach, M.C. McEwen, R.A. Weitenhagen, V.R. Wilmarth

AAS 74-125 Not Assigned

AAS 74-126 Design and Operation of the Skylab Attitude and Pointing Control System, J.F. Applegate, J.M. McMillion, R.E. Smith, Jr.

AAS 74-127 Design and Performance of Skylab Thermal/Environmental Control System, G.D. Hopson, J.W. Littles, W.C. Patterson

AAS 74-128 Performance of Solar Shields, R.J. Schwinghamer

AAS 74-129 Skylab Technology - Power Systems, A.P. Woosley, O.B. Smith, H.S. Nassen

AAS 74-130 Skylab Electronic Technological Advancements, G.L. Hornback

AAS 74-131 Data Management and Analysis (Abstract), C. Houston

AAS 74-132 Not Assigned

AAS 74-133 Skylab Experiment M487: Habitability/Crew Quarters, C.C. Johnson

AAS 74-134 Skylab Experiment M516: Crew Activities/Maintenance Study, R.L. Bond

AAS 74-135 An Evaluation of Skylab Habitability Hardware, J.W. Stokes

AAS 74-136 Design, Development, and Operation of a Zero Gravity Shower, R.L. Middleton, A.C. Krupnick, B.J. Schrick, J.C. Reily

AAS 74-137 Skylab Experiment M509: Astronaut Maneuvering Equipment - Orbital Test Results and Future Applications, C.E. Whitsett, Jr., B. McCandless II

AAS 74-138 Skylab Experiment T-020: Preliminary Results Concerning a Foot-Controlled Maneuvering Unit, D.E. Hewes

AAS 74-139 Investigation of Crew Motion Disturbances on Skylab Experiment T-013, B.A. Conway

AAS 74-140 Not Assigned

AAS 74-141 Flight Performance of the Skylab Earth Resources Experiment Package, A.E. Potter, A.L. Grandfield, C.K. Williams

AAS 74-142 Use of Skylab S-190A and S190B Photographs in Wyoming Earth Resources Studies, R.S. Houston, R.W. Marrs

AAS 74-143 Evaluation of Skylab Data for Land Use Mapping (Abstract & Microfiche), L.L. Biehl, L.F. Silva

AAS 74-144 Recent Processed Results from the Skylab S-192 Multispectral Scanner (Abstract & Microfiche), F.J. Thomson, R.K. Vincent, R.F. Nalepta

AAS 74-145 Automatic Categorization of Land-Water Cover Types of the Green Swamp, Florida, Using Skylab Multispectral Scanner (S-192) Data (Abstract & Microfiche), A.E. Coker, A.L. Higer, R.H. Rogers, N.J. Shah, L. Reed, S. Walker

AAS 74-146 Remote Sensing of Soil Moisture by Skylab Radiometer and Scatterometer Sensors (Abstract & Microfiche), J.R. Eagleman, F.T. Ulaby

AAS 74-147 The Kennedy Space Center Role in Skylab, W.H. Rock

AAS 74-148 Skylab Facilities Modifications at KSC, J.D. Phillips

AAS 74-149 Skylab Checkout and Launch Operations at KSC, R.E. Moser

AAS 74-150 Skylab Experiments Processing at KSC, U.R. Barnett

AAS 74-151 Introduction to the Session on Skylab Student Science (Abstract & Microfiche), H.B. Floyd

AAS 74-152 Development Innovations for Skylab Student Experiments, R.L. Gause

AAS 74-153 Not Available

AAS 74-154 Earth Orbital Neutron Analysis: A Skylab Student Project (Abstract & Microfiche), T. Quist

AAS 74-155 The Skylab Student Project - A Science Educator's Appraisal, L. Summerlin

AAS 74-156 The Skylab Student Science Program from a Student Investigator's Point of View, J. Leventhal

AAS 74-157 Not Assigned

AAS 74-158 Skylab Experiment Integration Overview, S.R. Schrock, B.P. Woods

AAS 74-159 Specimen Analysis of the Skylab M553 Metals Melting and Solidification Experiment (Abstract), D. Larson, Jr.

AAS 74-160 Not Available

AAS 74-161 Ultraviolet Stellar Spectra Obtained with Skylab Experiment S019, K.G. Henize, J.D. Wray, S.B. Parsons, G.F. Benedict, Y. Kondo, F.G. O'Callaghan

AAS 74-162 Magnetospheric Particle Composition Experiment, J. Geiss, D.L. Lind, W. Stettler

AAS 74-163 Composition of Cosmic Ray Nuclei with $Z \geq 70$ (Abstract), P.B. Price, E.K. Shirk

AAS 74-164 Comet Kohoutek Observations from Skylab, W.C. Snoddy, R.J. Barry

AAS 74-165 Not Assigned

AAS 74-166 A Discussion of the Development of the ATM Solar Instruments, V.L. Easterly, D.A. Roalstad, M.W. Frank, D.B. Hicks

AAS 74-167 Evolution of the Apollo Telescope Mount Control and Display Console, C.P. Fazio

AAS 74-168 ATM Test and Integration, J.W. Moore, J.R. Mitchell

AAS 74-169 A Solar Observatory in Space: Initial Results and Mission Assessment, E.M. Reeves

AAS 74-170 Solar Environment Support for Skylab, J.W. Hirman

AAS 74-171 Skylab Apollo Telescope Mount - Calibration Rocket Project, M.W. Frank, E.E. Bissell, Jr., C.M. Spencer

AAS 74-172 Not Assigned

AAS 74-173 Skylab Food System, W.H. Bush

AAS 74-174 Skylab Biomedical Hardware Development, W.J. Huffstetler, Jr., J.D. Lem

AAS 74-175 Skylab Medical Technology Utilization, J.C. Stonesifer

AAS 74-176 Evaluation of Life in Skylab from a Medical Viewpoint, J.R. Hordinsky

AAS 74-177 Skylab Medical Operational Support, G.R. Primeaux, F.R. Spross

AAS 74-178 to -179 Not Assigned

VOLUME 37, SCIENCE AND TECHNOLOGY
SPACE RESCUE AND RECOVERY, 1975
(International Astronautical Federation, September 30 - October 5, 1974, Amsterdam)

AAS 74-180 The Concept of Flight Safety, P.H. Bolger

AAS 74-181 An Emergency Astronaut Reentry Parachute System, J.J. Murray, F.R. DeJarnette

AAS 74-182 Survey of Space Flight Safety Systems, R. Fleisig, H.D. Wolf

AAS 74-183 Safety of Spacecraft Lightweight Pressure Vessels, P.H. Bolger, W. Boyes

AAS 74-184 Space Shuttle Personnel Rescue Provisions and Techniques, N.E. Brown

AAS 74-185 Predicting the Probability that Earth-Orbiting Spacecraft will Collide with Man-Made Objects in Space, D.R. Brooks, G.G. Gibson, T.D. Bess

AAS 74-186 System Safety as Applied to Skylab, K.S. Kleinknecht, B.J. Miller

AAS 74-187 Possibility of Astronaut Emergency Recovery by Fishing Vessels, M. Nagatomo

AAS 74-188 1974 Annual Survey of Recovery Capabilities, G.W. Heath

AAS 74-189 Engineering & Designing for Spacecraft Reliability, F.X. Kane

# AAS TECHNICAL PAPERS
# 1975

VOLUME 40, SCIENCE & TECHNOLOGY
FUTURE SPACE ACTIVITIES, 1976
(Thirteenth Goddard Memorial Symposium, April 1975, Washington, D.C.)

AAS75-001 The Space Transportation System, C.J. Donlan, E.J. Brazill

AAS75-002 Spacelab, D.J. Shapland

AAS75-003 Planetary Missions in the Shuttle Era, R.S. Kraemer

AAS75-004 Automated Missions of the Future, J. Purcell

AAS75-005 Military Space Shuttle Participation, H. Stelling, C. Tringali

AAS75-006 Not Available

AAS75-007 The Status of the Satellite Solar Power Station, P.E. Glaser

AAS75-008 Not Available

AAS75-009 Advanced Mars Mission Concepts: Mobile Lander and Surface Sample Return, W.T. Scofield

AAS75-010 Unified Mars Detection System, J.P. Martin, R.D. Johnson, B. Kok, R. Radmer

AAS75-011 to -019 Not Assigned

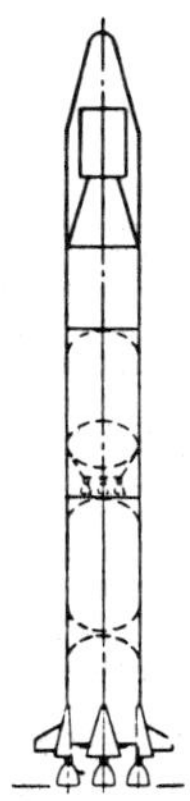

1975

VOL. 33 ADVANCES IN THE ASTRONAUTICAL SCIENCES ASTRODYNAMICS 1975, 1976
VOL. 26 AAS MICROFICHE SERIES, Supplement to Vol. 33 Advances
AAS/AIAA ASTRODYNAMICS CONFERENCE 1975, 1976
(AAS/AIAA Astrodynamics Conference, July 28-30, 1975, Nassau, Bahamas)

AAS 75-020 The Relativistic Dynamics of a Sub-Light Speed Interstellar Ramjet Probe, W.B.Roberts (v26 Micro)*

AAS 75-021 A Uniformly Valid Solution for Motion about the Interior Libration Point of the Perturbed Elliptic-Restricted Problem, D.L.Richardson, N.D.Cary (v26 Micro)

AAS 75-022 The Motion of a Satellite in Resonance with the Longitude-Dependent Harmonics, S.S.Dallas (v26 Micro)

AAS 75-023 Lunar & Solar Perturbations on the Orbit of a Geosynchronous Satellite, O.F.Graf (v26 Micro)

AAS 75-024 Solar Pressure Induced Perturbations of Orbital Elements for the Communications Technology Satellite, K.C.Pande, V.J. Modi, A.K.Misra (v33 Adv)*

AAS 75-025 Drag Perturbations of Near-Circular Orbits in an Oblate Diurnal Atmosphere, F.Santora, (v26 Micro)

AAS 75-026 Attitude Translational Coupling in a Rotating Drag-Free Satellite, S.S. Fernandez de Cordova, (v26 Micro)

AAS 75-027 Not Available

AAS 75-028 NASA/LTV Scout Orbital Accuracy Improvement Through Incorporation of a Fourth Stage Spin-Stabilized Attitude and Velocity Control System, J.J.Banchetti, G.W.Kreiter (v26 Micro)

AAS 75-029 Attitude Capture for the Geodynamics Experimental Ocean Satellite (GEOS)-3, G.M.Lerner, K.P.Coriell (v33 Adv)

AAS 75-030 Passive Three-Axis Stabilization of the Long-Duration Exposure Facility, E.K.Huckins,III, W.J.Breedlove,JR., J.D.Heinbockel (v26 Micro)

AAS 75-031 Automatic Control of a Momentum-Biased Observatory in Equatorial Orbit, T.M.Spencer (v33 Adv)

AAS 75-032 Not Available

AAS 75-033 On the Calculation of Low-Thrust Fail-Safe Trajectories, C.G.Sauer, Jr. (v26 Micro)

---

* Micro = AAS Microfiche Series
Adv = Advances in the Astronautical Sciences

AAS 75-034 Three-Dimensional G-Constrained Shuttle Boost Trajectory Optimization Using a Min-H/Secant Hybrid Algorithm, R.G.Gottlieb, W.T.Fowler (v26 Micro)

AAS 75-035 A Combined Gradient/Neighboring Extremal Algorithm for the Calculation of Optimal Transfer Trajectories Between Non-coplanar Orbits Using a Constant Low-Thrust Rocket, G.M.Anderson, E.A.Smith (v26 Micro)

AAS 75-036 Solution of an Optimal Control Lifting Body Entry Problem by an Improved Method of Perturbation Functions, F.Garcia,Jr. (v26 Micro)

AAS 75-037 Introduction to the Restricted Jupiter Orbiter Problem, T.A.Heppenheimer (v26 Micro)

AAS 75-038 Jupiter Orbiter Lifetime, the Hazard of Galilean Satellite Collision, A.L.Friedlander (v26 Micro)

AAS 75-039 Station-Keeping Requirements, J.B.Eades,Jr. (v26 Micro)

AAS 75-040 Satellite Disintegration Dynamics, R.R.Dasenbrock, B.Kaufman, W.B.Heard (v33 Adv)

AAS 75-041 Relativity Mission with Two Counter-Orbiting Polar Satellites, R.A.Van Patten, C.W.F.Everitt (v33 Adv)

AAS 75-042 Collision Avoidance for Two Counter-Orbiting Polar Satellites, D.B.Schaechter, J.V.Breakwell, R.A.Van Patten, C.W.F.Everitt (v26 Micro)

AAS 75-043 Satellite Attitude Acquisition by Momentum Transfer, P.M.Barba, J.N.Aubrun (v26 Micro)

AAS 75-044 Perturbation Solution for the Flat Spin Recovery of a Dual-Spin Spacecraft, J.R.Gebman, D.L.Mingori (v26 Micro)

AAS 75-045 Variable Spin Rate, Two-Degrees-of-Freedom, Nutation Damper Dynamics, C.J.Clouter (v33 Adv)

AAS 75-046 Nutational Stability of an Asymmetric Dual-Spin Spacecraft, G.T.Tseng (v26 Micro)

AAS 75-047 Geometrical Procedures for the Analysis of Spacecraft Attitude & Bias Determinability, J.R.Wertz, L.C.Chen (v26 Micro)

AAS 75-048 Not Available

AAS 75-049 A Study of Orbit Determination Accuracies for Future Earth Observatory Missions, W.C.Bryant,Jr., C.C.Goad (v33 Adv)

AAS 75-050 Maximum Confidence Estimation of Missile Flight Errors, C.W.Johnson, R.E.Maxwell, J.J.Horvath (v26 Micro)

AAS 75-051 Preliminary Mission Planning Using Interactive Graphics, F.G.Rea (v26 Micro)

AAS 75-052 Atmosphere Explorer (AE-c) Adaptive Mission Design and Trajectory Performance, R.W.Barbieri, J.C.Cooley, K.G.Nickerson (v33 Adv)

AAS 75-053 Not Available

AAS 75-054 A Two-Level Trajectory Decomposition Algorithm Featuring Optimal Intermediate Target Selection, F.M.Petersen, D.E.Cornick, G.L.Brauer, J.R.Rehder (v26 Micro)

AAS 75-055 Sensitivity Analysis of Short-Arc Station Coordinate Determinations from Range Data, B.E.Schutz, B.D.Tapley (v26 Micro)

AAS 75-056 The Use of Series-Solutions for Batch and Sequential Estimation, T.Feagin, R.P.Mikkilineni (v26 Micro)

AAS 75-057 The Orbit Determination System of Telesat Canada, U.Domb, A.J.Grisé, F.C.Kes (v26 Micro)

AAS 75-058 Orbit and Attitude State Recoveries from Landmark Data, A.J.Fuchs, C.E.Velez, C.C.Goad (v26 Micro)

AAS 75-059 GEOS-C Orbit Determination with Satellite to Satellite Tracking, P.Argentiero, R.Garza-Robles (v26 Micro)

AAS 75-060 Not Available

AAS 75-061 Optimal Rendezvous in the Neighborhood of a Circular Orbit, J.B.Jones (v26 Micro)

AAS 75-062 A Method for Handling Coast Arcs in Low-Thrust Interplanetary Trajectory Optimization, P.M.O'Neill, W.T.Fowler (v26 Micro)

AAS 75-063 Real-Time Orbiter Abort Guidance, V.S.Sohoni (v26 Micro)

AAS 75-064 Not Available

AAS 75-065 A Linear Feedback Guidance Law for an Aeromaneuvering Orbit-to-Orbit Shuttle, J.J.Rehder (v26 Micro)

AAS 75-066 Limited Variance Control in Statistical Low-Thrust Guidance Analysis, R.A.Jacobson (v33 Adv)

AAS 75-067 The Earth-Venus-Mercury Orbiter 1978 Mission, R. G. Brusch (v26 Micro)

AAS 75-068 Trajectory, Atmosphere, & Wind Reconstruction from Viking Measurements, F.W.Hopper (v26 Micro)

AAS 75-069 Mars Surface Sample Return--Some Mission Aspects of an Improved Design with Dual Launch, S.K.Asnin (v26 Micro)

AAS 75-070 Autonomous Rendezvous and Docking in Mars Orbit, F.Vandenberg (v33 Adv)

AAS 75-071 Preliminary Mariner Jupiter-Saturn 1977 Launch/Encounter Strategy, R.A.Wallace (v33 Adv)

AAS 75-072 The Navigation of a Satellite Re-Encounter in a Jupiter Orbiter Mission, G.N.Sherman, J.B.Jones (v26 Micro)

AAS 75-073 Not Available

AAS 75-074 The Theory of Enceladus and Dione--An Application of Computerized Algebra in Dynamical Astronomy, W.H.Jefferys, L.M.Ries (v26 Micro)

AAS 75-075 Not Available

AAS 75-076 Interactions between Positional and Dynamical Astronomy, C.A.Williams (v33 Adv)

AAS 75-077 Not Available

AAS 75-078 An Investigation of Finite Element Representations of the Geopotential, J.L.Junkins (v26 Micro)

AAS 75-079 Resolution of the Equations of Condition for Runge-Kutta-Nystrom Formulas through Eighth Order, R.H.Battin (v26 Micro)

AAS 75-080 Optimal Runge-Kutta Methods, D.G.Hull, D.G.Bettis (v26 Micro)

AAS 75-081 Not Available

AAS 75-082 A Variable-Metric Algorithm Employing Linear & Quadratic Penalties, H.J.Kelley, I.L.Johnson,Jr. (v26 Micro)

AAS 75-083 Satellite Research Applications of Digital Image Generation, M.B.Faintich (v26 Micro)

AAS 75-084 The Powered Swingby and Its Applications to Jupiter Orbit Mission Design, C.Uphoff (v26 Micro)

AAS 75-085 Shuttle-Launched Multi-Comet Mission 1985, R.W.Farquhar, D.P.Muhonen, F.I.Mann, W.H.Wooden, D.K.Yeomans (v33 Adv)

AAS 75-086 Multi-Asteroid Flyby Trajectories Using Venus-Earth Gravity Assists, D.F.Bender, A.L.Friedlander (v26 Micro)

AAS 75-087 New Flight Techniques for Outer Planet Missions, G.R.Hollenbeck (v26 Micro)

AAS 75-088 Mariner-Venus-Mercury Optical Navigation Demonstration: Results and Implications for Future Missions, C.H.Acton,Jr., H.Ohtakay (v26 Micro)

AAS 75-089 Accuracy of Estimating the Masses of Phobos and Deimos from Multiple Viking Orbiter Encounters, R.H.Tolson, M.L.Mason (v26 Micro)

AAS 75-090 A Stability Criterion for Passive Magnetically Anchored Rate Dampers, H.F.Foulke, S.H.Siegel, A.Das (v26 Micro)

AAS 75-091 Dynamic Behavior of an Extremely Flexible Gravity-Gradient Dipole Satellite, R.M.Davis, K.Yong (v26 Micro)

AAS 75-092 On the Natural Modes of Oscillation of Rotating Flexible Structures about Nontrivial Equilibrium, L.Meirovitch, J.N.Juang (v26 Micro)

AAS 75-093 Spin Stability of a Satellite Equipped with Four Booms, D.A.Levinson, T.R.Kane (v26 Micro)

AAS 75-094 Not Available

AAS 75-095 Model for Solar Torque Effects on DSCS II, T.E.Suttles, R.E.Beverly (v26 Micro)

AAS 75-096 Poinsot's Rolling & Sliding Cone Revisited, J.N.Blanton (v26 Micro)

AAS 75-097 The Mission of the German Aeronomy Satellite Aeros B Goals and Achievements, H.Wanke (v33 Adv)

AAS 75-098 Positioning of the German-French Communication Satellite Symphonie, N.Engersbach, A.Leibold, K.Well (v33 Adv)

AAS 75-099 In-Flight Performance of the French German Three-Axis Stabilized Telecommunication Satellite Symphonie, B.Ebel (v26 Micro)

AAS 75-100 Special Problems of Orbit Determination, M.C.Eckstein (v33 Adv)

AAS 75-101 Mission Analysis & Operations of the US-German Solar-Probe Helios, H.J.Panitz (v26 Micro)

AAS 75-102 Relativity Experiment on Helios: A Status Report, J.D.Anderson, W.Kundt (v26 Micro)

AAS 75-103 Application of Lie Transform Based Canonical Perturbation Methods to the Optimal Control of Bilinear Systems, M.K.Ozgoren, R.W.Longman, C.A.Cooper (v26 Micro)

AAS 75-104 Temporal Registration of Multispectral Digital Satellite Images Using Their Edge Images, M.L.Nack (v26 Micro)

AAS 75-105 Temperature Control for Solar Heating and Cooling of Buildings, C.B.Winn, D.Hull (v33 Adv)

AAS 75-106 Not Available

AAS 75-107 Quantization & Symmetry in Periodic Coverage Patterns with Applications to Earth Observation, J.C.King (v26 Micro)

AAS 75-108 Applications of Optimization Techniques to Solar Heating and Cooling, L.B.Anderson, H.E.Rauch (v33 Adv)

AAS 75-109 to -119 Not Assigned

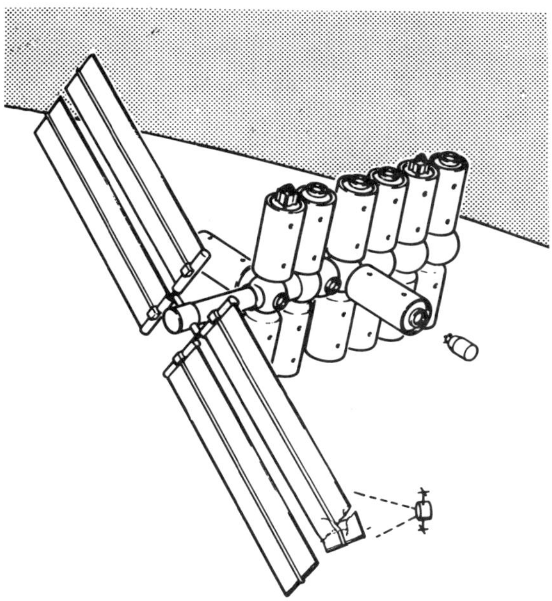

VOL. 32 ADVANCES IN THE ASTRONAUTICAL SCIENCES
SPACE SHUTTLE MISSIONS OF THE 80's, 1976

VOL. 25 AAS MICROFICHE SERIES
SPACE SHUTTLE MISSIONS OF THE 80's
SUPPLEMENT TO VOL. 32 ADVANCES
(Space Shuttle Missions of the 80's, 21st AAS Annual Meeting, August 26-28, 1975, Denver, Colorado)

AAS 75-120 Not Available

AAS 75-121 Keynote Address -- The Space Transportation System and Its Payloads, J.Naugle (v32I Adv)*

AAS 75-122 Department of Defense and Shuttle, H.B.Stelling,Jr.(v32I Adv)

AAS 75-123 Spacelab Status/Systems Capability, H.E.W.Hoffman(v32I Adv)

AAS 75-124 Spacelab Programme - Status Review, B.Deloffre (v32I Adv)

AAS 75-125 Spacelab Payloads, G.W.Sharp (v32I Adv)

AAS 75-126 Shuttle Payload Development, J.T.Murphy (v32I Adv)

AAS 75-127 Applications Payloads, C.W.Mathews (v32I Adv)

AAS 75-128 Potential for Advancement of Space Astronomy, A.D.Code (v32I Adv)

AAS 75-129 Concepts of LST Operation, C.R.O'Dell (v32I Adv)

AAS 75-130 Large Space Telescope Program Status, J.A.Downey III (v32I Adv)

AAS 75-131 The LST Scientific Instruments, G.M.Levin (v32I Adv)

AAS 75-132 Challenges in Space Physics, B.M.McCormac (v32I Adv)

AAS 75-133 Challenges in the Atmospheric Sciences, R.E.Dickinson (v32I Adv)

AAS 75-134 AMPS Science Objectives and Philosophy, E.R.Schmerling (v32I Adv)

AAS 75-135 AMPS Program Status, R.Ise (v32I Adv)

AAS 75-136 The Exploitation of Space for the Benefit of Mankind, F.E.Moss (v32I Adv)

---

* Micro - AAS Microfiche Series
Adv = Advances in the Astronautical Sciences

AAS 75-137 Space Tug Requirements and Planning, J.Wild, M.D.Kitchens (v32II Adv)

AAS 75-138 Space Tug Baseline Description and Status. W.Teir, A.G.Orillion (v32II Adv)

AAS 75-139 Not Available

AAS 75-140 Alternate Tug/IUS Approaches, A.O.Tischler (v32II Adv)

AAS 75-141 Orbital Transportation in the 1980's and Beyond, H.P.Davis (v32II Adv)

++AAS 75-142 Tug Avionics System Overview, M.T.Raaberg, J.I.Newcomb (v25 Micro)*

++AAS 75-143 Lightweight Fuel Cell Powerplant for Tug, L.M.Handley (v25 Micro)

AAS 75-144 Space Tug Laser Gyro IMU, R.Morrison,B.Walls(v32II Adv & v25 Micro)

++AAS 75-145 Interferometric Landmark Tracker Applied to Precise Space Tug and Payload Navigation, D.H.Aldrich, W.F.Hubbarth (v25 Micro)

++AAS 75-146 LSI Computer for 1980's Space Mission, W.A.Clapp, J.E.Saultz, J.B.White (v25 Micro)

++AAS 75-147 The Space Tug Economic Analysis Study--What We Learned, C.V.Hopkins (v25 Micro)

AAS 75-148 Not Available

AAS 75-149 A Survey of the Ecomonics of Materials Processing in Space, B.P.Miller (v25 Micro & v32II Adv)

++AAS 75-150 The Economics of Satellite Maintenance, W.L.Derocher,Jr., R.G.Sosnay (v25 Micro)

AAS 75-151 Not Available

++AAS 75-152 An Integrated Approach to Tug/IUS Mission Operations, B.S.King, R.A.Chamberlain (v25 Micro)

++AAS 75-153 Tug Ground Operations - An Approach to Early Identification of Operational Impacts, J.L.Best (v25 Micro)

++AAS 75-154 Space Tug Mission Operations, K.C.Nuss (v25 Micro)

++AAS 75-155 RL10 Derivatives for IUS/Tug, J.P.B.Cuffe (v25 Micro)

---

++ Summary in v32 II Advances, in full in v25 Micro

++AAS 75-156 Advances Space Engine Component Technology Status, A.T.Zachary (v25 Micro)

AAS 75-157 Aerospike Development Status, J.Campbell, H.G.Diem (v32II Adv)

AAS 75-158 Not Available

++AAS 75-159 Solid Rocket Technology Advancements for Space Tug and IUS Applications, W.Ascher,R.L.Bailey, J.W.Behm, W.Gin (v25 Micro)

++AAS 75-160 Requirements and Considerations in Selecting Space Tug Propulsion Systems, C.J.Cohan (v25 Micro)

++AAS 75-161 Space Tug Propulsion Systems--Storable Versus Cryogenic, W.E.Pipes (v25 Micro)

++AAS 75-162 Mixed-Mode Propulsion Systems for Full Capability Space Tugs, R.Salkeld, R.Beichel (v25 Micro)

AAS 75-163 A Candidate Mission Using the Shuttle and Solar Electric Propulsion, J.H.Duxbury, R.C.Finke (v25 Micro & v32II Adv)

++AAS 75-164 Status Report on Nuclear Electric Propulsion Systems, J.W.Stearns (v25 Micro)

++AAS 75-165 Tug and Payload-to-Orbiter Interface Requirements, E.H.Bock (v25 Micro)

++AAS 75-166 Tug Payload Intefaces, F.Runge (v25 Micro)

++AAS 75-167 Tug Rendezvous and Docking with a Spacecraft--A Remote, Manned Approach, M.J.Hurley (v25 Micro)

++AAS 75-168 Space Tug Thermal Control, T.L.Ward (v25 Micro)

AAS 75-169 NDT for Space Tug Thin Gage Materials, W.D.Rummel (v32II Adv)

AAS 75-170 Burner II Interim Upper Stage, H.Kudish (v32II Adv)

++AAS 75-171 Transtage Interim Upper Stage, P.B.Teets (v25 Micro)

++AAS 75-172 Centaur as an Initial Upper Stage for the Space Shuttle, D.J.Jones, D.A.Heald (v25 Micro)

++AAS 75-173 Delta as an Interim Upper Stage (IUS), R.P.Dawson, J.F.Meyers (v25 Micro)

++AAS 75-174 Spinning Solid Perigee Stage, H.A.Rosen, C.R.Jones, L.M.Bronstein (v25 Micro)

---

+ Summary in v32 I Advances, in full in v25 Micro
++ Summary in v32 II Advances, in full in v25 Micro

+AAS 75-175 Mirror Substrate Material and Manufacturing for the Large Space Telescope, W.C.Lewis (v25 Micro)

+AAS 75-176 Fabrication and Test of 1.8-Meter-Diameter, High Quality ULE Mirror, R.J.Wollensak, C.A.Rose (v25 Micro)

+AAS 75-177 Design and Testing with a Reflective Null System, L.Montagnino, A.Offner (v25 Micro)

+AAS 75-178 Test Results on Homogeneity of Expansion for a 1.8-M ULE Lightweight Mirror, G.Friedman, G.Gasser (v25 Micro)

+AAS 75-179 Large Space Telescope External Interfaces, R.E.Collart (v25 Micro)

+AAS 75-180 LST Refurbishment and Support, J.M.Henschke (v25 Micro)

+AAS 75-181 Simulation of On-Orbit Maintenance Cycle for LST, J.A.Donnelly (v25 Micro)

+AAS 75-182 LST Power System Long-Life Design Techniques, O.B.Smith, R.L.Donovan, J.L.Oberg (v25 Micro)

+AAS 75-183 Large Space Telescope--Orbital Crew EV Maintenance Operations, H.T.Fisher (v25 Micro)

+AAS 75-184 An Analytical and Experimental Evaluation of Actuator Vibration of LST Image Distortion, A.D.Houston, L.W.Hodge,Jr., T.J.Kertesz (v25 Micro)

+AAS 75-185 Development of a Large-Inertia Fine-Pointing and Dimensional Stability Simulator, R.L.Gates, D.H.Wine, R.W.Seiferth, N.A.Osborne (v25 Micro)

+AAS 75-186 Evaluation of Communication Antenna Drive System Design Requirements to Allow TDRS Tracking During LST Fine Pointing, A.J.Besonis, C.J.Chang (v25 Micro)

+AAS 75-187 A Small Instrument Pointing System for Shuttle Sortie Missions, C.H.Henrikson, E.E.Schmidt (v25 Micro)

+AAS 75-188 LST Interferometric Fine Guidance Sensor, A.B.Wissinger, R.H.Carricato (v25 Micro)

+AAS 75-189 Prismatic Grating Star Tracker, A.H.Greenleaf (v25 Micro)

+AAS 75-190 Thermo/Structural Design Considerations to Achieve the Large Space Telescope Line-of-Sight Requirements, D.J.Tenerelli (v25 Micro)

---

\+ Summary in v32 I Advances, in full in v25 Micro

+AAS 75-191 Design of Low-Thermal-Distortion LST Metering Structure, J.R.Lager (v25 Micro)

+AAS 75-192 3-Axis Simulation of the LST Pointing Control Subsystem--A Multi-Discipline Activity, W.W.Emsley, T.D.Fehr, D.C.Fosth, D.L.Knobbs (v25 Micro)

+AAS 75-193 LST Data Management and Mission Operations Concept, F.Walker, F.Hudson, L.Murphy (v25 Micro)

+AAS 75-194 Data Management for Large Space Telescope, G.R.Hope, T.J.Rasser (v25 Micro)

+AAS 75-195 A Cost Effective Data Management Subsystem for the LST, J.A.Dougherty, T.D.Patterson, A.E.Cole (v25 Micro)

AAS 75-196 Not Available

+AAS 75-197 System Consideration, Design Approach, and Test of a Low-Gain Spherical Coverage Antenna for Large Space Vehicles, R.E.Ferguson, T.D.Patterson, M.R.Moreno (v25 Micro)

AAS 75-198 A Scientific Operations Plan for the Large Space Telescope, D.K.West (v32I Adv)

AAS 75-199 LST Operations, A Typical Day, W.J.Pragluski, R.H.Brown (v32I Adv)

+AAS 75-200 Large Space Telescope Mission Analysis, F.M.Friedlaender (v25 Micro)

+AAS 75-201 Automation of the LST, W.W.Warnock, C.W.Case (v25 Micro)

+AAS 75-202 Science Performance Considerations for the Design of the LST, D.D.Ostrander, J.C.Tuttle (v25 Micro)

+AAS 75-203 Optical Performance Control, T.A.Facey (v25 Micro)

+AAS 75-204 Not Available

+AAS 75-205 Stray Light from Out-of-Field Sources, R.J.Noll (v25 Micro)

+AAS 75-206 Design of Highly Stable Optical Support Structure, M.H.Krim (v25 Micro)

+AAS 75-207 Large Format SEC-Orthicon Integrating Television Sensor for LST, J.L.Lowrance (v25 Micro)

+AAS 75-208 The Intensified Charge Coupled Device as a Photon Counting Imager, J.T.Williams (v25 Micro)

---

+ Summary in v32 I Advances, in full in v25 Micro

AAS 75-209 The Infrared Capabilities of the Large Space Telescope, D.E.Kleinmann (v32I Adv)

+AAS 75-210 Development of an Infrared Spectroradiometer, W.H.Alff, J.C.Thunen (v25 Micro)

+AAS 75-211 Faint Object Spectrograph, W.P.Devereux (v25 Micro)

+AAS 75-212 High-Speed Area Photometer: Conceptual Design and Integration, W.M.Bloomquist, F.J.Steputis, Jr. (v25 Micro)

+AAS 75-213 LST High Resolution Spectrograph, K.Peacock (v25 Micro)

AAS 75-214 Spaceflight Tracking and Data Networks Support in the Space Tug Era, L.R.Stelter, R.D.Godfrey (v25 Micro & v32II Adv)

AAS 75-215 Some Experiments with Energetic Particle Injectors, W.Bernstein, J.R.Winckler (v32I Adv)

AAS 75-216 Not Available

+AAS 75-217 AMPS Experiments Involving Gas Releases, L.M.Linson (v25 Micro)

AAS 75-218 LIDAR Systems for AMPS, R.D.Hake,Jr. (v32I Adv)

AAS 75-219 Not Available

AAS 75-220 Observation of Spacecraft-Generated Electrostatic Fields in the Vicinity of the ATS 6 Satellite, E.C.Whipple,Jr. (v25 Micro & v32I Adv)

AAS 75-221 Not Available

AAS 75-222 Measurement of Static Electric Fields at the Surface of Satellites, J.E.Nanevicz (v25 Micro & v32I Adv)

AAS 75-223 to -227 Not Available

+AAS 75-228 Preliminary Shuttle Payload Contamination Assessment, E.B.Ress, R.O.Rantanen, L.E.Bareiss (v25 Micro)

+AAS 75-229 AMPS Data Management Concepts, P.N.Metzelaar (v25 Micro)

AAS 75-230 Experiment Integration of AMPS, R.Witholder (v32I Adv)

---

\+ Summary in v32 I Advances, in full in v25 Micro
++ Summary in v32 II Advances, in full in v25 Micro

AAS 75-231 Ultraviolet Remote Sensing of Atmospheric Ozone from Payloads Using Shuttle Capabilities, A.J.Krueger, D.F.Heath, H.A.Roeder (v25 Micro & v32I Adv)

+AAS 75-232 Design of High-Voltage Insulation Systems for Aerospace Equipment, W.G.Dunbar (v25 Micro)

AAS 75-233 Not Available

++AAS 75-234 A Large Cooled Infrared Telescope Facility for Spacelab, S.G.McCarthy, L.S.Young, F.C.Witteborn (v25 Micro)

AAS 75-235 The Multimission Modular Spacecraft for the 80's, R.O.Bartlett, F.J.Cepollina (v25 Micro & v32II Adv)

++AAS 75-236 Spacelab Task Allocation--A Preliminary Determining of Onboard Versus Ground Operational Priorities, W.J.Harris, R.E.Holmen (v25 Micro)

++AAS 75-237 NASA/ESA CV-990 Airborne Simulation of Spacelab, D.Mulholland, C.Neel, J.DeWaard, R.Lovelett, L.Weaver, R.Parker (v25 Micro)

++AAS 75-238 Spacelab Payload Accommodation, D.M.Waltz (v25 Micro)

AAS 75-239 Concept Verification Test: Evaluation of Spacelab/Payload Operation Concepts, R.O.McBrayer, H.H.Watters (v32II Adv)

++AAS 75-240 The Integration of Commercial Payloads into Spacelab, H.L.Bloom, K.R.Taylor (v25 Micro)

++AAS 75-241 Low-Cost Integration Techniques for Spacelab Payloads, C.A.Braunwarth, T.C.Aepli(v25 Micro)

AAS 75-242 Not Available

AAS 75-243 Spacelab Payload and Program Planning in Germany,H.Schreiber (v25 Micro & v32II Adv)

++AAS 75-244 Space Transportation System Payloads Data and Analysis, J.D.Peterson, H.G.Craft,Jr. (v25 Micro)

AAS 75-245 Programmatic Aspects of German Shuttle/Spacelab Utilization, G.Greger (v25 Micro & v32II Adv)

++AAS 75-246 Future Payload Technology Requirements, H.M.Ikerd, L.R.Alton (v25 Micro)

++AAS 75-247 Computer-Aided Scheduling of Spacelab Ground Operations, J.K.Willoughby (v25 Micro)

---

\+ Summary in v32 I Advances, in full in v25 Micro
++ Summary in v32 II Advances, in full in v25 Micro

++AAS 75-248 An Orbiting Molecular Shield Vacuum Facility: A Materials Laboratory in Space, J.W.Youngblood, R.A.Outlaw, L.T.Melfi,Jr. J.R.McIlhaney (v25 Micro)

++AAS 75-249 Advanced Extravehicular Mobility Unit Technology Experiments, G.Wengrow (v25 Micro)

++AAS 75-250 The Advanced Technology Laboratory, C.Llewellyn, R.J.Milliken (v25 Micro)

++AAS 75-251 Shuttle Entry Technology Payloads, P.M.Siemers III (v25 Micro)

++AAS 75-252 Shuttle Communication Experiments, J.J.Woodruff, D.R.Peters (v25 Micro)

++AAS 75-253 Large Deployable Antenna Shuttle Experiment, R.E.Freeland, J.G.Smith, J.C.Springett, K.E.Woo (v25 Micro)

++AAS 75-254 Availability of a Communications Satellite, Requirement and Feasibility, G.D.Gordon (v25 Micro)

++AAS 75-255 Application of Space Shuttle to Fundamental and Applied Microbiological Research, J.V.Mayeux (v25 Micro)

++AAS 75-256 Life Sciences Manned Payloads for Shuttle/Spacelab, D.B.Heppner, G.L.Drake, C.B.May (v25 Micro)

++AAS 75-257 Shuttle Bioresearch Laboratory Breadboard Simulations, S.T.Taketa (v25 Micro)

AAS 75-258 A Life Sciences Spacelab Mission Simulation, J.A.Mason, F.S.Musgrave, D.R.Morrison (v32II Adv)

++AAS 75-259 Space Shuttle Trace Gas Analyzer, W.Dencker (v25 Micro)

++AAS 75-260 Astronomy Spacelab Payloads, R.Ott, G.Wengrow (v25 Micro)

++AAS 75-261 Atmospheric X-Ray Emission Experiment for Shuttle, R.A.Goldberg, K.L.Hallam, J.G.Emming (v25 Micro)

++AAS 75-262 Spacelab Ultraviolet-Optical Telescope Facility, M.Bottema (v25 Micro)

++AAS 75-263 Adaptation of an Existing Cosmic Ray Ionization Spectrometer Experiment to Spacelab, U.R.Alvarado, J.F.Ormes, C.V.Stahle (v25 Micro)

++AAS 75-264 Automation of Space Processing Applications Shuttle Payloads, W.E.Crosmer, O.T.Neau, J.Poe (v25 Micro)

---

++ Summary in v32 II Advances, in full in v25 Micro

AAS 75-265 Shuttle Benefits for Automated Retrievable Cryogenically Cooled Payloads J.O.Simpson, T.M.Spencer, W.H.Follett,Jr. (v25 Micro & v32II Adv)

AAS 75-266 Not Available

++AAS 75-267 Retrieval of the HEAO-C Spacecraft with Space Shuttle, D.H.Mitchell (v25 Micro)

++AAS 75-268 Mission Considerations for Multidiscipline Applications Payloads, J.M.Macdonald (v25 Micro)

++AAS 75-269 The Long-Duration Exposure Facility--A Shuttle Transported Low-Cost Technology Experiment Carrier, J.D.DiBattista (v25 Micro)

++AAS 75-270 Payload Planning for the First Spacelab Mission--A European View, A.V.Breitenstein, D.Davidts (v25 Micro)

AAS 75-271 Spacelab Utilization in Different Fields of Science and Applications, J.Collet (v25 Micro & v32II Adv)

AAS 75-272 Not Available

++AAS 75-273 Shuttle/IUS Performance for Planetary Missions, M.J.Cork, J.M.Driver, J.L.Wright (v25 Micro)

++AAS 75-274 Impact of Space Transportation System on Planetary Spacecraft and Missions Design, P.M.Barnett (v25 Micro)

AAS 75-275 Using the Shuttle for Future Advanced Planetary Missions, D.Friedman, W.Scofield (v32II Adv)

+AAS 75-276 A Fine-Pointing Facility for Spacelab Experiments in the 1980's--The Instrument Pointing Subsystem, H.Heusmann, J.C.Jones (v25 Micro)

AAS 75-277 Not Available

++AAS 75-278 Landmark Tracking Technology, J.D.Welch, W.E.Silverton, R.G.Wilson (v25 Micro)

++AAS 75-279 Possibilities for Reducing High-Energy Performance Requirements, G.R.Hollenbeck (v25 Micro)

AAS 75-280 Energy Research Overview/Alternatives for Energy Development, T.J.Vogenthaler (v32II Adv)

AAS 75-281 The Satellite Solar Power Station--A Focus for Future Space Shuttle Missions, P.E.Glaser (v32II Adv)

---

\+ Summary in v32 I Advances, in full in v25 Micro
++ Summary in v32 II Advances, in full in v25 Micro

AAS 75-282 Not Available

AAS 75-283 Nuclear Power in the Shuttle Era, S.R.Ross (v32II Adv)

AAS 75-284 Not Available

++AAS 75-285 Availability and Variability of Solar Energy in the Rocky Mountain Region, R.C.Burriss (v25 Micro)

AAS 75-286 Modeling the Western Coal Industry, W.Ganter, C.McMillan, F.Glover (v32II Adv)

AAS 75-287 Minimum Cost Solar Thermal Electric Power Systems: A Dynamic Programming Based Approach, W.S.Duff (v32II Adv)

AAS 75-288 100 MWe Solar Power Plant Design Configuration and Performance, F.A.Blake (v32II Adv & v25 Micro)

++AAS 75-289 Minimization of Overtime Shift Hours with an L-P Model, G.L.Kaes (v25 Micro)

AAS 75-290 Deposited Today--Consumable Tomorrow Sewage and Sold Waste Hydrogenation, C.W.LaGrone (v32II Adv)

++AAS 75-291 Early Space Station User Accommodations, D.R.Saxton, H.L.Wolbers (v25 Micro)

AAS 75-292 Agena Interim Upper Stage, J.H.Guill(v32II Adv & v25 Micro)

AAS 75-293 A Methodology for Assessing Reliability of Coal Conversion Plants, W.T.Long (v25 Micro)

AAS 75-294 The European Space Agency Study of Photon Counting Imaging for LST, R.J.Laurance (v32I Adv)

++AAS 75-295 Analysis of Extended--Duration Sortie Missions, R.C.Ring, W.C.Lide (v25 Micro)

AAS 75-296 System applications of the Fault-Tolerant Memory, L.J.Murphy (v32I Adv & v25 Micro)

++AAS 75-297 Spacelab Resources, H.M.Kappler (v25 Micro)

AAS 75-298 Defining Shuttle and Tug Requirements for Large Space Facility Construction, G.W.Driggers (v25 Micro)

AAS 75-299 Large Space Telescope, Image Quality Analysis, L.H.Solomon, W.B.Wetherell (v25 Micro)

---

++ Summary v32 II Advances, in full in v25 Micro

AAS 75-300 Ultraviolet and Visible Scattered Light Effects on the Optical Performance of the Large Space Telescope (LST), W.B.Wetherell (v25 Micro)

VOL. 41 SCIENCE AND TECHNOLOGY SPACE RESCUE AND SAFETY 1975, 1976 (International Academy of Astronautics, 26th International Astronautical Congress, September 21-27, 1975, Lisbon, Portugal)

IAF 75-001 to -012 Not Assigned

IAF 75-013 Survey of Space Flight Systems (Fifth Supplement: July 1974-June 1975), H.D.Wolf, R.Fleisig

IAF 75-014 Not Available

IAF 75-015 Medical Considerations on the Participation of Scientists in Future Space Flight, E.C.Burchard

IAF 75-016 The Space Shuttle--New Thresholds of Safety in Space, C.O.Baker, G.S.Canetti, W.E.Dean

IAF 75-017 Not Available

IAF 75-018 Safety Considerations in the Design of Satellites Flown by Manned Boosters, F.X.Kane

IAF 75-019 Space Tug Manned Safety, N.E.Tipton

IAF 75-020 Spacelab Environmental Control/Life Support System--Design Safety, R.Gartner

IAF 75-021 Not Available

IAF 75-022 1975 Annual Survey of Recovery Capabilities--Toward Normalization of Space Recovery, G.W.Heath

IAF 75-023 Emergency Supply and Return Capability of a Space Station Support Vehicle, A.I.Sibila

IAF 75-024 Space Shuttle Crew Safety Provisions, N.E.Brown

# AAS TECHNICAL PAPERS
# 1976

AAS/AAAS VIKING MARS SCIENCE EXPERIMENTS: EXPECTATIONS
(Bicentennial Meeting, American Association for the Advancement of Science, February 18-24, 1976, Boston)

No proceedings volume published

AAS 76-001 to -019 Not Available

VOLUME 44, SCIENCE AND TECHNOLOGY
SATELLITE COMMUNICATIONS IN THE
NEXT DECADE, 1977
(Fourteenth AAS Goddard Memorial Symposium, March 12, 1976, Washington, D.C.)

AAS76-020 Satellite Communication Uses in the 80's, J. Stenbit

AAS76-021 Satellite Communication Uses in the 1980's, P. Goldmark

AAS76-022 Satellite Communications in the 1980's, J.P. Witherspoon

AAS76-023 Satellite Communication Uses in the 1980's, R.F. Garbarini

AAS76-024 Communication Satellite Systems in the 1980's, W.E. Morrow, Jr.

AAS76-025 Satellite Communications Technology, B.I. Edelson

AAS76-026 Space Transportation for Satellite Communications, W. Pritchard

AAS76-027 The Challenge of a Changing Telecommunications Market (Summary), B.A. Johnson

AAS76-028 Advance in Non-Satellite Communications (Summary), S. J. Buchsbaum

AAS76-029 Frequency and Orbit needs for Satellite Communications, S.E. Probst

AAS76-030 Not available

AAS76-031 Communications Satellites - Contrasting Policy Approaches, B. Strassburg

AAS76-032 National and International Issues for Satellite Communications (Summary), D. Jansky

AAS76-033 The Internationalization of Communication Satellite Space Technology, C.L. Cuccia

VOL. 35 ADVANCES IN THE ASTRONAUTICAL SCIENCES, BICENTENNIAL SPACE SYMPOSIUM, New Themes in Space: Mankind's Future Needs and Aspirations, 1977
(Proceedings of 22nd AAS Annual Meeting, October 6-8, 1976, Washington, D.C.)

AAS 76-034 to -036 Not Assigned

AAS 76-037 Planetary Exploration Options for the Future--Abstract, J.Nichoff, L.Friedman

AAS 76-038 Climate and the Changing Sun, J.A.Eddy

AAS 76-039 The Solar-Terrestrial Observatory as a Major Module of a Space Station, C.R.Chappell

AAS 76-040 Strange Objects in Astrophysics--Abstract, G.B.Field

AAS 76-041 Designing to Meet Astronomy Requirements--Space Telescope--Abstract, K.P.Timmons

AAS 76-042 Not Available

AAS 76-043 Military Satellite Communications, W.E.Morrow,Jr.

AAS 76-044 Not Available

AAS 76-045 Satellite Communications for the Mobile Service, W.Luksch

AAS 76-046 Satellite Communication for Public Services, R.S.Cooper, W.N.Redisch

AAS 76-047 Space Industrialization, J.H.Disher

AAS 76-048 Prospects for Space Manufacturing, D.M.Waltz

AAS 76-049 Not Available

AAS 76-050 The Space Station and Space Industrialization, R.L.Kline

AAS 76-051 Transportation Concepts for Space Industrialization, J.T.Murphy

AAS 76-052, -053 Not Available

AAS 76-054 Monitoring Crops and Forests from Space, H.L.Hill, F.P.Weber, G.F.Hart

AAS 76-055 Europe's Need for Better Information on Natural Resources and Environment, J.Bodechtel, J.Plevin

AAS 76-056 Not Available

AAS 76-057 Quo Vadis, Remote Sensing, F.C.Billingsley

AAS 76-058 Prospects for Satellite Power Stations, R.A.Summers

AAS 76-059 Not Assigned

VOL. 34 ADVANCES IN THE ASTRONAUTICAL SCIENCES, APOLLO SOYUZ MISSION REPORT, 1977
Published by authorization of NASA Johnson Space Center, Houston, Texas

AAS 76-060 Apollo Soyuz Mission Report, ed. C.M.Lee

AAS 76-061 Mission Summary

AAS 76-062 Trajectory

AAS 76-063 Spacecraft Systems Performance, Docking Module

AAS 76-064 Spacecraft Systems Performance, Command and Service Modules

AAS 76-065 Spacecraft Systems Performance, Command and Service Module Consumables

AAS 76-066 Crew Station

AAS 76-067 Experiments, Space Sciences

AAS 76-068 Experiments, Life Sciences

AAS 76-069 Experiments, Technology

AAS 76-070 Inflight Demonstrations

AAS 76-071 Joint Flight Activities

AAS 76-072 Biomedical Evaluation

AAS 76-073 Pilot's Report

AAS 76-074 Mission Support Performance

AAS 76-075 Assessment of Mission Objectives and Technical Investigations

AAS 76-076 Launch Phase Summary

AAS 76-077 Anomaly Summary, Command and Service Module, Docking Module, and Government Furnished Equipment

AAS 76-078 Anomaly Summary, Experiments

AAS 76-079 Systems and Equipment Description, Docking Module

AAS 76-080 Systems and Equipment Description, Soyuz Spacecraft

AAS 76-081 Systems and Equipment Description, Apollo Spacecraft, Crew Equipment, and Stowage Configurations

AAS 76-082 Systems and Equipment Description, Applications Technology Satellite 6

AAS 76-083 Systems and Equipment Description, Experiment Hardware

AAS 76-084 ASTP Summary As-Flown Flight Plan

AAS 76-085 Spacecraft History

AAS 76-086 Spacecraft Mass Properties

AAS 76-087 Postflight Testing

AAS 76-088 Data Availability

AAS 76-089 Conversion Scales

AAS 76-090 Appendix I - Mission Report Supplement

AAS 76-091 to -099 Not Assigned

VOL. 43 SCIENCE AND TECHNOLOGY, THE EAGLE HAS RETURNED, 1976 (Proceedings of the Dedication Conference of the International Space Hall of Fame, October 5-9, 1976, Alamogordo, New Mexico)

AAS 76-100 Survey Paper "Engineering in Space Flight", E.A.Steinhoff

AAS 76-101 Assessment Paper for "Engineering in Space Flight", R.Beichel

AAS 76-102 Engineering in Space, M.A.Faget

AAS 76-103 Response to Survey Paper "Engineering in Space Flight", M.H.Kaplan, H.Maeda

AAS 76-104 Ignition Delay Studies of Hybrid Rocket Propellants, N.L.Munjal

AAS 76-105 Assessment Paper for "Engineering in Space Flight", R.A.Nau, R.Salkeld

AAS 76-106 Response to "Engineering in Space Flight", T.M.Tabanera

AAS 76-107 Assessment for "Engineering in Space Flight", R.Beichel

AAS 76-108 Second Response to "Engineering in Space Flight", M.H.Kaplan

AAS 76-109 Assessment Paper for "Engineering in Space Flight", R.A.Nau, R.Salkeld

AAS 76-110 Perspective in Space Communications, S.K.Sarkar

AAS 76-111 Remarks of the Program Director of the Roundtable "Engineering in Spaceflight"

AAS 76-112 Survey of the Pioneer 10 and 11 Jupiter Mission, J.H.Wolfe

AAS 76-113 Electric Propulsion: A High Energy Capability for Solar System Exploration, K.L.Atkins

AAS 76-114 Some Aspects of Planetary Atmospheres, K.S.W.Champion

AAS 76-115 Viking Survey Paper, G.Soffen

AAS 76-116 Planetary Missions--Possibilities for the Future, L.D.Friedman

AAS 76-117 Advancements in Astronomy During the Space Age, A.D.Code

AAS 76-118 Remarks on Dr. Arthur Code's Survey Paper, E.W.Silvertooth, J.D.Strong

AAS 76-119 The Current Status of Man's Response to the Space Environment, C.A.Berry

AAS 76-120 Proposed Comments for the Roundtable, P.A.Campbell

AAS 76-121 How Good is Exercise for Astronauts?, U.C.Luft

AAS 76-122 Proposed Comments for the Roundtable, C.S.White

AAS 76-123 Development Anomalies in Biological Systems of Animals Due to Heavy Nuclei of Cosmic Radiation, E.H.Graul, W.Ruther

AAS 76-124 Developments in Space Law: An Impressive Record for the Hall of Fame, S.Gorove

AAS 76-125 Is There a General International Law of Original Ownership? L.F.E.Goldie

AAS 76-126 Assessment of Space Law Development, H.DeSaussure

AAS 76-127 Comment on the Introductory Paper of Prof. Gorove, I.H.P.Diederiks-Verschoor

AAS 76-128 Response to "Developments in Space Law", E.Fasan, E.R.Finch,Jr.

AAS 76-129 Response to Survey Paper "Developments in Space Law", E.Galloway

AAS 76-130 Response to Survey Paper "Developments in Space Law", I.Herczeg, S.H.Lay, M.Menter

AAS 76-131 Comments on the Responses Received on the Survey Paper, S.Gorove

AAS 76-132 Response to Participants' Comments, H.DeSaussure, I.H.P.Diederiks-Verschoor, E.R.Finch,Jr., E.Galloway

AAS 76-133 Response to Survey Paper "Developments in Space Law" S.H.Lay, M.Menter

AAS 76-134 Comments on Second Responses Received in Connection with Dr. Gorove's Survey Paper, S.Gorove

AAS 76-135 Response to Participants' Comments, S.H.Lay

AAS 76-136 On the Evolution of Rocket Navigation, Guidance and Control (NGC) Systems, W.Haeussermann

AAS 76-137 The History of Apollo On-Board Guidance, Navigation and Control, D.G.Hoag

AAS 76-138 Guidance and Control Symposium, H.Hoeizer

AAS 76-139 Testing of the Lunar Excursion Module About Sensor Assembly at the Central Inertial Guidance Test Facility (CIGTF) M.G.Jaenke, N.L.Ingold

AAS 76-140 Goals and Needs of World Development, B.H.Chatel

AAS 76-141 Role of Space Stations in the Assembly and Maintenance of Large Space Structure, R.L.Kline

VOL. 1 AAS HISTORY SERIES, TWO HUNDRED YEARS OF FLIGHT IN AMERICA, A BICENTENNIAL SURVEY, 1977
(Proceedings of the first history symposium held in the National Air and Space Museum, November 4, 1976, Washington, D.C.)

AAS 76-142 Perspectives on Flight in America, E.M.Emme

AAS 76-143 Ballooning, 1783-1976, R.Pineau

AAS 76-144 The Airship, 1904-1976, R.K.Smith

AAS 76-145 General Aviation, 1919-1976, T.D.Crouch

AAS 76-146 Military Aviation, 1908-1976, R.L.Perry

AAS 76-147 Commercial Aviation, 1919-1976, R.P.Hallion

AAS 76-148 Instrumented Utilization of Space, R.C.Hall

AAS 76-149 Manned Space Flight, E.C.Ezell

AAS 76-150 Perspectives of a Historian of Technology, T.P.Hughes

VOL. 45 SCIENCE AND TECHNOLOGY, THE EAGLE HAS RETURNED, SECOND PART, 1977 (Second Part of Proceedings of the Dedication Conference of the International Space Hall of Fame, October 5-9, 1976, Alamogordo, New Mexico)

AAS 76-151 Keynote Address: The Future of Space, H.G.Stever

AAS 76-152 On the Evolution of Accurate Inertial Guidance Instrumentation, C.S.Draper

AAS 76-153 Assessment for Engineering in Space Flight, T.M.Tabanera

AAS 76-154 Assessment for Engineering in Space Flight, S.Sanz-Aranguez

AAS 76-155 Airbreathing and Oxygen Collecting HTOHL Launch Vehicle, R.A.Nau

AAS 76-156 Questions and Answers for Engineering in Space Flight, R.Beichel

AAS 76-157 Overview of the Viking Project and Most Recent Conclusions, J.Martin

AAS 76-158 Assessment of Solar System by Optical Astronomy On-Site Observations and Crust Samples, A.Dollfus

AAS 76-159 Assessment of Future Tasks of Astronomy, A.Code

AAS 76-160 Optical Astronomy, L.Perek

AAS 76-161 Remarks on Professor Code's Survey Paper, Amendment, J.Strong, E.W.Silvertooth

AAS 76-162 Past Interpretation of Mars in Retrospect, C.W.Tombaugh

AAS 76-163 Jupiter--A Laboratory for Meteorologists, R.F.Beebe

AAS 76-164 Unsolved Problems in the Solar System, J.Evans

AAS 76-165 The Large Space Telescope, an Introduction into the Future of Spaceborne Optical Astronomy, M.Hunter II

AAS 76-166 Very Long Baseline Interferometry (VLBI) in Modern Radio Astronomy, B.Clark

AAS 76-167 The Very Large Array at Magdalena, New Mexico, V.Herrero

AAS 76-168 Introduction to Life Sciences in Space, C.Berry

AAS 76-169 First Response to Dr. Berry's Survey Paper, H.Saiki

AAS 76-170 Life Sciences in Space Flight, H.J. von Beckh

AAS 76-171 The Development of the International Law of Outer Space, C.Q.Christol

AAS 76-172 Review of Space Law Developments, S.Gorove

AAS 76-173 Issues Pending Before the United Nations Legal Subcommittee, E.Galloway

AAS 76-174 The Utility of Morphology to Space Law, E.Fasan

AAS 76-175 Remote Sensing and Direct Broadcast with the United Nations' System, S.K.Sarkar

AAS 76-176 The Evolution of Flight Control of the Apollo Mission, M.A.Faget

AAS 76-177 Space Stations-Symbols and Tools of New Growth in an Open World, K.A.Ehricke

AAS 76-178 Space Stations and Their Prologue, J.F.Madewell

AAS 76-179 Astropolis and Androcell--The Psychology and Technology of Space Utilization and Extraterrestrialization, K.A.Ehricke

AAS 76-180 Contributions of Spaceflight and Space Stations to the Solution of Social and Environmental Problems, L.Jaffe

AAS 76-181 About the Need for Social Institutions to Change to Accomodate Technological Advances and Change. Comments on Dr. Jaffe's and Dr. Ehricke's Keynotes, W.Armstrong

AAS 76-182 The European Space Agency and the Spacelab Programme, W.J.Mellors

AAS 76-183 The Space Shuttle Orbiter, G.Merrick

AAS 76-184 Trends in Space Communications, G.Harter

AAS 76-185 From Ballistic Missiles to Space Operations--An Exercise in Perspectives, G.Hansen

AAS 76-186 The Role of Space Technology in Society, R.L.Gervais

AAS 76-187 The State and Promise of Space, T.O.Paine

UTILIZATION OF SPACE SHUTTLE AND SPACELAB

Proceedings published by DGLR, 1976. Distributed by Univelt, Inc.
P.O. Box 28130, San Diego, California 92128

International Meeting sponsored by Deutsche Gesellschaft für Luft- und Raumfahrt and American Astronautical Society, June 24, 1976, Bonn, West Germany

No AAS numbers assigned

DGLR 1.1 Utilization of STS-Spacelab Transportation System for Technology Applications to Meet Economic Requirements, G.Greger

DGLR 1.2 ESA Spacelab Utilization, A.Lebeau

DGLR 1.3 Perspektiven der Weltraumtechnik im Hinblick auf industriell nutzbare Innovationen (Perspectives if Industrial Innovation by Means of Space Technology), W.Dettmering

DGLR 2.1 Die Bedeutung der Spacelab-Nutzung für die Werkstoff-Wissenschaft, H.Ahlborn

DGLR 2.2 Space Processing Applications--Designing the Initial Space Transportation System Payloads Capability, R.L.Hammel

DGLR 2.3 Space Processing of Gasturbine Blades--A Possible Industrial Application, J.E.Feustel, P.Leven

DGLR 2.4 Deployable Radiators for Waste Heat Dissipation from Shuttle Payloads, R.L.Cox, J.B.Dietz, J.W.Leach

DGLR 2.5 Looking Homeward: Uses of the STS/Spacelab to view the Earth, C.E.Cheeseman

DGLR 2.6 Fernerkundungstechniken und ihre Nutzung aus europäischer Sicht (Remote Sensing Techniques and its Economical Applications from European Point of View), P.Hartl, J.Bodechtel, D.Davidts, R.Mühlfeld, H.Sax, S.Schneider, E.Velten

DGLR 2.7 Investigation of New Communication Technologies and its Operational Application, HJ Albrecht, D.Rother

DGLR 3.1 Spacelab Science Missions of the 1980's, C.R.Chappell

DGLR 3.2 Spacelab Science Utilization, J.Collet

DGLR 3.3 Scientific Objectives of SL-Borne Lidar Systems, W.Renger

DGLR 3.4 LIDAR Payloads for Atmospheric Research, J.C.Poher

DGLR 3.5 Das Konzept eines multidisziplinären Lidar Ergebnis einer BMFT-Studie (The Technical Concept of a Multidisciplinary Lidar), E.Achtermann, W.Englisch

DGLR 3.6 Remote Sensing of Atmospheric Constituents of Interest in the Photochemistry of the Ozone Layer, D.G.Murcray

DGLR 3.7 Meteorological Observations from Space and Spacelab, HJ Bolle

DGLR 3.8 The Spacelab Instrument Pointing Subsystem (IPS) and its Utilization, J.C.Jones, H.Siemann

DGLR 3.9 LIRTS: A Large Infrared Telescope for Spacelab, A.F.M.Moorwood

DGLR 4.1 Long-Duration Exposure Facility--a Multipurpose Free-Flying Experiment Carrier, W.H.Kinard

DGLR 4.2 Space Telescope Science Program, C.R.O'Dell

DGLR 4.3 Cooled Infrared Telescope Development, L.S.Young

DGLR 4.4 The Preliminary Design of an Orbiting Observatory--The Space Telescope, K.P.Timmons

DGLR 4.5 The ATS-6 Experience and the Future of Broadcast Satellites, H.Dornbrand, W.Johnston,Jr.

DGLR 4.6 Remote Manipulator System and Satellite Servicing Experiment for Space Shuttle, G.M.Lindberg, J.D.MacNaughton

DGLR 4.7 Progress in Extraterrestrial Research through Shuttle and Spacelab, K.Pinkau

DGLR 5.1 Further Development of Spacelab to Meet Presumed Increasing Economic Requirements, H.E.W.Hoffmann

DGLR 5.2 Space Stations for the International Future, J.F.Madewell, R.E.Sexton

DGLR 5.3 Space Power Demonstration Stations, R.F.Freitag

DGLR 5.4 Potential for Future Space Solar Power Systems, R.H.Nansen

DGLR 5.5 Entwurf und Leistungsfähigkeit von Sonnenenergie-Satelliten ("Design and Performance of Solar Power Plants"), D.Köhn, K.Jesche, J.Rath, D.Teichmann, G.Wirths

DGLR 5.6 Nutzungsvorschläge aus dem Bereich Biologie/Medizin--Stand der vorbereitenden Planung ("Proposals for Life Sciences Experiments--Status of Planning") K.O.Pfeiffer

DGLR 5.7 Microwave Transmission System for Space Power, R.M.Dickinson

DGLR 5.8 GroBe Entfaltantennen und Solargeneratoren in extremer Leichtbauweise ("Large Ultralight Deployable Antennas and Solar Generators") W.Schafer

DGLR 5.9 The Two-Frequencies Microwave Scatterometer for Measuring of Ocean Waves, W.Alpers

DGLR 6.1 Informing the Public of STS-Spacelab Benefits: A Case History, C.E.Carnahan

DGLR 6.2 STS-Spacelab Utilization Relative to Past and Future Manned Space Flight, E.G.Gibson

DGLR 6.3 NASA Presentation and Final NASA Statements on International Usage of the Space Transportation System, W.P.Murphy

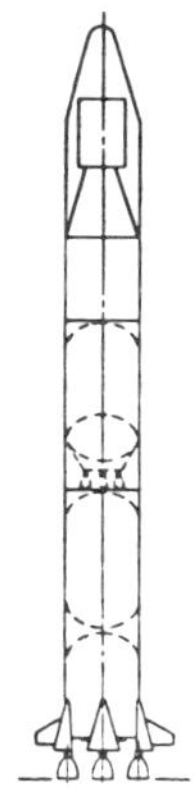

# AAS TECHNICAL PAPERS
# 1977

AAS/AAAS SOLAR ELECTRIC POWER CONFERENCE, February 20-26, 1977, Denver, Colorado
(Held in conjunction with the Annual Meeting of the American Association for the Advancement of Science)

No proceedings published.

AAS 77-001 to -049 Not Available

VOL. 46 SCIENCE AND TECHNOLOGY EXPORT OF AEROSPACE TECHNOLOGY, 15TH GODDARD MEMORIAL SYMPOSIUM, 1978
(Export of Aerospace Technology, 15th Goddard Memorial Symposium, March 31-April 1, 1977, Washington, D.C.)

AAS 77-050 Not Assigned

AAS 77-051 Aerospace Technology Transfer: A State Department Perspective, O.H.Ganley

AAS 77-052 Not Available

AAS 77-053 The NASA Program, E.Z.Gray

AAS 77-054 On the Implementation of the Defense Science Board Report, O.Lomacky

AAS 77-055 Not Available

AAS 77-056 National Security Implications of Technology Transfer, J.Thompson

AAS 77-057 to -059 Not Available

AAS 77-060 The Economic Impact of Aerospace Technology Transfer, B.A. Sharmon

AAS 77-061, -062 Not Available

AAS 77-063 Jobs, Technology Export, and U.S. Policy, W.C.Norris

AAS 77-064 Technology Suitable for Export, D.J.Fink

AAS 77-065 Technology Suitable for Export, H.G.McDonell

AAS 77-066 Technology Suitable for Export, R.L.Johnson



AAS 77-067 Examination of Our Obligations, S.Ross

AAS 77-068 to -099 Not Assigned

VOL. 27 AAS MICROFICHE SERIES, AAS/AIAA ASTRODYNAMICS 1977
(1977 AAS/AIAA Astrodynamics Conference, September 7-9, 1977, Jackson, Wyoming)

AAS 77-100 Not Assigned

AAS 77-101 Jupiter Orbiter and Probe Project: Synthesis of the Mission Design, J.C.Beckman, P.H.Roberts,Jr.

AAS 77-102 Trajectory Design Aspects of a Proposed NASA/ESA Out-of-Ecliptic Mission, R.A.Wallace

AAS 77-103 Ion Drive Multi-Asteroid Rendezvous Trajectories, D.B.Bender

AAS 77-104 Comparison of Solar Sail and Ion Drive Trajectories for a Halley's Comet Rendezvous, C.G.Sauer,Jr.

AAS 77-105 Mission Design for the Halley Comet Rendezvous Using Ion Drive, R.J.Boain

AAS 77-106 Mars Orbiter/Rover/Penetrator Mission for the 1984 Opportunity, R.Hastrup, J.Driver, R.Nagorski

AAS 77-107 1983 and 1984 Multirevolution Opportunities to Mars, P.H.Roberts, Jr.

AAS 77-108 Lunar Far-Side Sample Return Missions Using the Soviet Lunar System, P.H.Roberts,Jr.

AAS 77-109 Outer Planets Exploration: A Mission Strategy for the 1980's, J.W.Moore, J.C.Beckman

AAS 77-110 Velocity Matching Technique for Three-Dimensional Orbit Transfer in Conceptual Mission Design, S.T.Chu, T.J.Lang, B.E.Winn

AAS 77-111 Spinning Tandem Solid Motor Mission Flexibility, L.N.Berge, D.D.Van Winkle

AAS 77-112 Not Available

AAS 77-113 Single-Axis Attitude Determination Accuracy, L.C.Chen, J.R.Wertz

AAS 77-114 Star Pattern Recognition for Real-Time Attitude Determination, J.L.Junkins, C.C.White III, J.D.Turner

AAS 77-115 Attitude Determination Techniques for Nimbus 6 and G, A.Das, S.H.Siegel, H.L.Stallings

AAS 77-116 Analytic Investigation of the AEM-A/HCMM Attitude Control System Performance, G.M.Lerner, W.Huang

AAS 77-117 Space Sextant Attitude Reference System, H.A.Garcia, W.J.Owen

AAS 77-118 Variation of Parameters Approach to the Arbitrarily Torqued, Asymmetric Rigid Body Problem, L.G.Kraige, S.B.Skaar

AAS 77-119 On the Hanson Coefficients, P.Cefola

AAS 77-120 New Transformation Invariant in the Orbital Boundary Value Problem, R.H.Battin, T.J.Fill, S.W.Shepperd

AAS 77-121 Critical Second Species Periodic Orbits in the Restricted Problem for $\mu > o$ D.L.Hitzl

AAS 77-122 Boundary Value Methods with Time Transformations for Encounter Trajectories, P.E.Nacozy

AAS 77-123 Not Available

AAS 77-124 Linear Multistep Methods with Variable Step Size Using Back-Corrections, T.Feagin, J.Peterson

AAS 77-125 Radiometric Direction Finding: A New Approach to Deep Space Navigation, W.G.Melbourne, D.W.Curkendall

AAS 77-126 Demonstration of Dual Spacecraft Tracking Conducted with the Viking Spacecraft During the Approach Phase, C.C.Chao

AAS 77-127 Not Available

AAS 77-128 Orbit Determination for Earth Survey Satellites, P.J.Dunn, A.J.Fuchs

AAS 77-129 Advanced Spacecraft Tracking Techniques Using the Tracking and Data Relay Satellite System (Tross), J.Teles, C.Ayres

AAS 77-130 Interpolation Schemes for Orbit Determination with the Global Positioning System, P.Argentiero, G.E.Morduch

AAS 77-131 Differential Altimetry for Satellite Orbit Determination, H.H.Hagar,Jr., D.H.Boggs

AAS 77-132 Design of Minimum Variance Filters Subject to Stability Constraints, S.P.Karatsinides, W.F.Powers

AAS 77-133 Algorithms for Isolating Worst Case Systematic Data Errors with Applications to a Very Long Baseline Interferometry, D.W.Curkendall

AAS 77-134 Spline Method for Non-Linear Parameter and State Estimation in the Presence of Model Uncertainties, H.Schneider, R.Ramachandran

AAS 77-135 Curvilinear Projection Developments, H.J.Kelley, L.Lefton, I.L.Johnson,Jr.

AAS 77-136 Singular Optimal Control Computation, W.F.Powers, B.D.Cheng, E.R.Edge

AAS 77-137 Stochastic Optimal Attitude Control of Spacecraft with Movable Appendages, R.Sellappan, P.M.Bainum

AAS 77-138 Voyager Navigation Strategy and Accuracy, J.B.Jones, J.P.McDanell, M.H.Bantell,Jr. et al

AAS 77-139 Navigation of the 1982 Jupiter Orbiter--Probe Mission, K.H.Rourke

AAS 77-140 Autonomous Satellite Navigation with the Global Positioning System, A.J.Fuchs, W.H.Wooden II, A.C.Long

AAS 77-141 On Achieving Sufficient Dual-Station Range Accuracy for Deep Space Navigation at Zero Declination, H.L.Siegel, C.S.Christensen, D.W.Green, F.B.Winn

AAS 77-142 Efficient Estimation Algorithms for a Satellite-Aided Search and Rescue Mission, P.Argentiero, R.Garcia-Robles

AAS 77-143 Epoch State Navigation Filter, R.H.Battin, S.R.Croopnick, J.A.Edwards

AAS 77-144 Design and Reconstruction of Viking Lander Descent Trajectories, E.A.Euler, G.L.Adams, F.W.Hopper

AAS 77-145 Geostationary Longitude Acquisition Planning Algorithm, C.J.Petruzzo, W.C.Bryant,Jr., K.G.Nickerson

AAS 77-146 Monte Carlo Analysis of Impulse Requirements for Injection Error Correction, P.P.Rao

AAS 77-147 Station Keeping of Space Shuttle in Vicinity of a Deployed Payload, W.A.Fey

AAS 77-148 Differential Dynamic Programming Feedback Control for a Pursuing Spacecraft with Limited Fuel, G.M.Anderson

AAS 77-149 Evaluation of Parity Equations for Gyro Failure Detection and Isolation (FDI) in a Skew-Redundant Octahedron Configuration, A.L.Satin, R.L.Gates

AAS 77-150 Venus Orbital Imaging Radar Mission Analysis, K.T.Nock

AAS 77-151 Orbit Design for Solar and Dual Satellite Occultation Measurements of Atmospheric Constituents, D.R.Brooks, E.F.Harrison

AAS 77-152 Earth Reencounter Probabilities for Aborted Space Disposal of Hazardous Nuclear Waste, A.L.Friedlander, H.Feingold

AAS 77-153 Lunar Mass Transport Problem in Space Colonization, T.A.Heppenheimer

AAS 77-154 Dynamics of a Flexible Manipulator Arm for the Space Shuttle, P.C.Hughes

AAS 77-155 Aerodynamic Heating of Tip Command Antenna, R.W.Newman

AAS 77-156 Not Available

AAS 77-157 Evaluation of an Infrared Horizon Scanner Bias Determination Algorithm for Earth-Oriented Spacecraft, S.G.Hotovy

AAS 77-158 Minimization of Reaction Wheel Momentum Storage with Magnetic Torquers, A.Wernli

AAS 77-159 In-Flight Gyro Drift Rate Calibration on Viking Orbiters, W.G.Breckenridge, A.J.Treder

AAS 77-160 Not Available

AAS 77-161 Optimal Space Shuttle Ascent Trajectory for First Orbital Flight Test, I.L.Johnson,Jr.

AAS 77-162 Unusual Maneuvers on Nimbus and Landsat Spacecraft, S.H.Siegel, A.Das

AAS 77-163 Application of Altitude Control Techniques for Low Altitude Earth Satellites, K.G.Nickerson, R.W.Herder, A.B.Class

AAS 77-164 Atmospheric Braking to Circularize an Elliptical Venus Orbit, A.D.McRonald

AAS 77-165 Halo Orbit Station Keeping for International Sun-Earth Explorer-C (ISEE-C), H.S.Heuberger

AAS 77-166 Not Available

AAS 77-167 Multibody Flexible Spacecraft Integrated Analysis: Structures, Dynamics, and Control, G.Margulies, J.N.Aubrun, D.Bushnell, J.Y.L.Ho

AAS 77-168 Effects of Meteroid Impacts on Spacecraft Attitude Motion, D.A.Levinson

AAS 77-169 Near-Resonant and Transition Attitude Motion of a Class of Dual-Spin Spacecraft, J.E.Cochran,Jr., J.R.Beaty

AAS 77-170 to -172 Not Available

AAS 77-173 Optimal Solar Sail Spiral to Escape, L.L.Sackett, T.N.Edelbaum

AAS 77-174 Elements of Solar Sail Spacecraft Navigation with Application to a 1986 Halley's Comet Rendezvous Mission, R.A.Jacobson, C.L.Thornton

AAS 77-175 Navigation Capability for an Ion Drive Rendezvous with Halley's Comet, C.L.Thornton, R.A.Jacobson

AAS 77-176, -177 Not Available

AAS 77-178 Long-Term Stability of Solar Storage Orbits, D.R.Davis, R.Greenberg

AAS 77-179 Recursive Analytical Solution Describing Artificial Satellite Motion Perturbed by an Arbitrary Number of Zonal Terms, A.C.Mueller

AAS 77-180 Not Available

AAS 77-181 Onboard Magnetic Field Modeling for Solar Maximum Mission (SMM), R.D.Headrick, F.L.Markley

AAS 77-182 Effects of Solar Radiation Pressure on the Orbit of the Solar Power Satellite, O.F.Graf,Jr.

AAS 77-183 Computationally Efficient Modeling for Long-Term Prediction of Global Positioning System Orbits, S.K.Collins, P.J.Cefola

AAS 77-184 to -199 Not Assigned

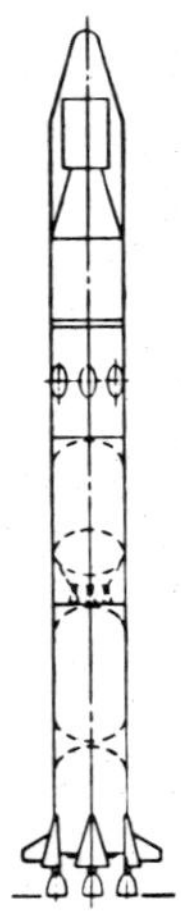

VOLUME 36 I & II, ADVANCES IN THE ASTRONAUTICAL SCIENCES, THE INDUSTRIALIZATION OF SPACE, 1978

VOLUME 28 AAS MICROFICHE SERIES, Supplement to Volume 36, Advances, 1978

(The Industrialization of Space, 23rd AAS Annual Meeting, October 18-20, 1977, San Francisco)

AAS 77-200 Freedoms from Constraints in Solar Power Satellite Design, R.Sperber, H.Zipursky (v36 I Adv)*

AAS 77-201 Automated Space Fabrication of Structural Elements, E.E.Engler, W.K.Muench (v36 I Adv)

AAS 77-202 A Near-Term Space Demonstration Program for Large Structures, C.A.Nathan (v36 I Adv)

AAS 77-203 Solar Power Satellite Construction Concepts, K.H.Miller (v36 I Adv)

AAS 77-204 Not Available

AAS 77-205 Structural and Assembly Concepts for Large Erectable Space Systems, E.Katz, E.T.Kruszewski, E.C.Naumann (v36 I Adv)

AAS 77-206 Technical Aspects of Large Space Structures, R.W.Johnson (v36 I Adv)

AAS 77-207 Not Available

AAS 77-208 Hidden Perils in Government Support of Space Activities, R.W.Poole,Jr (v36 II Adv)

AAS 77-209 Man and the Stars: A Summary, M.Gilson (v28 Micro)*

AAS 77-210 Government and Industrial Roles in the Initiation of Space Industrialization, G.H.Stine (v36 II Adv)

AAS 77-211 Expanding NASA's Charter to Facilitate Space Utilization, P.G.Grodzka, J.A.Tevepaugh (v36 II Adv)

AAS 77-212 Energy Crisis: A History Lesson, R.Sviedrys (v36 II Adv)

AAS 77-213 Two Lessons from the Past--An Analysis of Government's Role in Developing Super-Economies (Abstract)+, S.L.De Mandel (v36 II Adv)

---

* Adv = Advances in the Astronautical Sciences
Micro = AAS Microfiche Series

\+ In Abstract in Volume 36 Part II. In full in Volume 28 AAS Microfiche Series

AAS 77-214 The Integrated Use of Computers in Space Factories (Abstract)+, N.N.Matelan, L.W.Matelan (v36 II Adv)

AAS 77-215 Not Available

AAS 77-216 Space Shuttle Program Status and Utilization, R.F.Thompson, (v36 I Adv)

AAS 77-217 Advanced Launch Vehicle Systems and Technology, M.W.J.Bell (v36 I Adv)

AAS 77-218 Ground Operations and Concepts for Future Space Activities (Abstract)+, H.E.McCoy, T.A.Feaster (v36 II Adv)

AAS 77-219 A Hibrid Chemical Nuclear Space Freighter Concept, D.Kingsbury (v36 I Adv)

AAS 77-220 The Inertial Upper Stage, H.Kudish (v36 I Adv)

AAS 77-221 Solar Electric Propulsion and Interorbital Transportation, R.E.Austin (v28 Micro)

AAS 77-222 Mass Driver for Lunar Transport and as Reactor Engine, G.K.O'Neill, H.H.Kolm (v36 I Adv)

AAS 77-223 A Non-Synchronous Orbital Skyhook, H.Moravec (v28 Micro)

AAS 77-224 Space Processing of Turbine Blades by Means of Skin Technology, K.K.Schweitzer, J.Wortmann, A.Rossmann, W.Betz, H.Sprenger, E.Erben, H.Zeilinger (v36 I Adv)

AAS 77-225 Not Available

AAS 77-226 Space Industrialization and the Long-Term Prospects for Terrestrial Civilization, J.P.Vajk (v36 II Adv)

AAS 77-227 Not Available

AAS 77-228 Not Available

AAS 77-229 A Free Enterprise Model for the Operation of Industrial Facilities in Space (Abstract)+, T.W.Sanders (v36 II Adv)

AAS 77-230 Space Industrialization--The Long-Term View and the Near and Intermediate Steps, C.L.Gould (v36 II Adv)

AAS 77-231 Not Available

AAS 77-232 Marketing Techniques and Space Industrialization, G.H.Stine (v36 II Adv)

AAS 77-233 IDEF/Shuttle Capabilities for Environmental Testing in Space, LG.Clark, J.D.DiBattista (v36 I Adv)

AAS 77-233 (Suppl.) Long Duration Exposure Facility (LDEF) Guide for Experiment Accommodations (v28 Micro)

AAS 77-234 Not Available

AAS 77-235 Lunar Materials Processing, What, Where, Why? (Abstract), J.Green (v36 I Adv)

AAS 77-236 The Use of Outer Planet Satellites and Asteroids as Sources of Raw Materials for Life Support Systems, P.M.Molton, T.E.Divine (v36 I Adv)

AAS 77-237 A Baseline of Logistic and Power Requirements for Full-Scale Manufacturing of Metallic Materials in Earth Orbit, H. Bloom (v36 I Adv)

AAS 77-238 Materials Processing in Space: A Strategy for Commercialization, R.J.Naumann (v36 I Adv)

AAS 77-239 A Road Map to Space Products, R.L.Hammel. D.M.Waltz (v36 I Adv)

AAS 77-240 Satellite Mirror Systems for Providing Terrestrial Power: System Concept, K.W.Billman, W.B.Gilbreath, S.W.Bowen (v36 I Adv)

AAS 77-241 In-Orbit Manufacture of Solar Reflector Satellites, R.M.Miller (v36 I Adv)

AAS 77-242 Not Available

AAS 77-243 Interaction of Large, High-Power Systems with Operational Orbit Charged-Particle Environments, C.K.Purvis N.J.Stevens, F.D.Berkopec (v36 I Adv)

AAS 77-244 The Inevitability of Extraterrestrial Robotics in Space Industrialization (Abstract)+, R.W.Prehoda (v36 II Adv)

AAS 77-245 Isolation and Confinement: Considerations for Colonization, F.R.Akins (v36 II Adv)

AAS 77-246 Communication Requirements of a Space Settler: The Need to Keep in Touch, B.Copeland (v36 II Adv)

AAS 77-247 The Way in Space (Abstract)+, W.H.Cutler (v36 II Adv)

AAS 77-248 Woman's Future in Space (Abstract)+, C.Conley (v36 II Adv)

AAS 77-249 The Psychological Effects of High Orbital Migration: Summary, T.Leary (v36 II Adv)

AAS 77-250 An Analysis of a Space Community: Some Possible Areas of Diviance (Abstract)+, M.M.Cepin (v36 II Adv)

AAS 77-251 The Military Use of Outer Space, L.David (v36 II Adv)

AAS 77-252 Cultural Implications of Extraterrestrial Contact and the Colonization of Space, C.F.Urbanowicz (v36 II Adv)

AAS 77-253 to -255 Not Available

AAS 77-256 Public Service Satellite Communications, E.A.Wolff (v36 I Adv)

AAS 77-257 Maritime Satellite Communications--Where We Are and Where We are Going, D.W.Lipke (v36 I Adv)

AAS 77-258 Space Industrialization: Education, K.M.Joëls (v36 I Adv)

AAS 77-259 Assembly in Space of Large Communication Structures, F.Zylius, R.Donavan (v36 I Adv)

AAS 77-260 Legal Ties of a Space Colony to Earth, S.Gorove (v36 II Adv)

AAS 77-261 Not Available

AAS 77-262 The Impact of Treaties on Commercial Space Operations, M.Menter (v36 II Adv)

AAS 77-263 Information from Space: An Ethereal Resource for all Mankind, A.L.Moore (v36 II Adv)

AAS 77-264 Not Available

AAS 77-265 Extension of Terrestrial Law into Outer Space, H.DeSaussure (v36 II Adv)

AAS 77-266 Economical and Juridical Aspect of Industrialization of Activities in Outer Space, H.Safavi (v28 Micro)

AAS 77-267 Law and Policy in Operational Space (OpSpace), D.D.Smith (v36 II Adv)

AAS 77-268 Frontiers for Free Trade, M.Frazier (v36 II Adv)

AAS 77-269 The "Common Heritage of Mankind" Doctrine and Private Industrial Development of Outer Space, S.B.Rosenfeld (v36 II Adv)

AAS 77-270 Outer Space Treaty and the Great Deception: Civilian Industrialization or Military Outposts in Space?, G.S.Robinson (v36 II Adv)

AAS 77-271 The Politics of Law for Space Industrialization, J.D.Salmon (v36 II Adv)

AAS 77-272 Habitat Design--an Update, E.H.Bock, F.Lambrou,Jr., M.Simon (v36 I Adv)

AAS 77-273 Technology Requirements for Closed-Ecology Life Support Systems Applicable to Space Habitats, J.M.Spurlock, M.Modell (v36 I Adv)

AAS 77-274 Magnetic Radiation Shielding for Permanent Space Habitats, M.A.Paluszek (v36 I Adv)

AAS 77-275 Not Available

AAS 77-276 Requirements of Orbiting Facilities for Industrial Space Processing (Abstract)+, U.R.Alvarado, H.L.Bloom (v36 II Adv)

AAS 77-277 Concept for a Lunar/Orbital Logistics Support Station and Training Facility (Abstract)+, O.H.Feder (v36 II Adv)

AAS 77-278, -279 Not Available

AAS 77-280 Space Community Planning in a Down-to-Earth Context, R.Mack, T.Cullinan (v36 II Adv)

AAS 77-281 If We Can Get Along at Idaho State, We Can Get Along in Space (Abstract)+, B.Frentz, G.R.Mead, N.R.Solien (v36 II Adv)

AAS 77-282 Design Principles and Cultures, M.Maruyama (v36 II Adv)

AAS 77-283 Alternative Social Structures in a Vacuum, B.J.Bluth (v36 II Adv)

AAS 77-284 A Preliminary Investigation of Space Habitat Atmospheres, W.Ziegler (v36 II Adv)

AAS 77-285 Aesthetic Implications of the Crystal Palace Space Habitat, M.L.Stuart (v36 II Adv)

AAS 77-286 Drafting a Constitution for Orbis, The Business of Providing Community Services in Space, S.H.MacCallum (v36 II Adv)

AAS 77-287 Not Available

AAS 77-288 Space Sociology, A Terrestrial Perspective: Public Perceptions of the Space Program (Abstract)+, A.Rudoff (v36 II Adv)

AAS 77-289 Space Community Planning from a Viewpoint of Experience, T.A.Heppenheimer (v36 II Adv)

AAS 77-290 Space Industrialization, the Challenge to Private Enterprise Capitalism, C.O.Basler (v36 II Adv)

AAS 77-291 Pre-Production Problems in the Industrialization of Space (Abstract)+, E.W.Bjornen (v36 II Adv)

AAS 77-292 Cultural Futuristics: The New Role of Anthropology in Outer Space (Abstract)+, C.J.Motts (v36 II Adv)

AAS 77-293 A Methodology for Selecting Optimum Organizations for Space Communities, J.M.Ragusa (v36 II Adv)

AAS 77-294 Making it Work--Key Steps in the Development Program to Space Industrialization, M.A.Faget (v36 I Adv)

AAS 77-295 Solar Power Satellite Development Program, P.E.Glaser (v36 I Adv)

AAS 77-296 Remarks Delivered at the Panel Discussion on Making it Work, M.W.Hunter (v36 I Adv)

AAS 77-297 Statement for Panel on Making it Work, G.B.Merrick (v36 I Adv)

AAS 77-298 Remarks in Panel Discussion on Making it Work, B.M.Oliver (v36 I Adv)

AAS 77-299 Not Assigned

# AAS TECHNICAL PAPERS
# 1978

VOLUME 37 I & II, ADVANCES IN THE ASTRONAUTICAL SCIENCES, SPACE SHUTTLE AND SPACELAB UTILIZATION, Near-Term and Long-Term Benefits for Mankind, 1978

(Space Shuttle and Spacelab Utilization, 24th AAS Annual Meeting, 16th AAS Goddard Memorial Symposium, March 8-10, 1978, Washington, D.C.)

AAS 78-001 NASA Outlook, J.F.Yardley (Part I)

AAS 78-002 Outlook, H.A.Strub (Part I)

AAS 78-003 Shuttle/Spacelab System Program--Goals and Objectives, M.Bignier (Part I)

AAS 78-004 Space Shuttle Hardware Integration and Testing Status, W.E.Dean (Part I)

AAS 78-005 Spacelab Hardware Update, K.Berge (Part I)

AAS 78-006 Shuttle Mission Plans, J.T.Visentine, C.M.Lee (Part I)

AAS 78-007 Spacelab Mission Plans, G.Greger (Part I)

AAS 78-008 Advanced Missions for X-Ray Astronomy in the Shuttle Era, H.Tananbaum, R.Giacconi (Part I)

AAS 78-009 Prospects of Astronomy and Astrophysics from Spacelab, B.Aschenbach (Part I)

AAS 78-010 The NASA Life Sciences Experiment Program for Shuttle/Spacelab, D.L.Winter (Part I)

AAS 78-011 Life Sciences Laboratories in Spacelab, G.Wirths (Part I)

AAS 78-012 Not Available

AAS 78-013 German Materials Research and Space Processing Experiments in Microgravity and the Approach to Systematically Provide Mission Opportunities (Abstract), N.Kiehne (Part I)

AAS 78-013a Shuttle and Spacelab in the French Space Processing Programme, D.Sacotte, P.Pesenti (Part I)

AAS 78-014 The Department of Defense Role in the Shuttle/Spacelab System, K.L.Jordan,Jr. (Part I)

AAS 78-015 Serving the Public Via Platforms in Space, R.Fleisig, J.L.Bernstein (Part I)

AAS 78-016 A Spacelab Payload for Communication and Navigation Experiments, H.Öttl, H.Gössl, H.Haerter, D.Hounam (Part I) 1978

AAS 78-017, 018 Not Available

AAS 78-019 Landsat Missions, H.Mannheimer, S.Derdeyn (Part I)

AAS 78-020 Space Experiments Related to Earth Resources Technology (Abstract), P.Hart (Part I)

AAS 78-021 Objectives and Developments of Earth Observation Experiments for Spacelab, G.Rausch, E.H.Velten, W.Gilg (Part I)

AAS 78-022 Solar Power Satellite Developments, P.E.Glaser (Part I)

AAS 78-023 European Activities on Educational Physics and Young People's Experiments in Spacelab, G.Giampalmo (Part I)

AAS 78-024 Not Available

AAS 78-024a Advantages of Thin Silicon Solar Cells for Use in Space, O.S.Denman (Part I)

AAS 78-025 Preparatory Experiments in the Field of Metallurgy (Abstract), H.Ahlborn (Part I)

AAS 78-026 Preparatory Experiments for Crystal Growth, H.Weisz (Part I)

AAS 78-027 The German Space Processing Lab and Its Experiment Facilities P.Zimmermann (Part I)

AAS 78-027a Public Service Satellite Workship, R.Fleisig (Part I)

AAS 78-028 Medical and Psychological Selection and Training Criteria for European SL-Payload Specialists, K.E.Klein, J.R.Hordinsky (Part II)

AAS 78-028a The Space Shuttle: Some Legal Considerations, S.Gorove (Part II)

AAS 78-029 Space Sled--A Multi-Purpose Facility for Vestibular Research in Spacelab, P.Junk (Part II)

AAS 78-030, -031 Not Available

AAS 78-032 Radiation Sensitivity of Quartz Crystal Oscillators Experiment for the Long Duration Exposure Facility (LDEF), J.D.Venables, J.S.Ahearn (Part II)

AAS 78-033-I An Earth Oriented Multidisciplinary Mission Part 1: Experiment Objectives and Background, A.Drescher, A.Rossbach, F.Schlude (Part II)

AAS 78-033-II An Earth Oriented Multidisciplinary Mission (D 3) Part 2: Project Concept, H.W.K.Francois, H.Hirsch, H.Schwille, (Part II)

AAS 78-034 An Expanded Space Industrialization Program for the 1980's C.L.Gould (Part II)

AAS 78-035 Performance and Results of Apparatus and Equipment TEXUS I Payload, B.Franke, J.Geisel (Part II)

AAS 78-036 Skin Technology--A Shape Conserving Remelting Process in Space, H.Sprenger, E.Erben, H.Zeilinger, J.Wortmann, K.Schweitzer (Part II)

AAS 78-037 Teleoperator Retrieval System, R.J.Molloy, J.R.Tewell, R.A.Spencer (Part II)

AAS 78-038 Spacelab Demonstration Missions, G.Seibert (Part II)

AAS 78-039 Integration, Flight Operation, and Support Plan for Payloads Within the German Spacelab Utilization Program, H.J.Panitz (Part II)

AAS 78-040 European Spacelab Experiment Operations Philosophy, J.deWaard (Part II)

AAS 78-041 Not Available

AAS 78-042 Manned Maneuvering Units, C.E.Whitsett, Jr. (Part II)

AAS 78-043 NASA Standard Multimission Modular Spacecraft for Future Space Exploration, R.O.Bartlett (Part II)

AAS 78-044 Requirements for a Geosynchronous Information Services Platform, H.C.Ness, F.H.Shepphird, H.L.Wolbers (Part II)

AAS 78-045 Views on Spacelab Follow-On Development, G.Siebert (Part II)

AAS 78-046 Spas--The Shuttle-Dedicated Pallet Satellite Concept and Its Applications, D.E.Koelle, D.Davidts (Part II)

AAS 78-047 The Space Power Module: Utility for Shuttle/Spacelab--Tested for Technology Issues, R.W.Johnson (Part II)

AAS 78-048 Shuttle/Tethered Satellite System, C.C.Rupp, J.H.Laue (Part II)

AAS 78-049 The Shuttle Attached Remote Manipulator Program--A Status Review, J.MacNaughton (Part II)

AAS 78-050 Planning and Designing Payloads for the Space Transportation System, H.E.Emigh (Part II)

AAS 78-051 Not Available

AAS 78-052 Assembly of Large Space Structures, F.C.Runge (Part II)

AAS 78-053 Luncheon Address, D. Fuqua (Part II)

AAS 78-054, -055 Not Available

AAS 78-056 Current Topics and Prospects of Japanese Space Development, H.Uda (Part II)

AAS 78-057 to -099 Not Assigned

VOLUME 29 AAS MICROFICHE SERIES, GUIDANCE AND CONTROL CONFERENCE 1978 (Proceedings of Annual Rocky Mountain Guidance and Control Conference, March 10-13, 1978, Keystone, Colorado)

AAS 78-100 Navigation, Part I: Radio Navigation, Dead Reckoning, Accelerometers, Gyros, D.B.DeBra

AAS 78-100 Navigation, Part II: Inertial Navigation, Gravity Measuring Instruments, Tracking Propagation Corrections, D.B.DeBra

AAS 78-101 The New Generation of Dynamic Interaction Problems, P.Likins

AAS 78-102 Design of Digital Control Systems with State Feedback and Dynamic Output Feedback, B.C.Kuo

AAS 78-103 Some Limitations of Model Reduction and Controller Design Methods for Large, Uncertain Dynamical Systems, R.Skelton

AAS 78-104 A Method for Determining Overall Flexible Body Characteristics for a Series of Connected Substructures, S.C.Rybak

AAS 78-105 Optimal Digital Control of Large Space Structures, R.Gran, M.Rossi, H.G.Moyer

AAS 78-106 Some Approaches to the Control of Large Space Structures, M.J.Balas

AAS 78-107 to 109 Not Available

AAS 78-110 Transferring a Ground-Based Telescope to the Shuttle, R.B.Dunn

AAS 78-111 Not Available

AAS 78-112 The Annular Suspension and Pointing System (ASPS), W.W.Anderson, C.T.Woolley

AAS 78-113 Pointed Science Scan Platforms, R.S.Ward

AAS 78-114 Precision Pointing and Tracking System (PPTS), T.K.Brown, D.R.LeBlanc, E.Mettler, S.D.Gaalema

AAS 78-115 Earth Pointing with a Shuttle EPM, R.P.Iwens, R.A.Mayo, V.A.Spector, R.H.VanVooren

AAS 78-116 The Gimbalflex System, N.A.Osborne

AAS 78-117 Digital Pointing Control, P.A.Studer, R.J.Brown

AAS 78-118 A High Speed Pointing System, G.C.Winston

AAS 78-119 Not Available

AAS 78-120 HEAO Attitude Reference Design, D.P.Hoffman, T.T.McElroy

AAS 78-121 Payload Attitude Determination for the Small Satellites (S3) Project of the STP Office of SAMSO, A.K.Goodfellow

AAS 78-122 General Purpose Attitude Determination for Earth Orbiting Spinning Spacecraft, R.P.Zietz

AAS 78-123 A Decade of Attitude Sensors, R.L.Gutshall, R.A.Deters

AAS 78-124 Space Sextant Autonomous Navigation and Attitude Reference System, R.A.Booker

AAS 78-125 Digital Star Scanner Unit for Attitude Updates in Space: Design Flexibility and Performance, J.T.Tavolacci, C.Coscia, A.Lavaglia

AAS 78-126 to -129 Not Assigned

VOLUME 38 I & II ADVANCES IN THE ASTRONAUTICAL SCIENCES, THE FUTURE UNITED STATES SPACE PROGRAM, 1979

VOLUME 30 AAS MICROFICHE SERIES, Supplement to Volume 38, Advances, 1979

(The Future United States Space Program, 25th AAS Anniversary Conference, October 30-November 2, 1978, Houston, Texas)

AAS 78-130 Not Assigned

AAS 78-131 Youth in Space: Our Hope, Our Dilemma (Abstract), C.C.Hewitt (v38 I Adv)*

AAS 78-132 Space Experiments and the Small-Scale Investigator, R.L.Staehle (v38 I Adv)

AAS 78-133 The Space Program: Do Students Want to be Involved? J.J.Irons (v38 I Adv)

AAS 78-134 Development of Space Biology Experiments for the Space Shuttle, S. Walker (v38 I Adv)

AAS 78-135 Shuttle Small Self-Contained Payloads: "Getaway" to the Educational Opportunities of Space, M.K.Craig, C.A.Jacobson Sr., T.B.Murtagh (v38 I Adv)

AAS 78-136 Optimal Analytic Rendezvous Using Clohessy-Wiltshire Equations D.J.Jezewski, J.D.Donaldson (v38 I Adv)

AAS 78-137 Survey of Nonlinear Programming Algorithms and Software (Abstract)+ A.D.Waren, L.S.Lasdon (v30 Micro, v38 I Adv)*

AAS 78-138 Optimization Techniques Applied to Space Shuttle Explicit Re-Entry Guidance Law Design (Abstract)+ R.A.Metzler, W.F.Powers (v30 Micro, v38 I Adv)

AAS 78-139 Periodic Optimal Control Problems (Abstract), J.L.Speyer (v38 I Adv)

AAS 78-140 On the Possible Sociopolitical Nature of the Early L-5 Space Colonies: An Ecological Approach, R.D.McWilliams (v38 II Adv)

AAS 78-141 Constructing Space Communities: A Critical Look at the Paradigms, B.J.Bluth (v38 II Adv)

AAS 78-142 Not Available

---

* Adv = Advances in the Astronautical Sciences
Micro = AAS Microfiche Series

\+ In abstract in Volume 38 Advances.
In full in Volume 30, AAS Microfiche Series

AAS 78-143 Lunar Base Habitat Exploration Design Study (Abstract)+ G.Trotti, D.R.Criswell (v30 Micro, v38 II Adv)

AAS 78-144 Financial/Management Scenarios for a Satellite Power System Program, J.P.Vajk, R.D.Stutzke, R.Salkeld, G.H.Stine, M.S.Klan (v38 II Adv)

AAS 78-145 Developing Closed Life Support Systems for Large Space Habitats, J.M.Phillips, A.D.Harlan, K.C.Krumhar (v38 II Adv)

AAS 78-146 Space Shuttle Ascent Guidance, Navigation, and Control (Abstract), R.L.McHenry, A.D.Long, B.F.Cockell, J.R.Thibodeau, T.J.Brand (v38 I Adv)

AAS 78-147 Shuttle Entry Guidance (Abstract), J.C.Harpold, C.A.Graves,Jr. (v38 I Adv)

AAS 78-148 The Shuttle Central Computer Complex (Abstract), P.E.Sollock (v38 I Adv)

AAS 78-149 Space Shuttle Guidance, Navigation, and Control System Architecture (Abstract), W.L.Swingle (v38 I Adv)

AAS 78-150 Technology Requirements for an Orbital Navigation/Position Finding System, A.L.Lang,Jr., L.J.DeRyder,Jr. (v38 I Adv)

AAS 78-151 A Microwave Radiometer Spacecraft, U.M.Lovelace (v38 I Adv)

AAS 78-152 Thermally Stable Polyimide Components for Space and Commercial Applications, J. Gagliani, D.E.Supkis (v38 I Adv)

AAS 78-153 An Evolutionary Solar Power Satellite Program, G.M.Hanley, W.R.Rhote (v38 I Adv)

AAS 78-154 A Development Strategy for the Solar Power Satellite, D.L.Gregory (v38 I Adv)

AAS 78-155 Power Module Systems Concepts (Abstract), J.Craig, C.DeRos (v38 I Adv)

AAS 78-156 The Viking Results--The Case for Man on Mars, B.C.Clark (v38 I Adv)

AAS 78-157 Laser Aircraft Propulsion, K.C.Sun, A.Hertzberg, W.S.Jones (v38 I Adv)

AAS 78-158 International Cooperation and Satellite Telecommunications, R.R.Colino (v38 II Adv)

AAS 78-159 Not Available

AAS 78-160 Options for an Operational Remote-Sensing System, E.Galloway (v38 II Adv)

AAS 78-161 Not Available

AAS 78-162 Legal Aspects of Military Peaceful Uses of Space, W.D.Reed (v38 II Adv)

AAS 78-163 A Perception of Economics (Abstract), J.S.Greenberg (v38 II Adv)

AAS 78-164 Economic Parametrics--the Mature Industry Approach (Abstract)+ J.Gauger (v30 Micro, v38 II Adv)

AAS 78-165 The Value of Information as Applied to the Landsat Follow-On Benefit-Cost Analysis, D.B.Wood (v38 II Adv)

AAS 78-166 Costing the Satellite Power System, G.A.Hazelrigg, Jr. (v38 II Adv)

AAS 78-167 Defense and the Future of the U. S. Space Program, M.Sanborn (v38 I Adv)

AAS 78-168 Congress and the Future of the United States Program, G.P.Wilson (v38 I Adv)

AAS 78-169 A Technology Transfer Program for State and Local Governments (Abstract), W.Mooneyhan (v38 I Adv)

AAS 78-170 Alternative Futures for the U.S. Earth Resources Program, C.Sheffield (v38 I Adv)

AAS 78-171 The Role of Other Than Aerospace Industry (Abstract), H.Jensen (v38 II Adv)

AAS 78-171A Space Medicine/Introductory Remarks, H.S.Fuchs (v38 II Adv)

AAS 78-171A (Continued) Space Medicine/Conclusion, H.S.Fuchs (v38 II Adv)

AAS 78-172 -175 Not Available

AAS 78-176 Health Maintenance and Health Surveillance Considerations for an SPS Space Construction Base Community, J.P.Kornberg, P.K.Chapman, P.E.Glaser (v38 II Adv)

AAS 78-177 Satellite Communications for the 1980's, B.I.Edelson (v38 II Adv)

AAS 78-178 Space Shuttle: Providing for Man's Future in Space, G.B.Merrick (v38 II Adv)

AAS 78-179 Observing Planet Earth from Space - What's Next? D.S.Johnson (v38 II Adv)

AAS 78-180 Future Programs in Space, J.A.Snow (v38 II Adv)

AAS 78-181 Reaping the Space Investment, A.J.Calio (v38 II Adv)

AAS 78-182 Future Programs and Prospects for Resource Exploration from Space by the Year 2000, M.T.Halbouty (v38 II Adv)

AAS 78-183 Future Space Science Programs and Projects, (Abstract), A.Timothy (v38 II Adv)

AAS 78-184 Beyond the Shuttle, G.E.Mueller (v38 II Adv)

AAS 78-185 Planetary Exploration Objectives 1980-1990 (Abstract), D.Herman (v38 I Adv)

AAS 78-186 Radar Imagery and Future Exploration of Venus (Abstract) H.Masursky (v38 I Adv)

AAS 78-187 The Geological History of Mars--Implications for Understanding Earth History and Processes (Abstract), R.Greeley (v38 I Adv)

AAS 78-188 The Exploration of Mars--A Challenge for Science and Technology (Abstract), L.Friedman (v38 I Adv)

AAS 78-189 Understanding the Internal Dynamics of the Earth--The Space Connection (Abstract), E.A.Flinn (v38 I Adv)

AAS 78-190 Accessible Industrial Materials in Lunar Soil, W.N.Agosto (v38 I Adv)

AAS 78-191 The Polar Lunar Base--A Viable Alternative to L-5, J.Green (v38 I Adv)

AAS 78-192 Engineering in the 21st Century, J.F.McCarthy,Jr. (v38 II Adv)

AAS 78-193 Projected Aerospace Systems Development, F.Stuart (v38 II Adv)

AAS 78-194 Engineering in the 21st Century, W.T.Hamilton (v38 II Adv)

AAS 78-195 Technical Challenges of Large Space Systems in the 21st Century, T.J.Kelly (v38 II Adv)

AAS 78-196 Projected Progress in the Engineering State-of-the-Art, O.W.Nicks (v38 II Adv)

AAS 78-197 Engineering Education for the 21st Century, to Respond to the Needs of Space Systems Engineering, G.Bugliarello (v38 II Adv)

AAS 78-198 Utilization of Lunar Materials in Space (Abstract)+ D.R.Criswell, R.D.Waldron (v30 Micro, v38 II Adv)

AAS 78-199 Solar Power Space Stations: Some Issues of Law and Policy S. Gorove (v38 II Adv)

# AAS TECHNICAL PAPERS
# 1979 (PARTIAL)

VOLUME 39 ADVANCES IN THE ASTRONAUTICAL SCIENCES, 1979

VOLUME 30 AAS MICROFICHE SERIES,

(AAS Rocky Mountain Guidance and Control Conference, February 24-28, 1979, Keystone, Colorado)

*AAS 79-001 Space Navigation Using the Navstar Global Positioning System (GPS), J.E.Farr

AAS 79-002 Precision Correlation Tracking, J.M.Fitts

AAS 79-003 Thruster Select Logic Determination for Attitude Control of Skylab by Teleoperator Retrieval System, G.A.Cook

AAS 79-004 Pioneer Venus Star Sensor, R.L.Gutshall, G.Thomas

AAS 79-005 Autonomous Attitude Determination Systems, J.W.Lowrie

AAS 79-006 Agile Spacecraft Attitude Reference System Design, H.J. Dougherty, E.J.Pelka, J.J.Rodden

AAS 79-007 to -010 Not Assigned

AAS 79-011 Requirements and Opportunities for Autonomous Systems in Space, W.B.Gevarter, E.Heer

AAS 79-012 Advanced Guidance and Control Technology for Spacecraft Automation, R.T.Schappell

AAS 79-013 Autonomous Star Sensing and Attitude Estimation, J.L.Junkins, T.E.Strikwerda

AAS 79-014 Inertial Measurement Unit Redundancy Management, G.Quasius, J.Boutelle, D.E.Sanders, R.VanderVoort, S.P.Kau

AAS 79-015 The Fault Tolerant Spaceborne Computer (FTSC), G.Gilley

AAS 79-016 Spacecraft Automated Operations, T.H.Bird, B.L.Sharpe

AAS 79-017 to -020 Not Assigned

AAS 79-021 DRIRU II--The NASA Standard High Performance Inertial Reference Unit, R.B.Irvine, J.W.Ritter

AAS 79-022 A User's Guide for Standard Star Tracker, R.A.Deters

AAS 79-023 Standard Reaction Wheel (SRW), C.Sutter, J.Walberg

---

* Unless otherwise indicated all papers appear in Volume 39, Advances

AAS 79-024 NASA Standard Computers: A Description and Comparison, M.S.Ross

AAS 79-025 The NASA Multimission Spacecraft Modular Attitude Control System, J.W.Murrell

AAS 79-026 MMS--A Systems Level View of Standardization, S.Myers, H.Raymond

AAS 79-027 to -030 Not Assigned

AAS 79-031 Attitude Control for the TRS, F.E.Bikle

AAS 79-032 A Concept for Shifting the Reentry Point of Skylab, M.H. Kaplan, S.G.Alexander

AAS 79-033 Effects of Tether Attachments on the Shuttle/Tethered Satellite System Dynamics, L.L.Gresham, C.C.Rupp

AAS 79-034 Navigation and Flight Control of the Inertial Upper Stage, A.K.Goodfellow

AAS 79-035 Changing Inclination for Shuttle Payloads, T.M.Spencer, R.Glickman, G.Porcelli (v31 Mic)

AAS 79-036 Interim Upper Stage Guidance System Mission Success Analysis, W.G.McArthur, R.Baum

AAS 79-037 A Multivariable Control System Design Algorithm, L.L.Gresham, J.R.Mitchell, W.L.McDaniel,Jr. (v31 Mic)

AAS 79-038, -039 Not Assigned

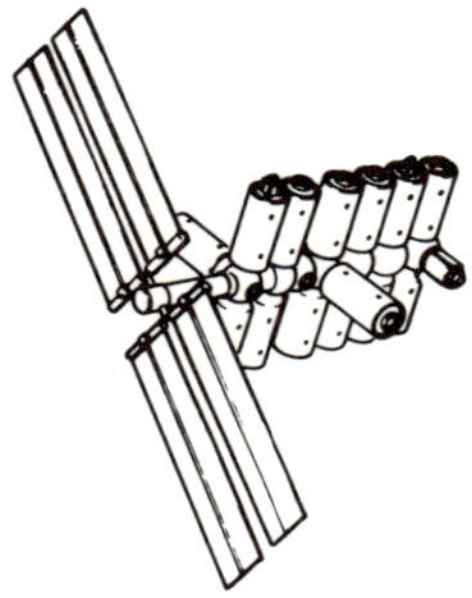

# CHRONOLOGICAL INDEX ASTRONAUTICAL SCIENCES REVIEW (ASR) 1959 - 1962

The following pages provide a chronological index of articles that appeared in *Astronautical Sciences Review* (1959-1962), a publication of the American Astronautical Society. Publication of the *Review* was discontinued in 1962. *The Astronautical Sciences Review* covers the early Sputnik era.

Most back issues of the *Review* may be obtained from the publishers for the American Astronautical Society, Univelt, Inc., P. O. Box 28130, San Diego, California 92128.

# CHRONOLOGICAL INDEX

## ASTRONAUTICAL SCIENCES REVIEW

## (1959-1962)

Following is a complete list of articles that appeared in the Astronautical Sciences Review (1959-1962). This quarterly journal was discontinued with the January-March 1962 issue.

Orbital Propulsion System for Space Maneuvering (PROFAC), S.T.Demetriades, pp. 17,18,26

Orbital Plane-Change Maneuver, D.B.DeBra, p. 19

VOLUME 2-1 (January-March 1960)

Some Aspects of Lunar Geology, J.Green, pp. 9-12,16

Lunik III Trajectory Predictions, J.E.Michaels, M.Wachman, A.Petty, pp. 13-16

VOLUME 2-2 (April-June 1960)

Parameter Transformation Tables for Two-Body Orbits, J.Morris, R.E.Roberson, pp. 11-15

Space Communications, J.A.Webb, pp. 16,17,24

VOLUME 2-3 (July-September 1960)

Personnel Requirements in Manned Space Vehicles, G.A.Peters, H.J.Dendl, pp. 9-12

A Life Support System for a Near Earth or Circumlunar Space Vehicle, G.B.Whisenhunt,Jr., pp. 13-20

VOLUME 2-4 (October-December 1960)

Some Radiation Problems of Space Conquest, W.H.Langham, pp. 9-18

Satellite Coverages, B.H.Gundel, pp. 19-23

VOLUME 3-1 (January-March 1961)

Radiation Shielding of Lunar Spacecraft, T.G.Barnes, E.M.Finkelman, A.L.Barazotti, pp. 11-18

A Family of Radioisotope-Fueled Auxiliary Power Systems for Lunar Exploration, R.J.Wilson, pp. 19-26

VOLUME 3-2 (April-June 1961)

Eclipses of Artificial Earth Satellites, L.G.Stoddard, pp. 9-14,16

Nomographs to Determine Velocity Requirements for Interplanetary Trips to Mars and Venus, J.V.Breakwell, R.W.Gillespie,S.Ross, pp. 15-16a

The Earth's Dust Belt, F.L.Whipple, pp. 17-20

VOLUME 3-3 (July-September 1961)

Nuclear Rockets for Space Flight, E.Sänger, G.E.Pieper, pp. 9-15

Temperature Control Considerations for Instrumented Lunar Packages, S.Caltabiano, pp. 16-19

VOLUME 3-4 (October-December 1961)

Adverse Effects Due to Space Vehicle Rotation, G.T.Beauchamp, pp. 9-11

Monitoring of Astronaut Brain Function by Means of Telemetered Electroencephalogram with Automatic Analysis, R.M.Morrell, pp. 13-19

A Space Vehicle Design for Unmanned Rendezvous, R.I.Johnson, G.R.Arthur, pp. 21-29

VOLUME 4-1 (January-March 1962)

Physiological Considerations Relevant to the Problem of Prolonged Weightlessness: A Review, R.W.Lawton, pp. 11-18,31-38

Drag-Life Curves for High Eccentricity, J.V.Breakwell, pp. 21,22,26

The Crater Contribution to Surface Roughness of the Moon, M.Kornhauser, pp. 23,24,26

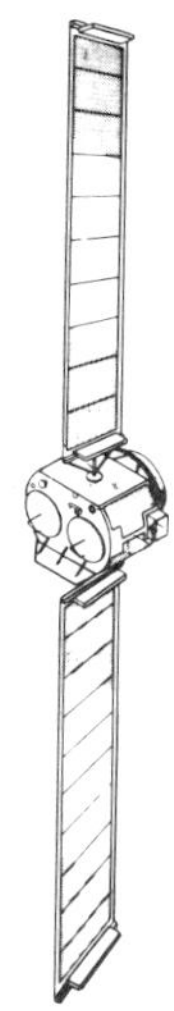

# CHRONOLOGICAL INDEX THE JOURNAL OF THE ASTRONAUTICAL SCIENCES (JAS) 1954 - 1978

The following pages provide a chronological index of articles that appeared in *The Journal of the Astronautical Sciences* (1954-1978) a publication of the American Astronautical Society. It should be noted that *The Journal of the Astronautical Sciences* was titled *Astronautics* in 1954-1955, then *The Journal of Astronautics* in 1956-57, and finally *The Journal of the Astronautical Sciences* from 1958 on.

Most back issues of the *Journal* may be obtained from the publishers for the American Astronautical Society, Univelt, Inc., P. O. Box 28130, San Diego, California 92128.

# CHRONOLOGICAL INDEX

# THE JOURNAL OF THE ASTRONAUTICAL SCIENCES*
(1954-1978)



* The first two issues were titled Astronautics, Journal of the American Astronautical Society, Inc. . The next three issues were titled The Journal of Astronautics . It was called The Journal of Astronautics until Volume 5 No. 1, 1958, and then became The Journal of the Astronautical Sciences .

Aspects of Nuclear Power Application for Jet Propulsion, Part I, H.J.Kaeppeler, pp. 50-56

Astronautics in the United States, Part II, F.I.Ordway III,H.E.Canney,Jr., pp. 57-61,77

VOLUME 2-3 (Fall 1955)

The MOUSE,A Mininum Orbital Unmanned Satellite of the Earth for Astrophysical Research, S.F.Singer, pp. 91-97

Power Transformer for the Moon, G.Camilli, pp. 98,99,118

Acceleration Force and the Space Pilot, J.J.Raffone, pp. 100-104

General Characteristics of Satellite Vehicles, Part II, N.V.Petersen, pp. 105-110

Aspects of Nuclear Power Application for Jet Propulsion, Part II, H.J.Kaeppeler, pp. 111-118

VOLUME 2-4 (Winter 1955)

On the Mechanics of Descent to a Celestial Body, K.A.Ehricke, pp. 137-144

Mercury--The Barren Plant, P.Moore, pp. 145-148

Electrical Propulsion System for Space Ships with Nuclear Power Source, Part I, E.Stuhlinger, pp. 149-152

Space Flight is International, F.C.Durant III, p. 153

VOLUME 3-1 (Spring 1956)

Satellite Ascent Guidance Requirements, J.Jensen, pp. 1-3,14

Thermal Control in a Space Vehicle, P.E.Sandorff, J.S.Prigg,Jr., pp 4-8,26

Fundamental Basis of Space Flight, W.O.Davis, pp. 9-10,25

Electrical Propulsion System for Space Ships with Nuclear Power Source, Part II, E.Stuhlinger, pp. 11-15

VOLUME 3-2 (Summer 1956)

The Oxygen Belt in the Planetary System, H.Strughold, pp. 27-28,52

Interplantary Communications, J.A.Webb, pp.29-30,36

Air for the Space Ship, R.P.Haviland, pp. 31-32,48

Electrical Propulsion System for Space Ships with Nuclear Power Source: Part III, E.Stuhlinger, pp. 33-36 JAS 1956

Present Status of Space Medicine, F.A.Hitchcok, pp.41-42,51-52 JAS 1957

The Physical and Chemical Fundamentals of Satellite Flight: Part I, P.F.Winternitz, pp.43-47

VOLUME 3, 3-4 (Autumn-Winter 1956)

Man Attempts to Understand his Environment, L.V.Berkner, pp. 53-58,76

Instrumentation for a Cosmic Ray Experiment for the Minimal Earth Satellite, G.H.Ludwig, J.A.Van Allen, pp. 59-60,80

Placing the Satellite in its Orbit, M.W.Rosen, pp. 61-64,82

The Physical and Chemical Fundamentals of Satellite Flight: Part II, P.F.Winternitz, pp. 65-71,80

VOLUME 4-1 (Spring 1957)

The Satellite and Meteorology, H.Wexler, pp.1-5,8

Accelerative Forces Associated with Leaving and Re-entering the Earth's Gravitational Field, R.R.Hessberg,Jr. pp. 6-8

Space Exploration: The New Challenge to the Electronics Industry, H.E.Prew, p.9-15

VOLUME 4-2 (Summer 1957)

Principal Directions of Current Research Activity in the Theory of Gravitation, B.S.DeWitt, pp.23-28

Education in the Astronautical Sciences, C.S.Draper, pp.29-30

Radio Tracking of IGY Satellite: The Mark II Minitrack System, R.L.Easton, pp.31-32,39

VOLUME 4-3 (Autumn 1957)

What the Future Holds for the Earth Satellite, R.P.Haviland, pp.41-45,151

Experimental Aspects of General Relativity Theory, L.Witten, pp.46-48

Space Vehicles as Tools for Research in Relativity, S.F.Singer, pp.49-51

Long-Term Human Confinement in Space Equivalent Vehicles, C.A.Dempsey, F.D.VanWart, J.H.Duddy, J.K.Hockenberry, pp.52-53,59

VOLUME 4-4 (Winter 1957)

Mechanoreceptors, Gravireceptors, H.O.Strughold, pp.61-63

Cosmic Ray Instrumentation for the IGY Program, R.C.Haymes, pp. 64-68

Meteoric Abrasion Studies Proposed for Vanguard, M.J.Swetnick, pp.69-71

Plants as a Means of Balancing a Closed Ecological System, J.G.Gaume, pp.72-75

VOLUME 5-1 (Spring 1958)

Fundamentals of Inertial Guidance and Navigation, W.E.Frye, pp.1-4,10

Plastic Balloons for Planetary Research, M.D.Ross, pp.5-10

Space Cabin Design, A.M.Mayo, pp.11-14

A Survey of Propulsion and Space Dynamics, D.B.Cross, J.Jensen, pp.15-19,30

VOLUME 5-2 (Summer 1958)

Minimum Time Interplanetary Orbits, D.M.Cole, pp.31-38

Plasma Motors, W.H.Bostick, pp.39-42

Synthesis of Human-Automatic Control Systems for High Performance Vehicles, W.L.Morris, R.C.Kaehler, pp.43-46

VOLUME 5, 3-4 (Autumn-Winter 1958)

Orientation of Research Needs Associated with Environment of Closed Spaces, W.T.Ingram, pp.51-57

Prediction of Cratering Caused by Meteoroid Impacts, M.Kornhauser, pp.58-63

Temperature Equilibria in Space Vehicles, R.A.Cornog, pp.64-67

VOLUME 6-1 (Spring 1959)

Artificial Modification of the Earth's Radiation Belt, S.F.Singer, pp. 1-9

Encke's Method and Variation of Parameters as Applied to Re-entry Trajectories, R.M.L.Baker,Jr., pp.10-13

Optimum Exhaust Velocity Programming and Propulsion Efficiency, R.Fox, pp.13-16

VOLUME 6-2 (Summer 1959)

Collision Trajectories in the Three-body Problem, E.T.Benedikt, pp.17-24 JAS 1958

A Review of the Current Status of Satellite Attitude Control, R.E.Roberson, pp.25-30 JAS 1959

Logical Approaches to Problems of Space Flight, G.W.Hoover, pp.31,32 JAS 1960

VOLUME 6-3 (Autumn 1959)

The Payload Capabilities of Ion Propulsion Rocket Systems, R.H.Fox, pp.33-39

The Effect of Aerodynamic Forces on Satellite Attitude, D.B.DeBra, pp.40-45

First Order Error Propagation in a Stagewise Smoothing Procedure for Satellite Observations, P.Swerling, pp.46-52

VOLUME 6-4 (Winter 1959)

Integration Methods for Differential Equations, S.Schlesinger, pp.53-57

Estimate of the Specific Ionization Caused by Heavy Cosmic Ray Primaries in Tissue or Water, T.Foelsche, pp.57-62

Some Aspects of the Electrical Properties of the Upper Atmosphere, P.A.Goldberg, pp.63-68

VOLUME 7-1 (Spring 1960)

Long Range Detection by Star Occultation, H.Dubner, pp.1-6

On the Problems of Re-entry into the Earth's Atmosphere, A.C.Robinson, A.J.Besonis, pp.7-20

TECHNICAL NOTES: Dual Burning Propulsion Systems for Satellite Stages, B.P.Martin, pp.21-23

TECHNICAL NOTES: Anisotrophy of Escape Velocity from the Moon, the Lunar Atmosphere, and the Origin of Craters, L.Gold, pp.23,24

VOLUME 7-2 (Summer 1960)

Limitations of a Rocket Propulsion System with Variable Exhaust Velocity Taking into Consideration Nuclear Data, T.Foelsche, pp.25-28

Attitude Drift of Space Vehicles, W.T.Thomson, G.S.Reiter, pp.29-34

Normal Lunar Impact Analysis, A.Liu, pp.35-39

Error Analysis of Satellite Orbits in the Presence of Drag, F.V.Pohle, pp.40-44

TECHNICAL NOTES: The Intersection of Coplanar, Confocal Conic Sections, G.S.Gedeon, pp.45,46

TECHNICAL NOTES: Two Maneuver Ascents to Circular Orbits, J.F.Wolfe, D.B.DeBra, pp.47,48

VOLUME 7-3 (Fall 1960)

Charge Neutralization in an Ionic Rocket, G.G.Cloutier, T.W.Johnston, pp.49-58

Ephemeris Time, T.P.Gabbard, pp.58-63

Biological Shielding for Radiation Belt Particles, D.L.Dye, J.C.Noyes, pp.64-70

The Influence of Heart Action and the Circulation of Blood on Manned Satellite Attitude Control, L.R.Zeitlin, pp.70-72

VOLUME 7-4 (Winter 1960)

Analytic Solutions to Several Optimum Orbit Transfer Problems, H.Munick, R.McGill, G.E.Taylor, pp.73-77

Orbital Patterns for Satellite Systems, L.G.Vargo, pp.78-86

Astronavigation-Guidance and Control in Space, R.E.Roberson, pp.87-95

VOLUME 8-1 (Spring 1961)

Analysis and Standardization of Astrodynamic Constants, M.W.Makemson, R.M.L.Baker,Jr., G.B.Westrom, pp.1-13

Rigid Body Attitude Stability and Natural Frequencies in a Circular Orbit, D.B.DeBra, R.H.Delp, pp.14-17

Analysis of Error Progression in Terminal Guidance for Lunar Landing, P.J.deFries, pp.18-27

VOLUME 8-2 (Summer 1961)

An Integrated Approach to the Determination and Selection of Lunar Trajectories, W.C.Nelson, pp.33-39

Application of Hill's Lunar Theory to the Motion of Satellites, P.Lanzano, pp.40-47

On Stromgren's Method of Special Perturbations, P.Munsen, pp.48-50

Goddard and His Early Rockets: 1882-1930, E.R.Hagermann, pp.51-59

TECHNICAL NOTES: Alignment Chart for Satellite Orbit Calculations, L.Holloway, pp.60,61

TECHNICAL NOTES: Radiation Shelters for Lunar Exploration, D.H.Robey, pp.62,63



VOLUME 8-3 (Fall 1961)

Measurements of Sheath Currents and Equilibrium Potential on the Explorer VIII Satellite, R.E.Bourdeau, J.L.Donley, G.P.Serbu, E.C.Whipple,Jr., pp.65-73

Furthering Basic Biological Knowledge Through Space Research, J.L.Biedler, pp.74-79

The Orbital Motion of Pellet Clouds, S.Ross, pp.79-83

Estimation of Longitudinal Variations in the Earth's Gravitational Field from Minitrack Observations, W.M.Kaula, pp.83-88

Bending Vibrations of a Disk Subjected to Gyroscopic Forces, L.Meirovitch, pp.88-93

VOLUME 8-4 (Winter 1961)

Green's Functions for Space Trajectory Perturbation Analysis, J.C.Dunn, pp.95-102

A Study of the Martian Upper Atmosphere and Ionosphere, G.Yanow, pp.103-109

Attitude Stability of an Elastic Body of Revolution in Space, L.Meirovitch, pp.110-113

Satellite Networks for Global Coverage, F.W.Gobetz, pp.114-126

VOLUME 9-1 (Spring 1962)

A Rapid Method for Selecting Interception Trajectories, G.S.Gedeon, pp.1-9

Review of Electromagnetic Effects on Space Vehicles, K.P.Chopra, pp.10-17

On the Use of Precision Frequency Signals in Space, W.E.Frye, pp.18-20

Variation of Parameters Equations of Motion for Space Vehicles in Predominantly Two-body Fields, K.C.Kochi, pp.21-30

TECHNICAL NOTES: Spin Stabilization of Attitude Against Gravity Torque, W.T.Thomson, pp.31-33

VOLUME 9-2 (Summer 1962)

Speculations on Space Observatories, B.F.Cohlan, pp.35-40

A New Model for Comets--The Cold, Partially Condensed, Magnetized Plasma, D.H.Robey, pp.41-48

The Use of High- and Low-Thrust Propulsion in Combination for Space Mission, T.N.Edelbaum, pp.49-60

TECHNICAL NOTES: A Limit to the Size of Solar Arrays which Can Supply their Own Stabilization Power, R.E.Roberson, pp.60-62

VOLUME 9-3 (Fall 1962)

Computer Calculations of Doses from Protons in Space, D.L.Dye, G.Butler,Jr. pp.63-70

Orbital Effects of Solar Radiation Pressure on an Earth Satellite, P.E.Koskela, pp.71-81

A Method for Determining Atmospheric Density from Short Arcs of an Almost Circular Satellite Orbit, M.Lecar, R.Morrison, pp.82-85

TECHNICAL NOTES: Attitude Reference as Established by a Horizon Scanner, R.E.Roberson, pp.86-88

TECHNICAL NOTES: A Rapid Method for Determining the Percentage of a Circular Orbit in the Shadow of the Earth, D.Pierce, pp.89-92

VOLUME 9-4 (Winter 1962)

Aerodynamic and Radiation Disturbance Torques on Satellites Having Complex Geometry, W.J.Evans, pp.93-99

Design for Adaptive Roll Control of a Missile, A.Gelb, T.C.Blaschke, pp.99-105

TECHNICAL NOTES: The Identity of Two Descriptions of Attitude Motions, R.E.Roberson, pp.106,107

TECHNICAL NOTES: Specific Heats of High Temperature Components of Air, S.Atallah, p.107

VOLUME 10-1 (Spring 1963)

A Method for Determining Approximate Initial Conditions for Interplanetary Trajectories, L.N.Rowell, pp.1-7

Least Squares Determination of Space Vehicle's Relative Range and Orientation from Film Data, G.W.Morgenthaler, pp.8-13

Hypersonic Viscid-Inviscid Interaction Solutions for Perfect Gas and Equilibrium Real Air Boundary Layer Flow, W.M.Mann,Jr., R.G.Bradley,Jr pp. 14-27

TECHNICAL NOTES: Comments on the Incorporation of Man Into the Attitude Dynamics of Spacecraft, R.E.Roberson, pp.27,28

JAS
1962
JAS
1963

VOLUME 10-2 (Summer 1963)

Pressure Estimation Techniques for Hypersonic Flows Over Blunt Bodies, L.G.Kaufman, pp.35-41

General Theory of Optimum Hypersonic Slender Bodies Including Frictional Effects, A.Miele, R.E.Pritchard, D.W.Hull, pp.41-54

Governing the Moon, H.G.Schwarz, pp.54-57

VOLUME 10-3 (Fall 1963)

Studies with a Simulated Martian Environment, T.L.Roberts, pp.65-74

Preliminary Report on the Calculation of Upper Atmosphere Models from First Principles; the Earth, G.Yanow, pp.75-81

Numerical Solutions of Non-coaxial Optimum Transfer Problems, K.Eckel, pp.82-91

TECHNICAL NOTES: On Meteoroids and Penetration, F.L.Whipple, pp.92-94

VOLUME 10-4 (Winter 1963)

Analysis of One- and Two-Gimbal Attitude Gyros in a Satellite Application, H.W.Evans, pp.103-106

An Improved Theory of the Use of High- and Low-Thrust Propulsion in Combination, W.R.Fimple, pp.107-113

Attitude Stability of a Spinning Unsymmetrical Satellite in a Circular Orbit, T.R.Kane, D.J.Shippy, pp.114-118

Rocket in Drag-Free Vertical Powered Flight Under Constant Thrust, C.H.Zee, pp.119-121

VOLUME 11-1 (Spring 1964)

Orbit Determination for a Lunar Satellite, J.Lorell, pp.1-7

Launch Windows for Orbital Missions, A.H.Milstead, pp.8-15

Feasibility of Electrostatic Systems for Space Vehicle Radiation Shielding, J.E.Felten, pp.16-22

VOLUME 11-2 (Summer 1964)

On Guidance Errors Induced by Thrust Decay in Vertical Powered Flight of Rockets, C.H.Zee, S.Suh, pp.33-36

Orbital Aspects of Development of Satellites, R.C.Nigam, pp.37-45

Vector Calculus for the Orbit Plane System, I.Sugai, pp.50-52

VOLUME 11-3 (Fall 1964)

Multidimensional Parameter Estimation by the Summed Weighted Least-Squares Minimization of Remainders, H.Schneider, pp.61-76

Addendum to Multidimensional Parameter Estimation, H.Schneider, A.Warwas, pp.77, 78

Inversion of Kepler's Equation and Herrick's Unified and Universal Time of Flight Equations, G.Gedeon, pp.79-84

Technical Papers Presented at the Second All-Union Congress on Theoretical and Applied Mechanics, J.L.Zygielbaum, pp.85-89

VOLUME 11-4 (Winter 1964)

The Minimum Time Problem, H.Heerman, P.Hretsinger, pp.93-107

Slavic Language Supplement, J.L.Zygielbaum, pp.108-112

Concerning the Change of the Mean Molecular Weight of the Nocturnal Atmosphere at Altitudes of from 100 to 210 km According to Mass Spectrometer Measurements, A.A.Pokhunkov, pp.113-117

Mass Spectrometer Research of Structural Parameter of the Earth's Atmosphere at Altitudes from 100 to 210 km, A.A.Pokhunkov, pp.118-124

VOLUME 12-1 (Spring 1965)

Powered Flight Trajectories of Rockets Under Constant Tangential Thrust, C.H.Zee, pp.1-6

Application of Lagrange's Planetary Equations to Orbits with Low Eccentricities, or Low Inclinations, or Both, A.G.Lubowe, pp.7-17

Stability of a Symmetrical Satellite in Attitudes Fixed in an Orbiting Reference Frame, P.W.Likins, pp.18-24

The Effect of Planetary Rotation on Aerodynamic Heating During a Glide, P.D.Arthur, pp.25,26

The Implications of Space Crew Physical Fitness Maintenance to System Design, I.Streimer, B.Wendrow, pp.27-29

VOLUME 12-2 (Summer 1965)

Three-dimensional Wings of Minimum Total Drag in Newtonian Flow, D.G.Hull, A.Miele, pp.39-45

Cones of Minimum Drag in Newtonian Flow, J.W.Reyn, pp.46-50

Electrical Surface Interaction in Hypersonic Flight I: The Aerodynamic Plasma Sheath, R.J.Heyman, S.J.Fenster, pp.51-58

(Technical Note Addendum): Time Derivatives of Unit Vectors in the Orbit Plane System, I.Sugai, pp.59,60

VOLUME 12-3 (Fall 1965)

A Linear Theory of Optimum Low-Thrust Rendezvous Trajectories, F.W.Gobetz, pp.69-76

Optimal Aerodynamics Steering for Anti-ICBM Missiles, H.Halkin, pp.77-82

Three-Dimensional Configuration of Minimum Total Drag in Newtonian Flow, D.G.Hull, pp.83-87

On the Logical Establishment of Global Surveillance and Communication Nets, B.H.Billik, H.L.Roth, pp.88-99

TECHNICAL NOTES: Effect of Lunar Inequality on the Anomalistic Year, R.C.Nigam, pp.100,101

TECHNICAL NOTES: Sufficient Conditions for Global Extrema in the Calculus of Variations, G.W.Ewing, pp.102-105

VOLUME 12-4 (Winter 1965)

Determination of Selenographic Latitude and Longitude of Star Observations from the Far Side of the Moon, M.W.Makemson, pp.117-122

The Dynamical Attitude Equations for n-Body Satellite, W.Hooker, G.Margulies, pp.123-128

On the High Order Effects in the Methods of Krylov-Bogoliubov and Poincare, P.Munsen, pp.129-134

Minimum Drag Bodies with Elliptical Cross Sections, J.Suddath, W.Oehman, pp.135-141

TECHNICAL NOTES: Long Term Behavior of Artificial Satellite Orbits Due to Third Body Perturbations, J.Lorell, pp.142-148

TECHNICAL NOTES: On the Inversion of the Universal Kepler Equations, E.T.Pitkin, pp.149-152

VOLUME 13-1 (January-February 1966)

Lift-to-Drag Ratios of Slender Bodies at Hypersonic Speeds, A.Miele, pp.1-6

Eclipse Observations from Orbiting Spacecraft, R.C.Moore, G.F.Schilling, pp.7-20

Low-Thrust Trajectories--A Bibliography, E.T.Pitkin, pp.21-28

Photographic Smear with Orbital Ground Reconnaissance Cameras, W.H.Tobey, J.R.Tewell, pp.29-36

The Soviet Program for Military Applications in Space, J.L.Zygielbaum, pp.37-41

VOLUME 13-2 (March-April 1966)

A Three-Mass Retrieval Study for the Gemini Tethered Astronaut, C.R.Poli, E.P.Hanavan,Jr., pp.61-71

Optimum Transfers Between Nearly Tangent Orbits, G.A.McCue, D.F.Bender, pp.72-75

Wake Analysis of an Orbiting Body, H.L.Crowson, pp.76-86

TECHNICAL NOTES: Uniqueness of Equilibrium Attitudes for Earth-Pointing Satellites, P.W.Likins, R.E.Roberson, pp.87,88

VOLUME 13-3 (May-June 1966)

A Solution to the Large ARC Angles--Only Orbit Determination Problem, P.R.Escobal, pp.101-105

An Investigation of the Effect of Total Simulation System Mass on Certain Human Force Outputs in Tractionless Environments, I.Streimer, D.P.W.Turner, K.Volkmer, pp.106-109

Invariant Imbedding and Nonlinear Filtering Theory, R.E.Bellman, H.H.Kagiwada, R.E.Kalaba, R.Sridhar, pp.110-115

Similarity Laws for Bodies Maximizing the Lift-to-Drag Ratio at Hypersonic Speeds, A.Miele, pp.116-120

TECHNICAL NOTES: Comparison of Simple Approximations to Elliptical Orbits, E.P.Wallner, pp.121-124

TECHNICAL NOTES: "Newar" Rendezvous by Two Satellites, J.Y.Miyamoto, pp.125,126



VOLUME 13-4 (July-August 1966)

Radiation Shielding Considerations for Interplanetary Spacecraft, C.F.Kottler,Jr., pp.133-152

Conditions for Optimal One-Impulse Transfer, D.F.Bender, G.A.McCue, pp.153-158

Universal Rendezvous Guidance Equations, G.S.Gedeon, pp.159-165

An Accelerated Gradient Method for Parameter Optimization with Non-Linear Constraints, H.J.Kelley, W.F.Denham, I.L.Johnson, P.J.Wheatley, pp. 166-169

TECHNICAL NOTES: Deboost from Circular Orbits, A.H.Milstead, pp.170,171

VOLUME 13-5 (September-October 1966)

Application of Krylov-Bogolubov Method to the Solution of the Stellar Three Body Problem, P.Musen, pp.177-189

Model for the Thermal Deformation of a Cylindrical Stabilization Boom in an Earth Orbital Environment, G.A.McCue, pp.190-198

A Method for Determining Satellite Illumination During an Eclipse by the Earth, R.H.Gersten, pp.199-201

TECHNICAL NOTES: Determination of an Orbit Between Two Postion Vectors with Specific Periapsis Radius, H.L.Roth, pp.202,203

TECHNICAL NOTES: Second Transition Partial Derivatives Via Universal Variables, E.T.Pitkin, pp.204-207

VOLUME 13-6 (November-December 1966)

Man's Future in Space, E.Z.Gray, pp.213-216

Some Operational Considerations of an Orbiting Space Station, F.S.Nyland, pp.217-225

Stability of a Nonsymmetrical Satellite with Elastically Connected Moving Parts, H.D.Nelson, L.Meirovitch, pp.226-234

Range of Hypersonic Glide Vehicles, S.Y.Chen, pp.235-247

Three-Dimensional Hypersonic Wings of Maximum Lift-to-Drag Ratio, A.Miele, D.W.Hull, pp.248-254

TECHNICAL NOTE: A Theorem of Inverse Extrema, A.H.Milstead, pp.260

VOLUME 14-1 (January-February 1967)

Photometric Investigations of Simulated Lunar Surfaces, J.D.Halajian, pp.1-12

High-Energy Trajectories from Earth to Mars and Return, S.S.Dallas, pp.13-26

Characteristics and Problems of Aerospace Company Management, G.Schuesel, pp.27-34

Some Operational Considerations of an Orbiting Space Station, F.S.Nyland, pp.35-46

TECHNICAL NOTE: Drag Coefficients of Hypersonic Optimum Bodies at Non-Optimum Mach Numbers, M.R.Fink, pp.47-48

VOLUME 14-2 (March-April 1967)

Some Geometric Aspects of Differential Games, G.Leitmann, G.MON, pp.56-65

Minimum Impulse Transfers in the Near Vicinity of a Circular Orbit, T.N.Edelbaum, pp.66-73

A Comparison of Point Performance and Gradient Technique, H.Heermann, P.Kretzinger, pp.74-78

Effect of Interval Size on Calculated Minimum Drag Coefficients and Optimum Body Shapes, M.R.Rink, pp.79-81

Flat-Top Body of Minimum Drag for Given Lift and Length in Hypersonic Flow, A.H.Lusty,Jr., pp.82-85

Two-Dimensional, Hypersonic Wings of Maximum Lift-to-Drag Ratio, D.G.Hull, pp.86-93

VOLUME 14-3 (May-June 1967)

Results of the Ranger, Lunar 9, and Surveyor 1 Missions, H.Behm, pp.101-111

Preliminary Orbit Determination for Lunar Satellites, B.C.Douglas, D.S.Ingram D.H.Lewis, pp.112-122

Attitude Instability Regions of a Spinning Unsymmetrical Satellite in a Circular Orbit, L.Meirovitch, F.Wallace, pp.123-133

TECHNICAL NOTES: Note on Exact Geodetic Sub-Latitude and Geodetic Altitude of Space Vehicle, I.Sugai, pp.134-137

TECHNICAL NOTES: Satellite Motion About an Oblate Planet for Low and Moderate Eccentricities, B.C.Douglas, D.S.Ingram, pp.137,138

VOLUMES 14-4 (July-August 1967)

Resonance Effects on Eccentric Satellite Orbits, G.S.Gedeon, B.C.Douglas, M.T.Palmiter, pp.147-157

The Attitude Stabilization of Spinning Satellites by Means of Gyroscopic Devices, D.L.Mingori, T.R.Kane, pp.158-166

On Liquid Sloshing in a Cylindrical Tank with a Flexible Bottom Under Strong Capillary and Weak Gravity Conditions, J.Siekmann, S.C.Chang, pp.167-172

Simultaneous Integration of N-Bodies by Analytic Continuation with Recursively Formed Derivatives, J.G.Hartwell, pp.173-177

TECHNICAL NOTES: Hypersonic Small-Disturbance Theory for a Plane in the Strong-Shock Approximation, G.S.Campbell, pp178,179

TECHNICAL NOTES: Rise and Set Time of an Interplanetary Space Vehicle, P.R.Escobal, D.A.Affatati, pp.179-182

VOLUME 14-5 (September-October 1967)

Future Interplanetary Unmanned Space Mission, M.W.Hunter II, R.L.Moll, pp.193-199

Future Manned Planetary Missions, R.L.Sohn, pp.200-211

Minimum Time Interplanetary Trajectories, P.R.Escobal, G.A.Chrichton, pp.212-217

Life Detection by Visual Imaging, G.A.Soffen, R.K.Sloan, pp.218-224

A Planetary Geodetic Laser Survey System, R.A.Fowler, V.Castellano, E.L.Coh, pp.225-229

Future Radar Developments, R.P.Cheetham, pp.230-236

Experimental Studies of Optimum Body Shapes at Hypersonic Speeds, B.Spenser,Jr., C.H.Fox,Jr. pp.237-240

TECHNICAL NOTES: Contensou's "Spool", G.A.McCue, D.F.Bender, p.241

TECHNICAL NOTES: The Effects of Tractionless Simulator Mass upon Manual Work, I.Streimer, D.P.W.Turner, K.Volkmer, pp.242-245

VOLUME 14-6 (November-December 1967)

The Developing Aerospace Law, M.Menter, pp.255-261

Engineering Considerations on the Transversal Contour of a Lifting Body, A.Miele, H.Huang, pp.262-269

Mechanical, Optical, Thermal and Electrical Properties, of the Surveyor 1 Landing Site, J.D.Halajian, pp.270-281

Effects of Atmosphere Rotation Rate on Orbits and Orbit Determination, R.B.Freund, pp.282-289

Forces and Moments on Cylinders and Cones in a Two-Phase Plume, B.K.Adler, P.M.Schwartz, K.D.Korkan, pp.290-299

TECHNICAL NOTES: Advanced Pioneer Launch Requirements, C. Poli, pp.300-301

TECHNICAL NOTES: Optimum Deorbit Positioning for Minimum Impulse Reentry with Variable Reentry Conditions, A.L.Green, G.S.Stern, pp.302-306

VOLUME 15-1 (January-February 1968)

Observation of Twenty-Seventh and Twenty-Eight Order Gravity Coefficients Based on Doppler Observations, R.J.Anderle, S.Smith, pp.1-4

Base Flow Characteristics and Thermal Environment of Launch Vehicles with Strap-On Solid Rocket Motors, G.A.Etemad, K.D.Korkan, pp.5-13

On the Effects of Higher-Order Inertia Integrals on the Attitude Stability of Earth-Pointing Satellites, L.Meirovitch, pp.14-18

A Computer Study of the Estimated Propagation of Errors in the Numerical Integration of Trajectory Equations, R.Zani, R.Hannen, pp.19-30

Observational Evidence of Anomalistic Phenomena, R.M.L.Baker, Jr.,pp.31-36

VOLUME 15-2 (March-April 1968)

Simulating the Lunar Environment, J.Seminara, R.Shavelson, pp.59-70

Jet-Shock Interactions, C.A.Whitehurst, J.Mourer, pp.71-79

Circular Orbits of Non-Infinitesimal Material Bodies in Inverse Square Fields, R.E.Roberson, pp.80-84

Conical Bodies of Given Length and Volume Having Maximum Lift-to-Drag Ratio at Hypersonic Speeds (Part I--Direct Methods), A.Miele, J.C.Heideman, R.E.Pritchard, pp.85-91

Establishing Observer Creditability: A Proposed Method, S.Walker, pp.92-96



VOLUME 15-3 (May-June 1968)

A Perturbation Solution for the Stability of the Triangular Points in the Elliptic Restricted Problem, K.T.Alfriend, R.H.Rand, pp.105-109

Optimum Three-Dimension Deorbit for Specific Reentry Angle from Circular Orbits, S.T.Chu, pp.110-117

Conical Bodies of Given Length and Volume Having Maximum Lift-to-Drag Ratio at Hypersonic Speeds (Part II--Variational Methods), H.Y.Huang, pp.118-123

On Some Possible Simplifications and Changes in Hansen's Lunar Theory, P.Musen, pp.124-135

An Approximate Solution for Normal Shocks in Imperfect Air, E.R.Thompson, pp.136,137

Application of Nonlinear Filter to Short Range Missile Guidance, L.Teng, P.L.Phipps, pp.138-147

A New Simulation Technique to Study Lunar Self-Locomotive Tasks, A.E.Wudell, D.A.Lewis, pp.148-152

VOLUME 15-4 (July-August 1968)

Indications of Possible Biological Substances in Carbonaceous Meteorites, B.Nagy, pp.161-168

Non-Equilibrium Flow About a Cone with an Expansion Corner, F.G.Gavolos, C.Studerus, I.Edelfelt, pp.169-176

Planetary Lander Re-Contamination Hazards and Spacecraft-Particle Interaction Physics, J.F.Vandrey, pp.177-182

Dynamic Analysis of Flexible Space Vehicle Systems Subjected to Arbitrary Force and Motion Inputs, Y.Y.Yu, pp.183-193

Extension of Linear Estimation Techniques to Nonlinear Problems, I.A.Gura, pp.194-206

TECHNICAL NOTES: Additional Conclusions on the Heat of Formation of CN, W.F.Rumpel, H.J.Deacon,Jr. pp.207-209

TECHNICAL NOTES: An Integral Solution of the Hypersonic Laminar Wake, K.D.Korkan, pp.210,211

VOLUME 15-5 (September-October 1968)

Ground Coverage of Spin Stabilized Satellites, P.R.Esobal, R.J.Stern, G.W.Canon, pp.223-230

National Programs and the Progress of Technological Societies, T.J.Gordon, A.L.Shef, pp.231-241

Equilibria of Orbiting Gyrostats, R.E.Roberson, pp.242-248

Orbit Selection for Planetary Mapping Missions, R.J.Stern, G.S.Stern, pp.249-256

Applications of an N-Body Reference Orbit, K.Stumpff, E.H.Weiss, pp.257-261

Preliminary Orbit Determination for High-Data-Rate Sensors, R.M.L.Baker,Jr. pp.262-269

A General Solution of the Lambert Problem, E.T.Pitkin, pp.270,271

VOLUME 15-6 (November-December 1968)

Application of Multivariable Search Techniques to Optimal Aerodynamic Shaping Problems, D.S.Hague, H.L.Rozendaal, F.A.Woodward, pp.283-296

Space Vehicles and the Distribution of Future Position, R.E.Beckwith, pp.297-302

Application of Hansen's Method to the Xth Satellite of Jupiter, P.Musen, M.Charnow, J.L.Maury, pp.303-312

Optimal Interstellar Trajectories with Acceleration--Limited Relativistic Rockets, G.M.Anderson, pp.313-318

TECHNICAL NOTES: Notes on the Thomson-Tait-Chetaev Stability Theorem, R.E.Roberson, pp.319-321

TECHNICAL NOTES: Application of the Method of Particular Solutions to Boundary-Layer Analyses, J.C.Heideman, pp.322-324

VOLUME 16-1 (January-February 1969)

Zero Mass-Shift Passive Damper for the Reduction of Precession in a Rotating Space Vehicle, F.Austin, G.A.Zetkov, pp.1-7

Attitude Stabilization of a Non-rotating Satellite by Means of a Gravitational Anchor, D.J.Shippy, P.L.Shome, pp.8-12

Mission Profile Requirements for Close Overflight of the Sun's Polar Regions, P.R.Odom, pp.13-21

On the Identification of Observable Orbit Parameters, with Application to Lunar Orbiter Tracking, C.G.Pfeiffer, pp.22-36

The Influence of Gravity Torque on Dual-Spin Satellite Attitude Stability, E.W.White, P.W.Likins, pp.32-37

TECHNICAL NOTES: A Formula for Hand Computation in Two-Stage Vehicle Optimization, H.B.Bosch, M.H.Skeer, pp.38-39

VOLUME 16-2 (March-April 1969)

General Solution for the Equilibria of Orbiting Gyrostats Subject to Gravitational Torques, R.W.Longman, R.E.Roberson, pp.49-58

Hypersonic Low Density Transitional Regime Flow Over Conical Vehicles, C.R.Ortloff, pp.59-67

A Unified Approach to Nonlinear Estimation, I.A.Gura, L.J.Henrikson, pp.68-78

Nutation Damper Design Principles for Dual-Spin Spacecraft, G.T.Cloutier, pp.79-87

A Method for Generating Coordinate Systems for the Analysis of Optimal Trajectories, W.F.Powers, pp.88-94

VOLUME 16-3 (May-June 1969)

The Hybrid Patched Conic Technique as Applied to the Lunar Trajectory Problem, P.R.Escobal, K.Forster, G.S.Stern, R.J.Stern, pp.97-109

Optimum Three-Dimensional Deorbit for Specific Reentry Angle from Elliptical Orbits, S.Chu, pp.110-120

A New Method for Computing the State Transition Matrix, E.Weiss, pp.121-125

Streamlining Selenodesy Simulations, W.Boyce, pp.126-135

TECHNICAL NOTE: On the Transition Processes in the Two-Dimensional Laminar Hypersonic Wake, K.D.Korkan, pp.136-139

VOLUME 17-1 (July-August 1969)

From Perigee to a Radial Distance in Given Time: A General Solution, R.G.Gottlieb, pp.1-7

Decision Policies for Midcourse Corrections To the Trajectory of a Space Vehicle, J.T.Wagner, pp.8-24

The Hybrid Patched Conic Applied to Lunar Return Trajectory Propagation, R.J.Stern, G.S.Stern, K.Forester, P.R.Escobal, pp.25-43

Deployment of a Tethered Orbiting Interferometer, C.J.Swet, J.J.Whisnant, pp.44-59

TECHNICAL NOTE: On Hypersonic Shapes of Minimum Heat Transfer, D.G.Hull, pp.60-62

Orbit Determination from Two Position Vectors and Periapsis Radius, E.R.Lancaster, R.H.Estes, pp.63-64

VOLUME 17-2 (September-October 1969)

Optimal Interstellar Trajectories with Thrust-Limited Relativistic Rockets, G.M.Anderson, pp.65-79

Bayes Theorem and Digital Realizations for Non-Linear Filters, R.S.Bucy, pp.80-04

Dynamics and Aerodynamics of Drop Sondes in the Lower Venus Atmosphere, J.F.Vandrey, pp.95-103

Attitude Stabilization of Spacecraft with Geomagnetic Rate Damping, V.L.Pisacane, pp.104-128

VOLUME 17-3 (November-December 1969)

Orbit Synthesis, H.K.Karrenberg, E.Levin, R.D.Lüders, pp.129-177

A First Order Solution for the Effects of a Small Displacement of the Center of Mass from the Origin on Artificial Satellite Orbits, J.G.Hartwell, pp.178-184

VOLUME 17-4 (January-February 1970)

Variational Equations Derived from the Herrick Variation of Parameters Method, A.S.Liu, pp.185-217

On Estimation of Maximum Wind Speeds in Tornadoes and Hurricanes, P.Dergarabedian, F.Fendell, pp.218-236

VOLUME 17-5 (March-April 1970)

The Effect of Longitudinal Vibration on Heat Pipe Performance, J.W.Richardson, C.A.Whitehurst, G.D.Whitehouse, pp.249-266

Limit-Cycle Analysis of a Controlled Libration-Point Satellite, R.W.Farquhar, pp.267-291

Passive Stabilization of a Symmetric Satellite on a Fly-By Mission, T.R.Kane, E.L.Marsh, pp.292-306

VOLUME 17-6 (May-June 1970)



An Analysis of Eccentric Deeply Resonant Orbits for Geodesy, B.C.Douglas, M.T.Palmiter, pp.313-336

Exact Relations of Optimum Switching in the Problem of Impulsive Transfer, N.X.Vinh, pp.337-364

VOLUME 18-1 (July-August 1970)

EVA Stabilization Shoes, E.H.Wrench, A.L.Greensite, pp.1-21

Guidance and Control Aspects of Abort from Manned Spacecraft, A.L.Greensite, pp.22-56

TECHNICAL NOTES: Some Aspects of Thermonuclear Propulsion, G.L.Matloff, H.H.Chiu, pp.57-62

VOLUME 18-2 (September-October 1970)

Stability of Deformable Gyrostats on a Circular Orbit, P.Y.Willems, pp.65-85

Optimal Attitude Control System for Spinning Aerospace Vehicles, V.S.Sohoni, D.R.Guild, pp.86-100

A Technique for Obtaining Optimal Interplanetary Transfers in the Real Case, J.P.Gravier, C.Marchal, R.D.Culp, pp.101-117

VOLUME 18-3 (November-December 1970)

Manned Exploration of Mars and Venus Using Chemical Propulsion, R.R.Titus, pp.129-149

Linear Coupling between Orbital and Attitude Motions of a Rigid Body, B.Lange, pp.150-167

TECHNICAL NOTES: A Power Series Inverse in Time of the Universal Kepler Equation, R.H.Estes, E.R.Lancaster, pp.168-172

TECHNICAL NOTES: Some Wake Observations Obtained on the Gemini/Agena Two-Body System, B.E.Troy,Jr., D.B.Medved, U.Samir, pp.173-183

VOLUME 18-4 (January-February 1971)

Simplified Performance Analysis of Space and Missile Guidance, J.T.Wagner, D.F.McAllister, pp.193-216

Transformation of a Potential Function Under Coordinate Rotations, S.L.Levie,Jr., pp.217-235

TECHNICAL NOTES: Subsurface Temperature of the Moon, W.W.Salisbury, D.L.Fernald, pp.236-243

A Criticism by T. R. Kane, p.53

VOLUME 18-5 (March-April 1971)

Interactions Between a Satellite and the Ionosphere and Applications to Drag and Probe-Measurement Problems, G.Fournier, pp.257-280

A Priori Techniques to Generate Partial Derivatives for Computing Interplanetary Trajectories, J.G.Purdy, L.P.McNamee, pp.281-305

VOLUME 18-6 (May-June 1971)

Centrifugally Obtained Artificial Gravity, R.E.McGaughy, D.B.Hoffman, pp.323-354

Some Predictions for Blunt Body Heat Transfer Illustrating the Effects of Slip and Real Gas Properties, W.A.NcNeill, J.P.Lamb, J.C.Hester, pp.355-367

Applications and Computer Implementation of Hansen's Planetary Theory, R.Broucke, E.Chapman, pp.368-382

FOREIGN TECHNICAL NOTES: Observations of Jupiter, Venus, and Source 3 C 273 at 2 and 8 mm Wavelengths, V.A.Yefanov, A.G.Kislyakov, I.G. Moseyev, A.I.Naumov, pp.383-389

FOREIGN TECHNICAL NOTES: Some Results of the Flight of the Soyuz 9 Spacecraft, A.G.Nikolayev, pp.390-396

FOREIGN TECHNICAL NOTES: Experience of Tracking of the Satellite Pageos with the AFU-75 Camera in Ulan-Bator, R.Radnaa, S.Sanzzav, V.T.Groschev, pp. 397-398

VOLUME 19-1 (July-August 1971)

Applications and Development of Space Electric Power Systems, J.D.Lafleur, F.Schulman, pp.1-39

Analysis of Multi-Stage Reaction Rate Controlled Systems, W.Engelmaier, E.C.Woodward,Jr., pp.40-54

Cold Gas Rocket Propulsion, G.Mouritzen, pp.55-70

FOREIGN TECHNICAL NOTES: Preliminary Results of Determination of the Nikolajev-Helwan Space Direction, A.Dinescu, pp.71-72

FOREIGN TECHNICAL NOTES: Extra-High Resolution Radiointerferometry, L.I.Matveyenko, pp.73-76

FOREIGN TECHNICAL NOTES: Study of Reactivity to Transverse Accelerations and Radioprotectants, V.V.Antipov, M.V.Vasin, B.I.Davydov, E.F. Panchenkova, P.P.Saksonov, pp.77-81

FOREIGN TECHNICAL NOTES: Oxidation of Metals in an Unsteady (Transient) Region at High Temperatures. III. Method of Calculating the Oxygen Diffusion Coefficient in a Metal (Titanium), V.I.Tikhomirov, V.I.D'yachkov, pp.82-88

VOLUME 19-2 (September-October 1971)

Underwater Range Tracking, G.A.Nussear, P.E.Seelinger, pp.97-115

Hypersonic Bodies of Maximum Drag for a Given Lift-to-Drag Ratio, W.McMillan III, D.G.Hull, pp.116-129

Minimum-Time Transfer Between Close Coplanar Orbits with a Thrust-Limited Rocket, G.M.Anderson, J.R.Hamlin, W.L.Othling, pp.130-137

FOREIGN TECHNICAL NOTES: Calculation of Minimum-Weight Beams and Frames Under Random Loading, A.M.Araslanov, pp.150-158

FOREIGN TECHNICAL NOTES: Social Factors in Controlling the Development of Scientific Teams, Y.V.Poshekhonov

VOLUME 19-3 (November-December 1971)

Some Computational Aspects of the Minimum Fuel Continuous Low Thrust Orbit Transfer Problem, J.L.Starr, R.D.Sugar, pp.169-204

Optimum Transfer Between Hyperbolic Asymptotes with Turn Angles Less than the Maximum Natural Turn Angle, J.M.Walton, C.Marchal, R.D.Culp, pp. 205-231

FOREIGN TECHNICAL NOTES: Processes of Formation of Coatings Deposited by the Method of Gas-Flame Spraying, L.D.Svirskiy, Y.A.Pirogov, pp.246-251

FOREIGN TECHNICAL NOTES: Air Density Distribution in the Stratosphere, K.T.Logvinov, pp.252-254

VOLUME 19-4 (January-February 1972)

Aerodynamic Optimization by Combining the Gradient Method with the Integral Flow Model, W.E.Simon, pp.257-285

Use of the Method of Particular Solutions in Determining Periodic Orbits of the Earth-Moon System, K.H.Well, pp.286-296

TECHNICAL NOTES: Modification of the Laplace and Leuschner Methods of Heliocentric Orbit Determination, Using Five Observations, H.H.Jacoby, Jr., pp.297-306

TECHNICAL NOTES: "Vulcan" Evidence Re-Examined, J.P.Bagby, pp.307-311

FOREIGN TECHNICAL NOTES: Photographic Observation of Satellites Using the K-24 Aerial Camera with Polaroid Film, Th.M.Vermeesch, L.J. Lantwaard, pp.312-314

FOREIGN TECHNICAL NOTES: Combustion of Magnesium Particles in Rarefield Air, V.A.Fedoseyev, G.A.Mokhov, pp.315-317

FOREIGN TECHNICAL NOTES: Calculation of Uniformly Dependable Minimum-Weight Structures, A.M.Araslanov, pp.318-325

VOLUME 19-5 (March-April 1972)

Longitudinal Dynamic Stability of a Shuttle Vehicle, N.X.Vinh, E.V.Laitone, pp.337-363

Attitude Stability and Damping of a Tethered Orbiting Interferometer Satellite System, P.M.Bainum, R.E.Harkness, W.Stuiver, pp.364-389

TECHNICAL NOTES: Calculation of Free-Fall Trajectories Using Numerical Optimization Methods, D.G.Hull, W.T.Fowler, R.G.Gottlieb, pp.390-398

FOREIGN TECHNICAL NOTES: The Equation of the Energy Balance for Turbulent Pulsations in the Theory of a Free Turbulent Boundary Layer, N.I.Akatnov, A.P.Kuzenetsov, pp.399-407

VOLUME 19-6 (May-June 1972)

On the Elliptic Restricted Three-Body Problem, R.Broucke, pp.417-432

Free Vibrations of Multilayer Sandwich Plates in the Presence of In-Plane Loads, R.M.Shahin, pp.433-447

Eclipse Design Limits for Circular Orbits, P.R.Escobal, R.H.Johnson, pp.448-461

TECHNICAL NOTES: Attitude Stability and Performance of a Dual-Spin Satellite with Nutation Damping, D.K.Anand, J.M.Whisnant, pp.462-469

VOLUME 20-1 (July-August 1972)

Coronal Polarization Observations, W.N.Arnquist, pp.1-8

The Surface Frictional Layer under a Hurricane Vortex, P.Dergarabedian, F.Fendell, pp.9-34

Space Shuttle Abort: Downrange Basing and Crossrange Capability Requirements, R.Salkeld, pp.35-38

Chemical and Nuclear Space Tugs in the Earth Orbital Shuttle Mission, R.Salkeld, pp.39-43

TECHNICAL NOTES: Estimation of Gravity Field Harmonics in the Presence of Spin-Axis Direction Error Using Radio Tracking Data, J.Lorell, pp.44-52

JAS 1972

VOLUME 20-2 (September-October 1972)

The Effects of Trajectory Characteristics on Scientific Objectives for Major Planetary Orbiters, C.K.Paul, pp.65-87

Optimal 3-Dimensional Minimum Time Turns for an Aircraft, R.P.Humphreys, G.R.Hennig, W.A.Bolding, L.A.Helgeson, pp.88-112

A Novel Method for Averaging Equations of Motion, F.Morrison, pp.113-119

VOLUME 20-3 (November-December 1972)

Singular Perturbation and Turbulent Shear Flow Near Walls, F.E.Fendell, pp.129-165

Attitude Dynamics of a Three-Axis Stabilized Satellite with a Large Flexible Solar Array, P.C.Huges, pp.166-189

VOLUME 20-4 (January-February 1973)

Extremization of Terminally Constrained Control Problems, H.Y.Huang, S.Naqvi, pp.193-222

Differential Transformation for Satellite Injection, I.Sugai, pp.223-232

Further "Vulcan" Evidence, J.P.Bagby, pp.233-241

An Argument Against Bagby's "Vulcan" Earth Satellites, V.J.Slabinski, pp.242-243

Reply to Slabinski's Criticism of "Vulcan" as Earth Satellites, J.P.Bagby

TECHNICAL NOTES: A New Proof of an Old Theorem of Geometry, N.F.Eraslan pp. 246-248

VOLUME 20-5 (March-April 1973)

The Solar Radiation Pressure Force and Torques Model, R.M.Georgevic, pp.257-274

Cartography and Orbit Determination Using TV Data from a Mars Orbiting Spacecraft, S.N.Mohan, G.A.Ransford, pp.275-301

Errata to "Attitude Dynamics of a Three-Axis Stabilized Satelitte with a Large Flexible Array, P.C.Hughes, pp.302-303

FOREIGN TECHNICAL NOTES: Effect of Secondary Emission of the Potential of a Metallic Body in the Electron Radiation Belts of the Earth, T.Y.Ryabova, K.A.Trukhanov, pp.304-309

VOLUME 20-6 (May-June 1973)

A Study of Planar Deployment Control and Libration Damping of a Tethered Orbiting Interferometer Satellite, W.Stuiver, P.M.Bainum, pp.321-346

Large Space Vehicles--Platforms for Second Generation in-situ Wake Observations, U.Samir, pp.347-369

VOLUME 21-1 (July-August 1973)

Eddy Current Nutation Dampers for Dual-Spin Satellites, G.A.Haines, C.T.Leondes, pp.1-25

One- and Two-cell Tornado Structure and Funnel-Cloud Shape, P.Dergarabedian, F.Fendell, pp.26-31

Optimum Altitude for Coasting Flight of a Hypervelocity Vehicle, A.Busemann, N.X.Vinh, R.D.Culp, pp.32-48

FOREIGN TECHNICAL NOTES: Analysis of the State of the Atmosphere from the Picture Taken by the Zond-7 Space Station at an Altitude of 70,000 km on August 8, 1969, V.A.Bugayev, T.P.Popova, pp.49-56

VOLUME 21-2 (September-October 1973)

The Geometry of Solar Reflections for an Orbiting Sensor, D.L.Hitzl, pp.65-98

Alfvén Waves Associated with Long Cylindrical Satellites, N.S.Venkataraman, W.A.Gustafson, pp.99-117

VOLUME 21-3 & 4 (November 1973-February 1974)

A Comparison of Cowell's Method and a Variation-of-Parameters Method for the Computation of Precision Satellite Orbits, S.S.Dallas, E.A.Rinderle, pp.129-153

The Tri-Lobed Azimuthal Symmetric Case of the Optimum Orientation of Accelerometers for Ballistic Vehicles, E.C.Hoy, C.T.Leondes, pp.154-184

VOLUME 21-5 & 6 (March-June 1974)

Some Necessary Conditions for Impulsive Thrust-Coast-Thrust Minimum-Time, Fixed-Fuel Transfers Between Coplanar Orbits, G.M.Anderson, R.C.Tubbs, pp.193-205

Multilateration: A Nondegenerate Method of Obtaining Station Coordinates and Satellite Ephemerides, K.M.Ong, P.R.Escobal, pp.206-229

The Motion of an Artificial Satellite in a Nonspherical Gravitational Field and an Atmosphere with a Quadratic Scale Height, R.E.Willey, V.L.Pisacane, pp.230-243

Closed-Form Solution for the State Equation for Dual-Spin and Spinning Spacecraft, V.Larson, P.W.Likins, pp.244-251

TECHNICAL NOTES: The Motion of a Particle in Orbit Around a Growing Mass, S.H.Dole, pp.252-254

VOLUME 22-1 (July-September 1974)

A Bang-Bang Solar Pressure Attitude Control System, V.J.Modi, K.C.Pande, pp.1-20

State Equations for an n-Body Spacecraft, V.Larson, pp.21-35

Aerodynamic-Solar Hybrid Attitude Control of Near-Earth Satellites, V.J.Modi, K.C.Pande, pp.36-54

TECHNICAL NOTE: Exact Geodetic Latitude of Subvehicle Point, I.Sugai pp.55-63

VOLUME 22-2 (October-December 1974)

The Crisis in Astronautics, F.E.Moss, pp.67-72

The Reliability of Radiation Protection System and Space Flight Safety, E.E.Kovalev, pp.73-83

Reliability of Life Support Systems as Related to General Space Flight Safety Requirements, B.A.Adamovich, G.G.Ter-Minasyan, pp.84-94

MSFC Skylab System Safety Program, E.M.McNail, pp.95-116

VOLUME 23-1 (January-March 1975)

Engineering Principles to Assure Compatible Docking Between Future Spacecraft of USA and USSR, C.C.Johnson, pp.1-18

A Time Series (*F* & *G*) Approach to Missile Trajectory Prediction, R.M.L. Baker, Jr., N.H.Jacoby, Jr., pp.19-60

Automatic Detection and Recognition of Faint Artificial Earth Satellites Using Ground Based Electro-Optical Sensors, A.J.Verderese, pp.61-74

TECHNICAL NOTE: On the Motion of a Particle in Orbit Around a Growing Mass, R.Greenberg, pp.75-78

VOLUME 23-2 (April-June 1975)

Unified Mars Life Detection System, J.P.Martin, R.D.Johnson, B.Kok, R.Radmer, pp.99-120

Evaluation of Skylab Data for Land Use Mapping, L.L.Biehl, L.F.Silva, pp.121-146

Remote Sensing of Soil Moisture by Skylab Radiometer and Scatterometer Sensors, J.R.Eagleman, F.T.Ulaby, pp.147-160

The Exterior Tidal Potential Acting on a Satellite, P.Musen, pp.161-178

VOLUME 23-3 (July-September 1975)

Recent Processed Results from the SKYLAB S-192 Multispectral Scanner, F.J.Thomson, R.F.Nalepta, R.K.Vincent, B.C.Salmon

Spacelab Environmental Control/Life Support System-Design Safety, R.Gartner, pp.205-224

A Combined Gradient/Neighboring Extremal Algorithum for the Calculation of Optimal Transfer Trajectories between Noncoplanar Orbits using a Constant Low Thrust Rocket, G.M.Anderson, E.A.Smith, pp.225-240

GEOS-C Orbit Determination with Satellite to Satellite Tracking, P.Argentiero, R.Garza-Robles, pp.241-256

The Orbit Determination System of Telesat Canada, U.Domb, A.J.Grise, F.C.Kes, pp.257-270

VOLUME 23-4 (October-December 1975)

Range-Safety Debris-Pattern Analysis, R.M.L.Baker,Jr., T.J.Mucha, D.R.Darby, N.H.Jacoby,Jr., A.W.Johsnon, R.E.Ryan, pp.287-324

Space Tug Manned Safety, N.E.Tipton, pp.325-348

Jupiter Orbiter Lifetime: The Hazard of Galilean Satellite Collision, A.L.Friedlander, pp.349-368

Orbit and Attitude State Recoveries from Landmark Data, A.F.Fuchs, C.E. Velez, C.C.Goad, pp.369-382

VOLUME 24-1 (January-March 1976)

Survey of Space Flight Safety Systems (Fifth Supplement: July 1974 to June 1975), H.D.Wolf,R.Fleisig, pp.1-34

A Method for Developing K/S Boundary Conditions, D.J.Jezewski, pp.35-54

Optimal Rendezvous in the Neighborhood of a Circular Orbit, J.B.Jones, pp.55-90

VOLUME 24-2 (April-June 1976)

The Motion of a Satellite in Resonance with the Longitude-Dependent Harmonics, S.S.Dallas, pp.97-110

Sensitivity Analysis of Short-Arc Station Coordinate Determinations from Range Data, B.E.Schutz, B.D.Tapley, pp.111-136

Collision Avoidance for Two Counter Orbiting Satellites, D.Schaechter, J.V.Breakwell, R.A.Van Patten, F.W.Everitt, pp.137-146

On the Nature of the Radial and Cross Track Errors for Artificial Earth Satellites, N.L.Bonavito, R.A.Gordon, J.G.Marsh, J.C.Foreman, pp.147-164

Model for Solar Torque Effects on DSCS II, T.E.Suttles, R.E.Beverly, pp. 165-184

VOLUME 24-3 (July-September 1976)

Satellite Communications Technology, B.I.Edelson, pp.193-220

Accuracy of Estimating the Masses of Phobos and Deimos from Multiple Viking Orbiter Encounters, M.L.Mason, R.H.Tolson, pp.221-242

Space Solar Power--An Available Energy Source, S.Ferdman, R.L.Kline, pp.243-256

Attitude Stability of a Flexible, Dual-Spin Spacecraft with Active Nutation Damping Using Products of Inertia, G.T.Tseng, K.J.Phillips, pp.257-272

Free Vibration Analysis of Spinning Flexible Space Structures, K.K.Gupta, pp.273-280

VOLUME 24-4 (October-December 1976)

Spin-Stabilized Satellites, V.A.Sarychev, V.V.Sazonov, pp.291-310

Results of Numerical Determination of Poles and Zeros of a Transfer Function for Flexible Spacecraft Attitude Control, Y.Ohkami, P.W.Likins, pp.311-328

Optimal Control of Spin-Stablized Spacecraft with Telescoping Appendages, P.M.Bainum, R.Sellappan, pp.329-346

Quantization and Symmetry in Periodic Coverage Patterns with Applications to Earth Observation, J.C.King, pp.347-364

Monitoring Stratospheric Ozone and Aerosols Using Orbiting Photoelectric Detectors, G.L.Matloff, pp.365-382

VOLUME 25-1 (January-March 1977)

Librational Dynamics of Satellites with Thermally Flexed Appendages, V.J.Modi, K.Kumar, pp.3-20

Nonlinear Stress Waves in a Perfectly Flexible String, D.N.Fan, J.F.McGarvey, pp.21-34

An Analytic Trajectory Algorithm Including Earth Oblateness, A.R.Morrison, pp.35-62

Partial Differentiations of Vis-Viva and Lambert Equations, I.Sugai, pp.63-74

TECHNICAL NOTE: Conversion of Optimal Control Problems with Free Initial State into Optimal Control Problems with Fixed Initial State, A.Miele, B.P.Mohanty, A.K.Wu, pp.75-85

VOLUME 25-2 (April-June 1977)

Lagrange Configurations in Space System Dynamics, W.Stuiver, pp.91-106, pp.91-106

A Numerical Method for Generating Three-Impulse Minimum-Time Orbital Transfers, G.M.Anderson, J.M.Basine, pp.107-122

Effects of Meteoroid Impacts on Spacecraft Attitude Motion, D.A.Levinson, pp.129-142

Shuttle Vehicle Safety Assurance in Unpowered Landing Through Adaptive Control, A.A.Zhevnin, V.I.Toloknov, O.V.Sheviakow, pp.143-150

ASTRONAUTICAL LIBRARY: Hugh Latimer Dryden, E.M.Emme, pp.151-172

VOLUME 25-3 (July-September 1977)

The Use of Gyrodampers in Passive Satellite Attitude Control Systems, V.A.Sarychev, S.A.Mirer, pp.179-206

A Variation of Parameters Approach to the Arbitrarily Torqued, Asymmetric Rigid Body Problem, L.G.Kraige, S.B.Skaar, pp.207-226

Multilaterating the GEOS-3 Satellite, P.R.Escobal, J.F.Gallagher, o.h. von Roos, pp.227-250

Star Pattern Recognition for Real Time Attitude Determination, J.L.Junkins, C.C.White, III, J.D.Turner, pp.251-270

TECHNICAL NOTE: Orbital Perturbations of Tethered Satellite Systems, V.J.Modi, A.K.Misra, pp.271-278

VOLUME 25-4 (October-December 1977)

Analytical Evaluation of Solar Radiation Induced Orbital Perturbations of Space Structures, J.C. Van der Ha, V.J.Modi, pp.283-306

A Non-Synchronous Orbital Skyhook, H.Moravec, pp.307-322

Stochastic Optimal Attitudes Control of Spacecraft with Movable Appendages, R.Sellappan, P.M.Bainum, pp.323-347

TECHNICAL NOTE: Mass Driver for Lunar Transport and as a Reaction Engine, G.K. O'Neill, H.H.Kolm, pp.349-363

TECHNICAL NOTE: Torque-Free Motion of a Simple Gyrostat Initially at Rest, T.R.Kane, pp.365-370

VOLUME 26-1 (January-March 1978)

Shuttle/Tethered Satellite System, C.C.Rupp, J.H.Laue, pp.1-17

Near-Resonant and Transition Attitude Motion of a Class of Dual-Spin Spacecraft, J.E.Cochran, J.R.Beaty, pp.19-45

Optimization Techniques Applied to Space Shuttle Explicit Re-Entry Guidance Law Design, R.A.Metzler, W.F.Powers, pp.47-68

The Inertial Upper Stage, H.Kudish, pp.69-99

VOLUME 26-2 (April-June 1978)

Solar Power Satellite Developments, P.E.Glaser, pp.101-127

Application of Altitude Control Techniques for Low Altitude Earth Satellites, K.G.Nickerson, R.W.Herder, A.B.Glass, J.L.Cooley, pp.129-148

Orbit Determination for Earth Survey Satellites, P.J.Dunn, A.J.Fuchs, pp.149-158

Geometric Theory of Single-Gimbal Control Moment Gyro Systems, G.Margulies, J.N.Aubrun, pp.159-191

VOLUME 26-3 (July-September 1978)

Navigation Capability for an Ion Drive Rendezvous with Halley's Comet, C.L.Thornton, R.A.Jacobson, pp.197-210

A Distortion-Free Map Projection for Analysis of Satellite Imagery, J.L.Junkins, J.D.Turner, pp.211-234

Multibaseline: A Precision Ground Based Geodetic Measurement Technique, P.R.Escobal, pp.235-256

Minimization of Reaction Wheel Momentum Storage with Magnetic Torquers, A.Wernli, pp.257-278

TECHNICAL NOTE: Optimum Two Dimensional Wings in Supersonic Flows, N.H.V.Dutt, A.K.Sreekanth, pp.279-287

VOLUME 26-4 (October-December 1978)

Computationally Efficient Modelling for Long Term Prediction of Global Positioning, S.K.Collins, P.J.Cefola, pp.293-314

Orbit Analysis for SEASAT-A, E.Cutting, G.H.Born, J.C.Frautnick, pp.315-342

A Velocity Matching Techniques for Three-Dimensional Orbit Transfer in Conceptual Mission Design, S.T.Chu, T.J.Lang, B.E.Winn, pp. 343-368

Shaping the Course of Aeronautics, Rocketry, and Astronautics, R.C.Hall, pp.369-386

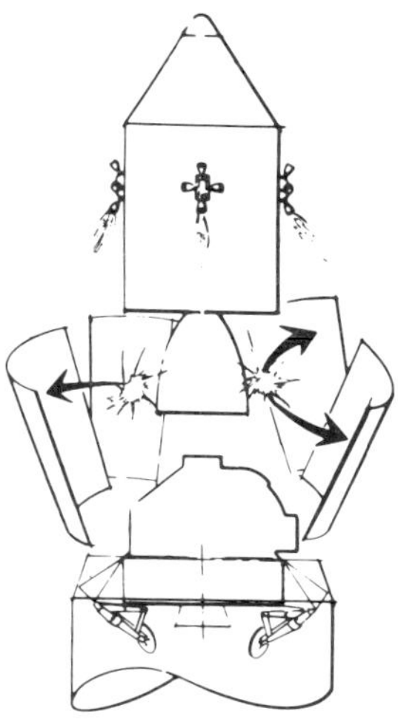

# AUTHOR INDEX
# 1954 - 1978

The Author Index includes the names of authors of AAS papers and articles presented and/or published in AAS publications from 1954 through 1978 and into early 1979. Where there are multiple authors for an article, all authors are included. Authors for unavailable papers or articles have been excluded. Also cited in this index are editors of AAS publications. In all cases, only the family name and initials of authors and editors are cited.

Following authors' names are the citations for articles and papers written by the authors. For the full citation, it is necessary to refer back to the Numerical/Chronological Index (pp 1-298). The following abbreviations are used:

| | |
|---|---|
| AAS | American Astronautical Society |
| AAS History | *AAS History Series* |
| Adv | *Advances in the Astronautical Sciences* |
| ASR | *Astronautical Sciences Review* (see pp 257-262) |
| Bonn Proc. | Bonn Proceedings (see pp 223-225) |
| DGLR | Deutsche Gesellschaft für Luft- and Raumfahrt (see pp 223-225) |
| IAF | International Astronautical Federation |
| JAS | *Journal of the Astronautical Sciences* |
| Mic | *AAS Microfiche Series* |
| Sp v | Special volume (see p 441) |
| S&T | *Science and Technology Series* |

# A

ABBOTT, A.S., AAS 66-306, Mic v9

ABDEL-GHANI, A.H., AAS 70-140, S&T v33; 73-018, S&T v31

ABDEL-GAWAD, M., AAS 67-171,Mic v5

ABEL, J., AAS 62-037, Adv v11

ABT, C.C., AAS 61-077, Adv v9

ACHTERMANN, E., DGLR 3.5 1976, Bonn Proc

ACTON, C.H., JR., AAS 75-088, Mic v26

ADAMOVICH, B.A., IAF 73-03, Mic v24-3/-4; JAS v22-2 pp84-94 1974

ADAMS, B.B., AAS 60-083. Adv v7

ADAMS, C.R., AAS 63-010, Adv v13

ADAMS, G.L., AAS 77-144, Mic v27

ADAMS, J.B., AAS 65-107, Adv v20I

ADAMS, J.D., AAS 71-369, Mic v20

ADAMS, M.C., AAS 66-031, S&T v10

ADAMS, R.M., AAS 70-057, Adv v27

ADAMS, W.M., AAS 68-036, S&T v25

ADELBERG, M., AAS 60-101, SP v1; 63-042, Adv v14

ADELMAN, I., AAS 69-424, Mic v13

ADLER, B.K., AAS 67-477, Mic v4; JAS v14-6, pp290-299, 1967

ADLER, I., AAS 68-197, Adv v25

AEPLI, T.C., AAS 75-241, Mic v25, Adv v32II

AFFATATI, D.A., JAS v14-4 pp179-182, 1967

AFSHAR, H.K., AAS 71-407, S&T v33

AGOSTO, W.N., AAS 78-190, Adv v38I

AHEARN, J.S., AAS 78-032, Adv v37II

AHLBORN, H., DGLR 2.1 1976, Bonn Proc; AAS 78-025, Adv v37I

AIKEN, E.G., AAS 57-016, Adv v2

AINTABLIAN, A.A., AAS 67-197, S&T v17

AKATNOV, N.I., JAS v19-5 pp399-407 1972

AKINS, F.R., AAS 77-245, Adv v36II

ALBRECHT, H.J., DLGR 2.7 1976 Bonn Proc

ALDRICH, D.H., AAS 75-145, Mic v25. Adv v32II

ALEXANDER, H.S., AAS 63-061, Adv v14; 69-433, Mic v18

ALEXANDER, J.K., AAS 71-108, Adv v29I

ALEXANDER, R., AAS 65-055, S&T v4

ALEXANDER, S.G., AAS 79-032, Adv v39

ALEXANDROV, I., AAS 59-023, Adv v5

ALEXIOU, A.G., AAS 67-539, Mic v6

ALFF, W.H., AAS 75-210, Mic v25, Adv v32I

# B

BECKMAN, J.C., AAS 73-231, Mic v21; 77-101, Mic v27; 77-109, Mic v27

BECKMANN, P., AAS 67-230, Mic v10

BECKWITH, R.E., JAS v15-6 pp297-302 1968

BEDDINGFIELD, S.T., AAS 70-043, Adv v28

BEEBE, R.F., AAS 76-163, S&T v45

BEERER, J.C., AAS 73-230, Mic v21

BEGGS, H.C., AAS 66-008, Adv v21

BEGGS, J.E., AAS 69-633, S&T v23

BEHM, H., JAS v14-3 pp101-111 1967

BEHM, J.W., AAS 75-159, Mic v25, Adv v32II

BEICHEL, R., AAS 75-162, Mic v25, Adv v32II; 76-101, S&T v43; 76-107, S&T v43; 76-156, S&T v45

BEILOCK, M., AAS 64-008, Adv v18

BELASCO, N., AAS 63-108, Adv v16I

BELCHER, D.J., AAS 69-573, Mic v12

BELL, A., AAS 68-296, Mic v11

BELL, B., AAS 61-010, Adv v8

BELL, J.H., AAS 65-167, S&T v6

BELL, J.H., JR., AAS 66-289, Mic v9; 67-400, Mic v4; 69-051, Mic v19

BELL, L.E., AAS 69-142, Mic v14

BELL, M.W.J., AAS 77-217, Adv v36I

BELLMAN, R.E., JAS v13-3 pp110-115 1966

BELLMORE, M., AAS 69-096, Mic v17

BELSKY, F.W., AAS 63-135, Adv v16II

BENDER, D.B., AAS 77-103, Mic v27

BENDER, D.F., AAS 63-146, Adv v16I; 66-124, S&T v11; JAS v13-2 pp72-75, 1966; v13-4 pp153-158 1966; v14-5 p240 1967; AAS 71-360, Mic v20; 73-204, Mic v21; 75-086, Mic v26

BENDER, R.L., AAS 66-222, Mic v9

BENDER, W.W., AAS 67-235, Mic v10

BENEDICT, G.F., AAS 74-161, Adv v31 II

BENEDIKT, E.T., ed. v14 Adv, 1963; AAS Sp v1, 1961; AAS 58-021, Adv v3; 58-040, Adv v4; 59-008, Adv v5; JAS v6-2 pp17-24 1959; 60-037, Adv v6; 60-093, Sp v1; 60-103a, Sp v1; 61-090, Adv v9

BENES, K., AAS 69-681, S&T v25

BENNETT, A., AAS 68-085, Mic v7

BENNETT, A.G., AAS 66-240, Mic v9; 66-250, Mic v9; 67-349, Mic v4

BENNETT, H.F., AAS 66-267, Mic v9

BENNETT, M., AAS 66-093, S&T v11, Mic v2

BENNION, J.L., AAS 58-013 (Abstract) Adv v3

BENSON, J.A., AAS 70-075, S&T v26

BENTLY, N.E., AAS 64-031, S&T v2

BOSSART, L.M., AAS 67-265, Mic v10

BOSTICK, W.H., AAS 57-002, Adv v2: JAS v5-2 pp39-42 1959

BOTLEY, C.M., AAS 69-011, S&T v25

BOTTEMA, M., AAS 72-021, S&T v28; 75-262, Mic v25, Adv v32II

BOURDEAU, R.E., JAS v8-3 pp65-73 1961

BOURKE, D.G., AAS 65-025, Adv v19

BOURKE, R.D., AAS 74-024, S&T v35

BOUTELLE, J., AAS 79-014, Adv v39

BOVEE, H.H., AAS 61-087, Adv v9

BOWEN, S.W., AAS 77-240, Adv v36I

BOWERS, J.L., AAS 67-303, Mic v4

BOYCE, W., JAS v16-3 pp126-135 1969

BOYER, K., AAS 63-066, Adv v15

BOYES, W., IAF 73-07, Mic v24-3/-4; 74-183, S&T v37

BOYLAND, R.E., AAS 67-273, Mic v10

BOYLE, E.J., JR., AAS 63-120, Adv v16I

BOYLE, W., AAS 64-005, Adv v18

BRACEWELL, R.N., AAS 58-041, Adv v4

BRADLEY, R.C., JR., AAS 61-019, Adv v8

BRADLEY, R.G., JR., JAS v10-1 pp14-27 1963

BRADLEY, R.H., IAF 69-10, Mic v23-2

BRADSHER, O.E., AAS 72-108, S&T v32

BRAND, T.J., AAS 78-146 (Abstract) Adv v38I

BRANNEN, J.P., AAS 67-220, Mic v10

BRASHER, W.E., AAS 66-241, Mic v9

BRASSE, A.W., AAS 66-250, Mic v9

BRASWELL, R.N., AAS 69-079, Mic v17; 69-198, Mic v17

BRATT, H.R., AAS 63-059, Adv v14

BRAVER, G.L., AAS 75-054, Mic v26

BRAUN, M.T., AAS 58-009, Adv v3

BRAUNWARTH, C.A., AAS 68-261, Mic v8; 69-494, Mic v19; 75-241, Mic v25, Adv v32II

BRAYSHAW, J.M., AAS 63-017, Adv v13

BRAZILL, E.J., AAS 75-001, S&T v40

BREAKWELL, J.V., AAS 58-034, Adv v3; 59-005, Adv v5; 60-076, Adv v7; ASR v3-2 pp15-16a 1961; v4-1 pp21-22,26 1962; AAS 66-094, S&T v11; 66-114, S&T v11; 68-094, Mic v7; 73-215, Mic v21; 73-239, Mic v21; 75-042, Mic v26; JAS v24-2 pp137-146 1976

BRECKENRIDGE, W.G., AAS 77-159, Mic v27

BREEDLOVE, W.J., JR., AAS 75-030, Mic v26

BREITENSTEIN, A.V., AAS 75-270, Mic v25, Adv v32II

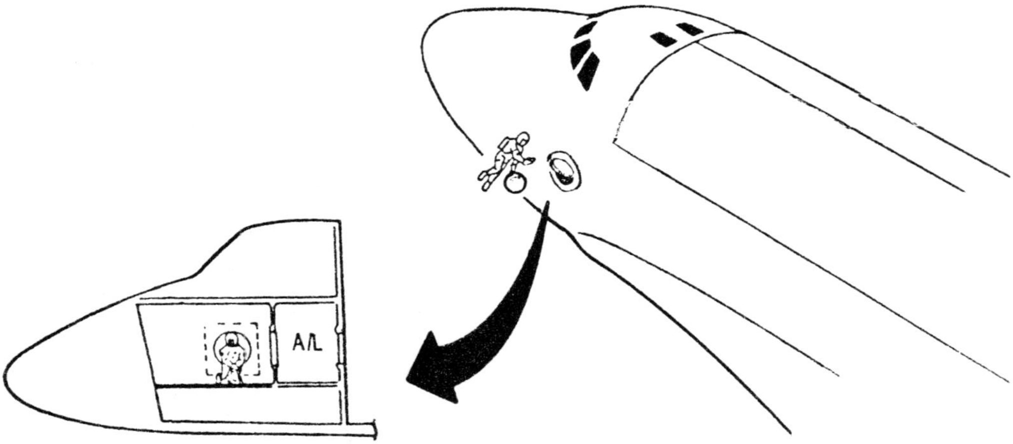
A/L

CABOTI, C.R., AAS 57-006, Adv v2

CADLE, R.D., AAS 69-523, S&T v22

CAGGIANO, G., AAS 63-104, Adv v16I

CAGLE, T.M., AAS 68-050, Adv v24

CAHILL, L.J., JR., AAS 68-212, Adv v25

CAIDIN, M., ed. JAS 1954; v1-1 pp8-9, 33, 1954

CAIN, C.C., AAS 63-060, Adv v14

CALDERON, O.H., AAS 67-276, Mic v10

CALDWELL, D.M., AAS 63-137, Adv v16I

CALDWELL, R.L., AAS 68-042, S&T v25

CALICO, R.A., AAS 71-345, Mic v20

CALIO, A.J., AAS 78-181, Adv v38II

CALTABIANO, S., ASR v3-3, pp16-19 1961

CAMERON, A.G.W., AAS 62-004, Adv v11

CAMERON, R.J., AAS 69-211, Mic v13

CAMERON, W.S., AAS 68-310, S&T v25; 69-007, S&T v25

CAMILLI, G., JAS v2-3 pp98-99, 118, 1955

CAMPBELL, D., AAS 74-057, S&T v38

CAMPBELL, G.S., JAS v14-4 pp178-179 1967

CAMPBELL, H.M., JR., AAS 66-253, Mic v9; 66-254, Mic v9

CAMPBELL, J., AAS 75-157, Adv v32II

CAMPBELL, J.O., AAS 69-606, S&T v23

CAMPBELL, P.A., AAS 67-135, Mic v3; IAF 72-00, Mic v24-2; 76-120, S&T v43

CAMPBELL, T.A., AAS 66-229, Mic v9

CAMPBELL, W.E., AAS 66-287, Mic v9

CANALE, R.P., AAS 73-224, Mic v21

CANETTI, G.S., IAF 75-016, S&T v41

CANNEY, H.E., JR., JAS v2-1 pp9-13, 1955; v2-2 pp57-61, 77, 1955

CANNON, G.W., JAS v15-5 pp223-230 1968

CANNON, L., AAS 68-259, Mic v8

CAPEHART, M.E., AAS 65-135, Adv v20II; 66-236, Mic v9

CAPELLARI, J.O., JR., AAS 66-131, S&T v11

CAPLINGER, E., AAS 65-150, S&T v6

CARAHAN, C.E., DGLR 6.1 1976, Bonn Proc.

CARDEN, A.E., AAS 66-309, Mic v9

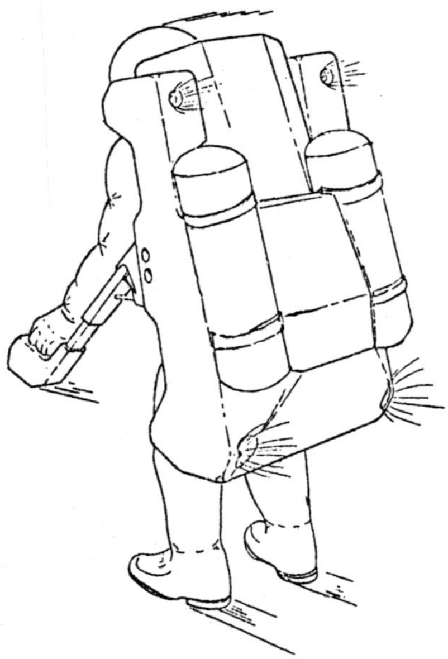

DACEY, M.F., AAS 65-055, S&T v4

DADDARIO, E.Q., AAS 66-038, S&T v10

DAEHLER, T.G., AAS 67-471, Mic v4

DAHL, P.R., AAS 67-458, Mic v4

DALKE, G.W., AAS 67-383, Mic v4

DALLAS, S.S., JAS v14-1 pp13-26 1967; v21-3/4 pp129-153 1974; AAS 75-022, Mic v26; JAS 24-2 pp97-110 1976

DALTON, C.C., AAS 67-334, Mic v4

D'AMARIO, L.A., AAS 73-217, Mic v21

DANIEL, J.N., JR., AAS 66-151, S&T v12

DANIELS, G.E., AAS 66-208, Mic v9

DANIELSON, R.E., AAS 72-013, S&T v28

DARBY, D.R., JAS v23-4 pp287-324 1975

DARWIN, C.R., AAS 67-511, Mic v4

DAS, A., AAS 75-090, Mic v26; 77-115, Mic v27; 77-162, Mic v27

DASENBROCK, R.R., AAS 75-040, Adv v33

DAVENPORT, P.B., AAS 71-389, Mic v20; 71-396, Mic v20

DAVENPORT, W.W., AAS 65-129, Adv v20II; 65-137, Adv v20II

DAVID, E.E., JR., AAS 72-035, S&T v29

DAVID, L., AAS 77-251, Adv v36II

DAVIDS, N., AAS 66-308, Mic v9

DAVIDSON, R.P., AAS 69-282, Mic v19

DAVIDTS, D., AAS 75-270, Mic v25, Adv v32II; DGLR 2.6 1976 Bonn Proc 78-046, Adv v37II

DAVIN, D.E., AAS 71-303, Mic v20

DAVIS, C.M., AAS 74-110, Adv v31 I

DAVIS, D.A., AAS 67-349, Mic v4

DAVIS, D.R., AAS 77-178, Mic v27

DAVIS, E.S., AAS 74-024, S&T v35

DAVIS, H.P., AAS 75-141, Adv v32II

DAVIS, J.B., AAS 69-549, Mic v17

DAVIS, L.I., AAS 67-048, S&T v15

DAVIS, L.P., AAS 67-459, Mic v4

DAVIS, N.S., AAS 57-031, Adv v2

DAVIS, R.E., AAS 67-052, S&T v15

DAVIS, R.J., ASR v1-1 pp9-12,14 1959

DAVIS, R.M., AAS 75-091 Mic v26

DAVIS, W.O., AAS 55-015, JAS v3-1 pp9-10 1956; AAS 67-104, Mic v3

DILLAWAY, R.B., AAS 63-001, Adv v13

DI MAGGIO, O.D., AAS 63-043, Adv v14; 66-318, Mic v9

DI NARDO, A., AAS 67-205, Mic v5

DINESCU, A., JAS v19-1 pp71-72 1971

DINKSESPILER, J.A., AAS 71-008, S&T v27

DISHER, J.H., AAS 74-075, S&T v36; 76-047, Adv v35

DIVEN, J.L, AAS 66-244, Mic v9

DIVINE, T.E., AAS 77-236, Adv v36I

DIXON, F.P., AAS 63-074, Adv v15; 66-144, S&T v12; 67-005, S&T v16

DIXON, J.F., AAS 73-247, Mic v21

DIXON, W.J., AAS 71-136, Adv v29II; 73-050, S&T v34

DOANE, G.B., III, AAS 66-310, Mic v9

DOBKINS, J.H., AAS 66-315, Mic v9

DOBNER, D.J., AAS 66-242, Mic v9

DOBROWALSKI, A., AAS 58-018, Adv v3

DOBRYSHMAN, E., IAF 71-07, Mic v24-1

DOBRZELECKI, A., AAS 69-244, Mic v18

DODDS, J.I., AAS 69-027, S&T v21

DOHERTY, L.R., AAS 72-007, S&T v28

DOHRMANN, R., AAS 67-216, Mic v10

DOLAN, J.E., AAS 61-032, Adv v8

DOLE, S.H., IAF 69-05, Mic v23-2; JAS v21-5/6 pp252-254 1974

DOLINSKY, E.L., AAS 69-456, Mic v13

DOLL, J.R., AAS 66-262, Mic v9; 68-095, Mic v7

DOLLFUS, A., AAS 76-158, S&T v45

DOLLMAN, T.S., AAS 67-325, Mic v4

DOMANOVSKY, P., AAS 72-115, S&T v32

DOMB, U., AAS 75-057, Mic v26; JAS v23-3 pp257-270 1975

DOMINY, F., AAS 69-115, S&T v22

DON, J.P., AAS 71-154, Adv v29II

DONALDSON, J.D., AAS 78-136, Adv v38II

DONAVAN, R., AAS 77-259, Adv v36I

DONDES, S., AAS 67-115, Mic v3

DONLAN, C.J., AAS 75-001, S&T v40

DONLEY, J.L., JAS v8-3 pp65-73 1961

DONNELLY, J.A., AAS 75-181, Mic v25, Adv v32I

DONOVAN, R.L., AAS 75-182, Mic v25, Adv v32I

DOOLIN, B.F., AAS 60-034, Adv v6

DORDICK, H.S., AAS 60-052, Adv v6

DORE, F.J., AAS 69-023, S&T v21

DORIAN, C., IAF 69-08, Mic v23-2

DORNBACH, J.E., AAS 65-105, Adv v20I

DORNBERGER, W.R., JAS v2-1 pp14-21 1955

DORNBRAND, H., DGLR 4.5 1976 Bonn Proc

DUNCAN, H.F., AAS 67-212, Mic v10

DUNCOMBE, R.L., AAS 71-106, Adv v29I; 71-312, Mic v20

DUNIN, S.E., AAS 66-104, S&T v11

DUNIPACE, K.R., AAS 67-451, Mic v4

DUNN, J.C., JAS v8-4 pp95-102 1961

DUNN, J.P., AAS 63-122, Adv v16I

DUNN, J.R., AAS 69-214, S&T v22

DUNN, P.J., AAS 77-128, Mic v27; JAS v26-2 pp149-158 1978

DUNN, R.B., AAS 78-110, Mic v29

DURANT, F.C., III, AAS 55-017, JAS v2-4, p153 1955

DUSKIN, F.H., AAS 63-126, Adv v16I

DUTT, H.N.V., JAS v26-3, pp279-287 1978

DUXBURY, J.H., AAS 75-163, Mic v25, Adv v32II

DUXBURY, T.C., AAS 71-372, Mic v20

DVORAK, J., IAF 68-06, Mic v23-1; 69-11, Mic v23-2; 71-11, Mic v24-1; 72-08, Mic v24-2; 73-10, Mic v24-3/-4

D'YACHKOV, V.I., JAS v19-1 pp82-88 1971

DYAL, P., AAS 68-206, Adv v25

DYE, D.L., JAS v7-3 pp64-70 1960; 9-3 pp63-70 1962

DYER, J.B., AAS 68-108, Mic v7

# E

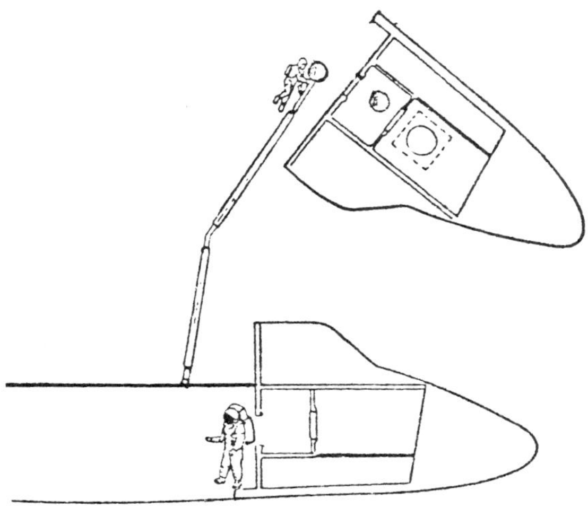

# F

F

# G



G

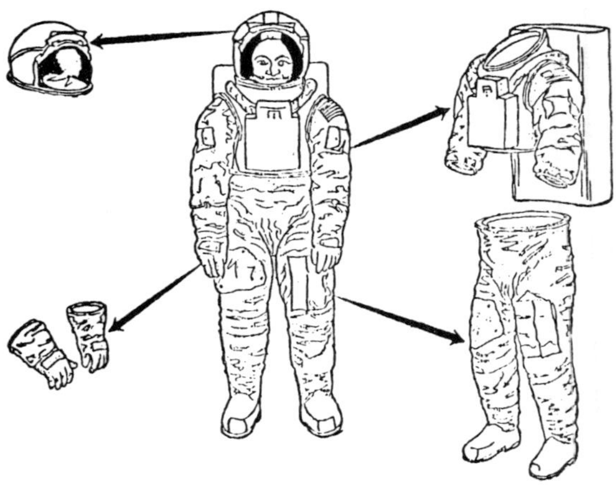

# H

HEATH, G.W., IAF 68-01, Mic v23-1; 69-01, Mic v23-2; 69-04, Mic v23-2; 70-01, Miç v23-3; 71-03, Mic v24-1; 71-04, Mic v24-1; 72-02, Mic v24-2; 73-01, Mic v24-3/-4; AAS 74-188, S&T v37; IAF 75-022, S&T v41

HEATON, R.A., AAS 67-524, Mic v4

HEBELER, H.K., AAS 58-008, Adv v3; ASR v1-2, pp11-15 1959; AAS 60-044, Adv v6

HEBERLIG, J.C., AAS 72-151, S&T v30

HECHT, E., AAS 62-056, Sp v5

HECKMAN, R.T., AAS 74-121, Adv v31 I

HEER, E., AAS 79-011, Adv v39

HEERMANN, H., JAS v11-4 pp93-107 1964; v14-2 pp74-78 1967

HEFFNER, W.A., AAS 69-275, Mic v17

HEFFRON, W.G., AAS 71-331, Mic v20

HEIDEMAN, J.C., JAS v15-2 pp85-91 1968; v15-6 pp322-324 1968

HEINBOCKEL, J.D., AAS 75-030, Mic v26

HEINE, W., AAS 68-094, Mic v7

HEISS, K.P., AAS 69-113, Mic v14; 72-163, S&T v30

HEITCHUE, R.D., AAS 63-107, Adv v16I

HELBERG, R.J., AAS 67-362, Mic v4

HELGESON, B.P., AAS 60-009, Adv v5

HELGESON, L.A., JAS v20-2 pp88-112 1972

HELLAND, J.C., AAS 69-366, Mic v19

HELLIWELL, R.A., AAS 68-211, Adv v25; 71-125, Adv v29I

HELVEY, T.C., AAS 58-026, Adv v3

HELVEY, W.M., AAS 65-063, S&T v4

HEMPHILL, W.R., AAS 65-105, Adv v20I

HENAULT, P., AAS 69-564, S&T v23

HENDERSON, C.W., AAS 65-115, Adv v20II

HENDERSON, G.C., AAS 67-321, Mic v4; 69-514, Mic v19

HENDRICK, J.J., AAS 71-161, Adv 29II

HENDRICKS, E.L., AAS 69-114, Mic v13

HENDRICKS, T., AAS 71-117, Adv v29I

HENDRICKS, T.C., AAS 73-245, Mic v21

HENF, G., AAS 63-150, Adv v16I

HENIZE, K.G., AAS 72-024, S&T v28; 74-161, Adv v31 II

HENNES, J.P., AAS 71-163, Adv v29II

HENNIG, G.R., JAS v20-2 pp88-112 1972

HENRARD, J., AAS 66-098, S&T v11; 68-087, Mic v7

HENRICI, J., AAS 73-015, S&T v31

HENRICKSON, C.H., AAS 75-187, Mic v25, Adv v32I

HENRIKSON, L.J., JAS v16-2 pp68-78 1969

HENRY, D.G., AAS 67-051, S&T v15

HENRY, J.P., JAS v1-1 pp12-19 1954; v2-1 pp22-26 1955

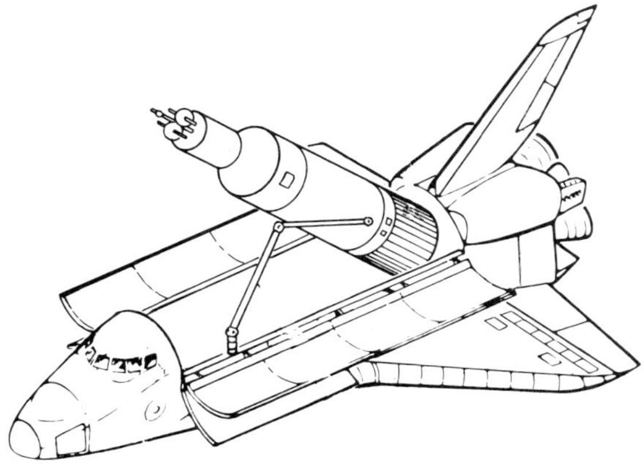

## I



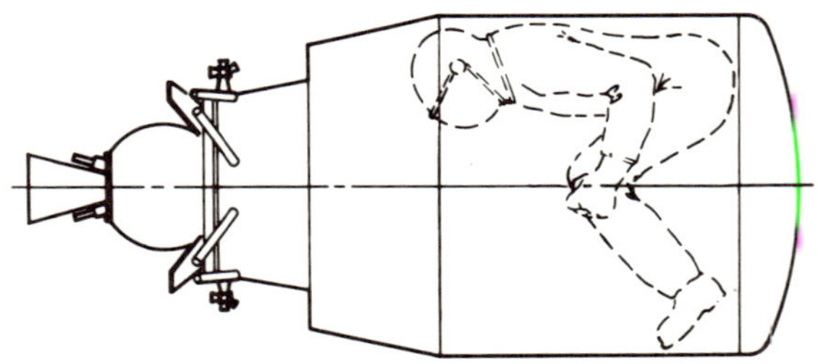

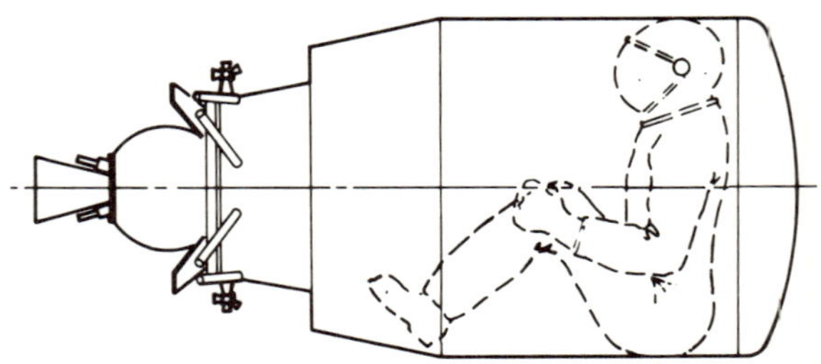

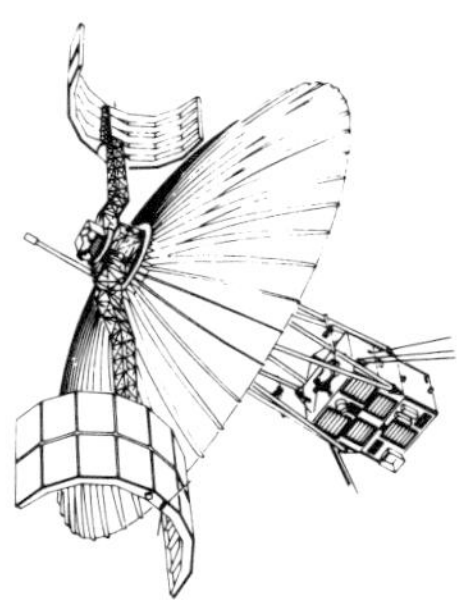

# K

KABAK, I.W., AAS 69-271, Mic v17

KAEHLER, R.C., AAS 57-020, Adv v2; JAS v5-2 pp43-46 1959; AAS 60-026, Adv v6

KAEPPELER, H.J., JAS v2-2 pp50-56 1955; v2-3 pp111-118 1955

KAERCHER, A., AAS 67-352, Mic v4

KAES, G.L., AAS 75-289, Mic v25, Adv v32II

KAFAFIAN, H., AAS 72-116, S&T v32

KAGIWADA, H.H., JAS v13-3 pp110-115 1966

KAHAN, A.M., AAS 69-123, Mic v19

KAIN, R.R., AAS 74-120, Adv v31 I

KALABA, R.E., JAS v13-3 pp110-115 1966

KALITINSKY, A., AAS 63-063, Adv v15

KALLAY, N., AAS 65-098, Adv v20I

KALTENBACH, M.C., AAS 74-124, Adv v31 I

KAMEL, A.A., AAS 73-239, Mic v21

KAMEN, S.K., AAS 63-148, Adv v16II

KAMM, J.L., AAS 71-328, Mic v20; 71-329, Mic v20

KANE, F.X., IAF 69-06, Mic v23-2; 72-06, Mic v24-2; 73-12, Mic v24-3/-4; 74-189, S&T v37; 75-018, S&T v41

KANE, T.R., JAS v10-4 pp114-118 1963 v14-4 pp158-166 1967; v17-5 pp292-306 1970; v18-4 p253 1971; AAS 75-093, Mic v26; JAS v25-4 pp365-370 1977

KAPLAN, C.J., ASR v1-4 pp11-12,21 1959

KAPLAN, E., AAS 70-086, S&T v26

KAPLAN, M.H., IAF 72-09, Mic v24-2; 73-09, Mic v24-3/-4; AAS 75-297, Mic v25, Adv v32II; 76-103, S&T v43; 76-108, S&T v43; 79-032, Adv v39

KARATSINIDES, S.P., AAS 77-132, Mic v27

KARAYALCIN, I.I., AAS 69-444, Mic v15; 69-445, Mic v15; 69-446, Mic v15

KAREN, B., AAS 61-107, Adv v10; 67-391, Mic v4

KARMIOL, E.D., AAS 58-010, Adv v3

KARNIK, K.S., AAS 70-145, S&T v33

KARRENBERG, H.K., AAS 68-136, Mic v7; JAS v17-3 pp129-177 1969

KARTH, J.E., AAS 66-044, S&T v10; 68-024, S&T v18

KASE, P.G., AAS 70-054, Adv v27

KATTERFELD, G.N., AAS 68-047, S&T v25; 69-676, S&T v25

KATZ, E., AAS 77-205, Adv v36I

KAU, S.P., AAS 79-014, Adv v39

K

KERTESZ, T.J., AAS 75-184, Mic v25, Adv v32I

KES, F.C., AAS 75-057, Mic v26; JAS v23-3 pp257-270 1975

KESSLER, D.J., AAS 67-333, Mic v4

KETCHUM, H.B., AAS 56-006, Adv v1

KEYS, G.T., AAS 67-503, Mic v4

KEYSER, H., AAS 63-069, Adv v15

KEVILLE, J.F., AAS 63-114, Adv v16I

KHODAREV, U.K., AAS 69-033, Adv v26; 73-170, Adv v30

KIEHNE, N., AAS 78-013(Abstract), Adv v37I

KIERSTEAD, F.H., ASS 59-006, Adv v5

KIMBLETON, S.R., AAS 69-535, Mic v17

KINARD, W.H., DGLR 4.1 1976 Bonn Proc

KING, B.S., AAS 75-152, Mic v25, Adv v32II

KING, J.C., AAS 71-353, Mic v20; 75-107, Mic v26; JAS v24-4 pp 347-364 1976

KING, J.M., AAS 69-389, Mic v13

KINGSBURY, D., AAS 77-219, Adv v36I

KINGSLAND, L., JR, AAS 69-553, Mic v19

KINNEY, W.D., AAS 66-131, S&T v11

KINZEY, S., AAS 74-057, S&T v38

KIRBY, F.M., AAS 73-005, S&T v31

KIRBY, R.C., AAS 67-078, Mic v3

KIRCHGESSNER, T.E., AAS 67-497, Mic v4

KIRKLAND, F.P., AAS 67-445, Mic v4

KIRKLAND, V.D., AAS 67-113, Mic v3

KIRPICH, A.S., AAS 67-491, Mic v4

KISLYAKOV, A.G., JAS v18-6 pp383-389 1971

KISNER, F.A., AAS 66-251, Mic v9

KITCHENS, M.D., AAS 75-137, Adv v32II

KLAESSIG, F.C., AAS 74-055, S&T v38

KLAN, M.S., AAS 78-144, Adv v38II

KLEIN, H.P., AAS 65-188, S&T v8; 68-286, Mic v8

KLEIN, K.E., AAS 78-028, Adv v37II

KLEIN, M., AAS 73-161, Adv v30

KLEINKNECHT, K.S., IAF 72-07, Mic v24-2; AAS 74-186, S&T v37

KLEINMANN, D.E., AAS 75-209, Adv v32I

KLEPCZYNSKI, W.J., AAS 71-106, Adv v29I; 71-312, Mic v20

KLICK, A.F., AAS 66-148, S&T v12

KLIGER, I.E., AAS 66-272, Mic v9

KLINE, R.L., JAS v24-3 pp243-256 1976; AAS 76-050, Adv v35; 76-141, S&T v43

KLIORE, A.J., AAS 68-185, Adv v25

KUBIS, J.F., AAS 74-123(Abstract), Adv v31 I

KUBOKAWA, C.C., AAS 69-438, Mic v18

KUCHNEGGER, W., AAS 59-003, Adv v5

KUDISH, H., AAS 75-170, Adv v32II; AAS 77-220, Adv v36I; JAS v26-1 pp69-99 1978

KUETTNER, J.P., AAS 66-013, Adv v21; 67-104, Mic v3

KUHN, K.G., AAS 60-110, Sp v2

KUHRT, W.A., AAS 66-030, S&T v10

KUIPER, G., AAS 63-090, Adv v15; 66-201, S&T v14

KULLAS, A.J., AAS 67-210, Mic v10

KUMAR, K., AAS 73-232, Mic v21; JAS v25-1 pp3-20 1977

KUNDT, W., AAS 75-102, Mic v26

KUNEN, A.E., AAS 58-006(Abstract), Adv v3; 59-013, Adv v5

KUNDT, W., AAS 75-102, Mic v26

KUO, B.C., AAS 78-102, Mic v29

KUPPERIAN, J.E., JR, AAS 62-089, Adv v12

KURTZ, H.F., JR, AAS 74-116, Adv v31 I

KURTZ, M., AAS 69-249, Mic v13

KURZHALS, P.R., AAS 63-004, Adv v13

KURZWEG, H.H., AAS 66-336, S&T v24

KUTCHMA, E.K., AAS 69-095, Mic v17

KUZNETSOV, A.P., JAS v19-5 pp399-407 1972

KYLSTRA, C.D., AAS 65-134, Adv v20II

KYNER, W.T., AAS 66-093, S&T v11 Mic v2

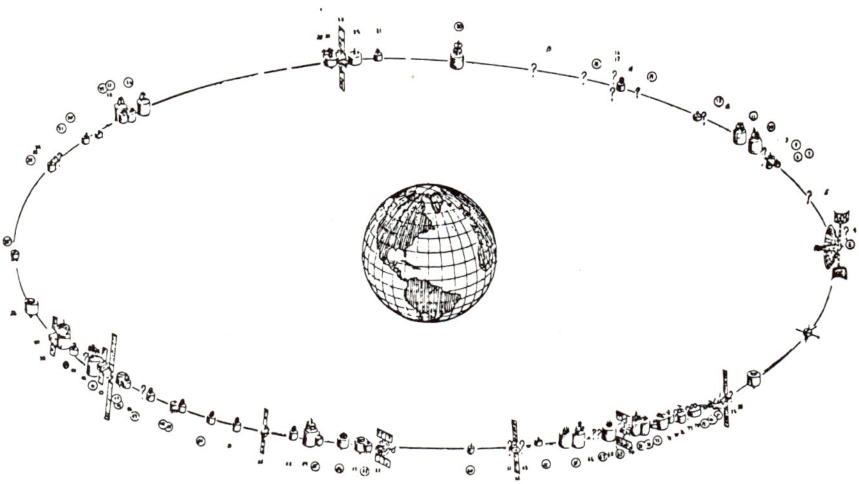

# L

LASSEN, H.A., AAS 65-022, Adv v19

LASTER, L., AAS 72-041, S&T v29

LATHAM, G.V., AAS 68-204, Adv v25

LAUE, J.H., JAS v26-1 pp1-17 1978; AAS 78-048, Adv v37II

LAURANCE, R.J., AAS 75-294, Adv v32I

LAUSCHNER, E.A., AAS 70-146, S&T v33

LAVAGLIA, A., AAS 78-125, Mic v29

LAVINE, M.C., AAS 74-053, S&T v38

LAWTON, R.W., ASR v4-1 pp11-18, 31-38 1962

LAY, S.H., AAS 76-130, S&T v43; 76-133, S&T v43; 76-135, S&T v43

LAYTON, D.M., AAS 71-121, Adv v29I

LAYTON, J.P., AAS 69-330, Mic v18

LAZARA, M., AAS 69-086, Mic v13

LEA, R.H., AAS 67-031, S&T v15; 69-021, S&T v21

LEACH, J.W., DGLR 2.4 1976, Bonn Proc

LEARY, T., AAS 77-249(Summary), Adv v36II

LEBEAU, A., DGLR 1.2 1976, Bonn Proc

LE BLANC, D.R., AAS 78-114, Mic v29

LEBOVITZ, R.M., AAS 72-117, S&T v32

LECAR, M., JAS v9-3 pp82-85 1962

LECK, D., AAS 67-261, Mic v10

LEDBETTER, K.W., AAS 71-358, Mic v20

LEDERBERG, J., AAS 65-187, S&T v8

LEE, C.M.., ed. Adv v34, 1977; AAS 78-006, Adv v37I

LEE, C.N., AAS 58-011, Adv v3

LEE, D.O., AAS 74-026, S&T v35

LEE, Y.C., AAS 58-038-11, Adv v3; 61-107, Adv v10

LEES, A.B., AAS 61-028, Adv v8

LEES, L., AAS 74-003, S&T v35

LEESTMA, R.A., AAS 65-060, S&T v4

LEFSTAD, S.M., AAS 70-071, S&T v26

LEFTON, L., AAS 77-135, Mic v27

LEHNER, J.W., AAS 66-295, Mic v9

LEIBOLD, A., AAS 75-098, Adv v33

LEIGHTON, G.S., AAS 74-006(Abstract), S&T v35

LEIMKUHLER, F.F., AAS 69-357, S&T v22

LEITMANN, G., AAS 61-030, Adv v8; JAS v14-2 pp56-65 1967

LEKUCK, H.W., AAS 64-023, S&T v1

LELIAKOR, I.P., AAS 73-205, Mic v21

LEM, J.D., AAS 74-174, Adv v31 II

LEMAY, J.L., AAS 68-129, Mic v7

LEMON, W.W., AAS 66-108, S&T v11 Mic v2

LENARD, M., AAS 69-314, Mic v16

LENOIR, W.B., AAS 74-124, Adv v31 I

LEONDES, C.T., JAS v21-1 pp1-25 1973; JAS v21-3/4 pp154-184 1974

LEPPER, R., AAS 60-103a, Sp v1; 63-049, Adv v14

LERNER, G.M., AAS 75-029, Adv v33; 77-116, Mic v27

LEUTHARD, J.J., AAS 67-282, Mic v10

LEVEN, P., DGLR 2.3 1976, Bonn Proc

LEVENTAHAL, E.L., AAS 65-145, S&T v6

LEVENTHAL, J., AAS 74-156, Adv v31 II

LEVIE, S.L., JR, JAS v18-4 pp217-235 1971

LEVIN, E., ASR v1-4 pp15-16,21 1959; AAS 68-136, Mic v7; JAS v17-3 pp129-177 1969

LEVIN, G.M., AAS 75-131, Adv v32I

LEVIN, G.V., AAS 65-047, Adv v19; 66-066, Adv v22; 67-576, S&T v19;

LEVINE, D.S., AAS 69-021, S&T v21

LEVINE, P., AAS 69-499, Mic v18

LEVINE, R.B., AAS 60-103, Sp v1

LEVINSON, D.A., JAS v25-2 pp129-142 1977; AAS 77-168, Mic v27; 75-093, Mic v26

LEVINSON, E., AAS 60-017, Adv v6

LEVITT, I.M., JAS v2-1, pp1-6,13 1955

LEVY, M., AAS 73-016, S&T v31

LEVY. E.Z., AAS 57-021, Adv v2

LEVY, S.L., AAS 67-079, Mic v3

LEWALLEN, J.M., AAS 68-099, Mic v7

LEWIS, A.E., AAS 74-017, S&T v35

LEWIS, D.A., JAS v15-3, pp148-152 1968

LEWIS, D.H., JAS v14-3 pp112-122 1967

LEWIS, W.C., AAS 75-175, Mic v25, Adv v32I

LEWYN, L., AAS 65-061, S&T v4

LEY, W., AAS 63-084, Adv v15

LIBBY, W.F., AAS 68-231, Mic v8

LICHT, B.W., AAS 67-414, Mic v4

LICHTENSTEIGER, M., AAS 74-053, S&T v38

LIDE, W.C., AAS 75-295, Mic v25, Adv v32II

LIEBELT, P.B., AAS 68-108, Mic v7

LIEBENOW, R.C., AAS 69-087, Mic v13

LIKINS, P., AAS 78-101, Mic v29

LIKINS, P.W., JAS v12-1 pp18-24 1965; v13-2 pp87-88 1966; JAS v16-1 pp32-37 1969; v21-5/6 pp244-251 1974; v24-4 pp311-328 1976

LILLESTRAND, R.L., AAS 60-061, Adv v6; 61-095, Adv v9

LILLY, J.C., AAS 66-077, Mic v1

# M

MA, E.C., AAS 67-417, Mic v4

MAAS, D.E., AAS 68-255, Mic v8

MAC CALLUM, S.H. AAS 77-286, Adv v36II

MAC DONALD, D., AAS 61-101, Adv v9

MAC DONALD, G.J., AAS 63-093, Adv v15

MAC DONALD, G.J.F., AAS 66-194, S&T v13

MAC DONALD, J.M., AAS 72-167, S&T v30; 75-268, Mic v25, Adv v32II

MAC DONALD, R.B., AAS 65-097, Adv v20I; AAS 67-513, Mic v4

MAC DOWALL, J., AAS 73-114, Adv v30

MACK, R., AAS 77-280, Adv v36II

MACKAY, D.B., AAS 58-052, Adv v4

MAC NAUGHTON, J., AAS 78-049, Adv v37II

MAC NAUGHTON, J.D., AAS 63-012, Adv v13; DGLR 4.6 1976, Bonn Proc

MAC QUEEN, R.M., AAS 68-217, Adv v25

MAC QUIVEY, D.R., AAS 66-017, Adv v21

MADAY, C.J., AAS 61-102, Adv v9; 65-133, Adv v20II

MADEWELL, J.F., AAS 76-178, S&T v45; DGLR 5.2 1976, Bonn Proc

MADEY, R., AAS 62-015, Adv v11

MADIGAN, R.J., AAS 69-401, Mic v18

MADLEY, R., AAS 62-022(Abstract), Adv v11

MAEDA, H., AAS 76-103, S&T v43

MAERSCHALK, J.C., AAS 67-428, Mic v4

MAES, W.J., AAS 68-275, Mic v8

MAGGIN, B., AAS 69-359, Mic v18

MAGNESS, T.A., AAS 58-057, Adv v4

MAGNO, P., AAS 70-150, S&T v33

MAHONEY, J.J., AAS 66-062, Adv v22

MAIR, C.C., AAS 63-140, Adv v16I

MAKEMSON, M.W., JAS v8-1 pp1-13 1961; AAS 61-027, Adv v8; JAS v12-4 pp117-122 1965

MALCHOW, H.L., AAS 71-116, Adv v29I

MALITSON, H.H., AAS 67-177, S&T v17

MALLON, J.A., AAS 66-307, Mic v9

MALONE, T.B., AAS 72-165, S&T v30

MANATT, S.L., AAS 68-040, S&T v25

MANGELSDORF, J.E., AAS 58-002, Adv v3; 65-075, S&T v 5

MATHESON, J.E., AAS 67-378, Mic v4

MATHEWS, C.W., AAS 67-004, S&T v16; 67-549, Mic v6; 69-020, S&T v21; 75-127, Adv v32I

MATLOFF, G.L., JAS v18-1, pp57-64 1970; v24-4 pp365-382 1976

MATSUO, M., IAF 72-10, Mic v24-2

MATSUMOTO, I., IAF 69-15, Mic v23-2

MATTHEWS, H.F., AAS 69-290, Adv v26; 71-167, Adv v29I

MATVEYENKO, L.I., JAS v19-1 pp73-76 1971

MATZENAUER, J.O., AAS 67-370, Mic v4

MAUL, G.A., AAS 73-146, Adv v30;

MAURITSEN, T.E., AAS 58-003, Adv v3

MAURY, J.L., JAS v15-6 pp303-312 1968

MAXWELL, J.C., AAS 69-323, S&T v22

MAXWELL, R.E., AAS 75-050, Mic v26

MAY, C.B., AAS 75-256, Mic v25, Adv v32II

MAYER, R.W., AAS 68-232, Mic v8

MAYEUX, J.V., AAS 75-255, Mic v25, Adv v32II

MAYO, A.M., AAS 57-030, Adv v2; JAS v5-1 pp11-14 1958; AAS 58-038-7, Adv v3; 61-013, Adv v8; 67-030, S&T v15; 67-074, Adv v23; ed. Adv v23, 1968 ; IAF 68-02, Mic v23-1; AAS 68-293, Mic v11

MAYO, R.A., AAS 78-115, Mic v29

MAYOR, R., AAS 72-112, S&T v32

MC ADAMS, H.T., AAS 67-168, S&T v17

MC ADOO, S.F., JR, AAS 70-039, Adv v28; 71-302, Mic v20

MC ALLISTER, D.F., AAS 61-038, Adv v8; 65-161, S&T v6; 69-404, Mic v18; JAS v18-4 pp193-216 1971

MC ARTHUR, W.G., AAS 79-036, Adv v39

MC BRAYER, R.O., AAS 75-239, Adv v32II

MC BRIDE, A.L., AAS 69-608, Mic v12

MC CABE, E.L., IAF 69-12, Mic v23-2

MC CALL, J., AAS 66-203, S&T v14

MC CALLY, M., AAS 63-161, Adv v17

MC CANDLESS, B., II, AAS 74-137, Adv v31 I

MC CARTHY, J.F., JR, AAS 78-192 Adv v38II

MC CARTHY, S.G., AAS 75-234, Mic v25, Adv v32II

MC CARTNEY, E.J., AAS 58-049, Adv v4

MC CLAIN, E.P., AAS 69-596, S&T v23

MC CLELLAN, H.J., AAS 69-026, S&T v21

MC CORMAC, B.M., AAS 67-369, Mic v4; 75-132, Adv v32I

MC CORMICK, R.A., AAS 69-620, S&T v23

MC COY, C.T., AAS 58-014(Abstract), Adv v3

MC COY, H.E., AAS 77-218(Abstract), Adv v36II

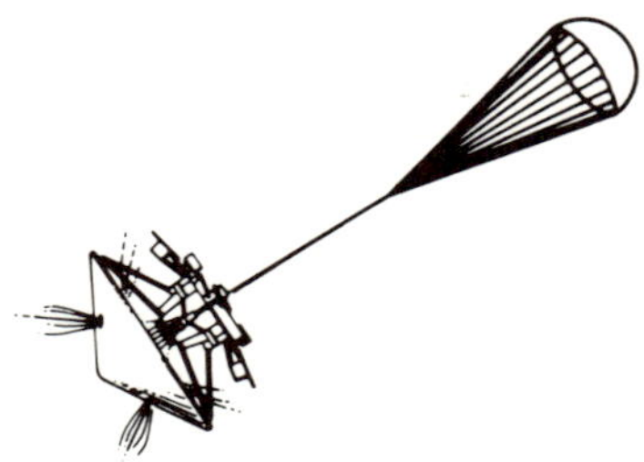

# N

NACHTSHEIM, P.R., AAS 71-147, Adv v29II

NACK, M.L., AAS 75-104, Mic v26

NACOZY, P.E., AAS 71-315, Mic v20; 73-220, Mic v21; 77-122, Mic v27

NAGATOMO, M., IAF 72-10, Mic v24-2; AAS 74-187, S&T v37

NAGORSKI, R., AAS 77-106, Mic v27

NAGORSKI, R.P., AAS 71-378, Mic v20

NAGVI, S., JAS v20-4 pp193-222 1973

NAGY, A.R., JR, IAF 69-03, Mic v23-2; 70-05, Mic v23-3

NAGY, B., AAS 66-054, Mic v1; JAS v15-4 pp161-168 1968

NALEPTA, R.F., AAS 74-144, Mic v22 JAS v23-3 pp193-204 1975

NANDA, V.P., AAS 69-042, Mic v19

NANEVICZ, J.E., AAS 75-222, Mic v25, Adv v32I

NANSEN, R.H., DGLR 5.4 1976 Bonn Proc

NAPOLITANO, L.G., AAS 58-047, Adv v4; 68-022, S&T v18

NARIN, F., AAS 65-124, Adv v20II; ed. Adv v20 1966

NARODNY, L., AAS 65-096, Adv v20I

NARODNY. L.H., AAS 67-181, S&T v17

NASON, M.L., AAS 61-040, Adv v8

NASSEN, H.S., AAS 74-129, Adv v31 I

NATHAN, C.A., AAS 77-202, Adv v36I

NAU, R.A., AAS 76-105, S&T v43; 76-109, S&T v43; 76-155, S&T v45

NAUGLE, J., AAS 75-121, Adv v32I

NAUGLE, J.E., AAS 72-025, S&T v28

NAUMANN, A.,JR, ed. Adv v38 1979

NAUMANN, E.C., AAS 77-205, Adv v36I

NAUMANN, R.J., AAS 62-046, Sp v5; 77-238, Adv v36I

NAUMCHEFF, M., AAS 74-116, Adv v31 I

NAUMOV, A.I., JAS v18-6 pp383-389 1971 N

NAYMARK, S., AAS 60-043, Adv v6

NEACY, W.P., AAS 65-034, Adv v19

NEAU, O.T., AAS 75-264, Mic v25, Adv v32II

NEEL, C., AAS 75-237, Mic v25, Adv v32II

NEGRO, A.G., AAS 63-001, Adv v13

NEKRASOV, P., AAS 68-064, Adv v24

NEKRASOV, P.S., AAS 68-067, Adv v24

NELSON, D.A., AAS 63-017, Adv v13

NELSON, H.D., JAS v13-6 pp226-234 1966

NELSON, R.N., AAS 69-395, Mic v18

NELSON, R.R., AAS 68-016, S&T v18

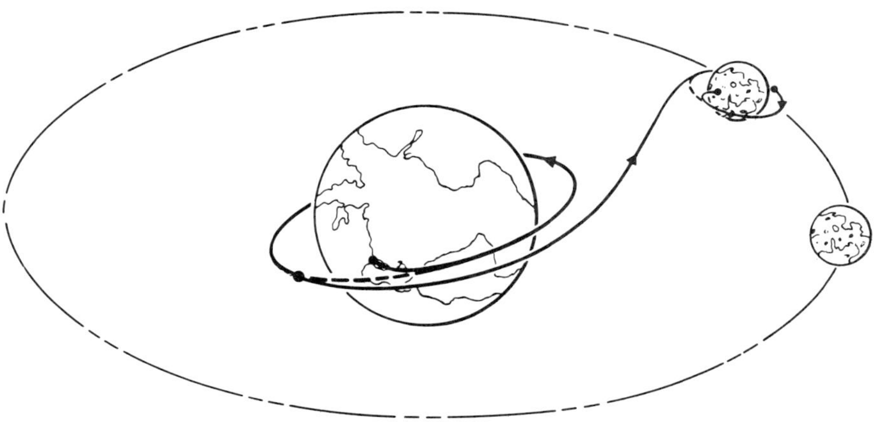

# O

OBERG, J.L., AAS 75-182, Mic v25, Adv v32I

O'BRIEN, B.J., AAS 62-084, Adv v12; 68-213, Adv v25

O'CALLAGHAN, F.G., AAS 74-161, Adv v31 II

O'DELL, C.R., AAS 75-129, Adv v32I; DGLR 4.2 1976, Bonn Proc

ODEN, J.T., AAS 67-416, Mic v4; 67-436, Mic v4

ODOM, P.R., AAS 66-263, Mic v 9; JAS v16-1 pp13-21 1969; AAS 71-322, Mic v20; 73-228, Mic v21

O'DONNELL, A.J., AAS 71-166, Adv v29II

O'DONNELL, C.A., AAS 70-032, Adv v27;

O'DWYER, W., AAS 67-352, Mic v4

OEHMAN, W., JAS v12-4 pp135-141 1965

OFFENHARTZ, E., AAS 66-220, Mic v9

OFFIELD, T.W., AAS 68-304, S&T v25

OFFIK, W.G., AAS 66-230, Mic v9; 67-426, Mic v4

OFFNER, A., AAS 75-177, Mic v25, Adv v32I

OGDEN, D.H., AAS 67-426, Mic v4

OGLETREE, E.G., AAS 62-007, Adv v11; 62-055, Sp v5

OHKAMI, T., JAS v24-4 pp311-328 1976

OHTAKAY, H., AAS 75-088, Mic v26

O'KEEFE, J.A., AAS 66-187, S&T v13

OLDEG, H.W., AAS 66-158, S&T v12

OLDFIELD, H.R., AAS 72-109, S&T v32

OLDS, R.H., SR, AAS 67-500, Mic v4

OLIVER, B.M., AAS 77-298, Adv v36I

OLIVER, C.C., AAS 66-218, Mic v9; 67-471, Mic v4; 67-472, Mic v4

O'LAUGHLIN, J.R., AAS 67-446, Mic v4

OLSON, R.N., AAS 67-427, Mic v4

O'MAHONY, M.S., AAS 66-240, Mic v9; 66-250, Mic v9; 67-348, Mic v4

O'MALLEY, J.P., AAS 65-030, Adv v19

OMAN, R.A., AAS 58-045, Adv v4

ONDRASIK, V.J., AAS 71-399, Mic v20; 73-244, Mic v24

O'NEIL, W.D., AAS 69-217, Mic v17

O'NEILL, G.K., JAS v25-4 pp349-363 1977; AAS 77-222, Adv v36I

O'NEILL, P.M., AAS 75-062, Mic v26

ONG, K.M., JAS v21-5/6 pp206-229 1974

ÖPIK, E.J., AAS 61-063, Sp v4

ORDWAY, F.I., III, ed. JAS 1955; JAS v2-1, pp9-13, 1955; v2-2 pp57-61,77 1955

ORILLION, A.G., AAS 75-138, Adv v32II

O'RIORDAN, P.A., AAS 71-160, Adv v29II

ORMES, J.F., AAS 75-263, Mic v25, Adv v32II

ORR, E.D., AAS 70-080, S&T v26

ORRALL, F.Q., AAS 72-002, S&T v28

ORTLOFF, C.R., JAS v16-2 pp49-67 1969

OSBORNE, N.A., AAS 75-185, Mic v25, Adv v32I; 78-116, Mic v29

OSGOOD, C.C., AAS 60-040, Adv v6; 60-068, Adv v7

OSTRANDER, D.D., AAS 75-202, Mic v25

OSTRANDER, D.R., AAS 65-007, Adv v19

OSTRIKER, J.P., AAS 72-012, S&T v28

OSWALT, F.W., AAS 67-223, Mic v10

OTHLING, W.L., JAS v19-2 pp130-137 1971

OTT, R., AAS 75-260, Mic v25, Adv v32II

OTTL, H., AAS 78-016, Adv v37I

OUTLAW, R.A., AAS 75-248, Mic v25, Adv v32II

OVERCAMP, J.J., AAS 66-253, Mic v9

OWEN, W.J., AAS 77-117, Mic v27

OWENS, A.T., AAS 65-160, S&T v6

OYAMA, V., AAS 69-352, Mic v19

OZGOREN, M.K., AAS 75-103, Mic v26

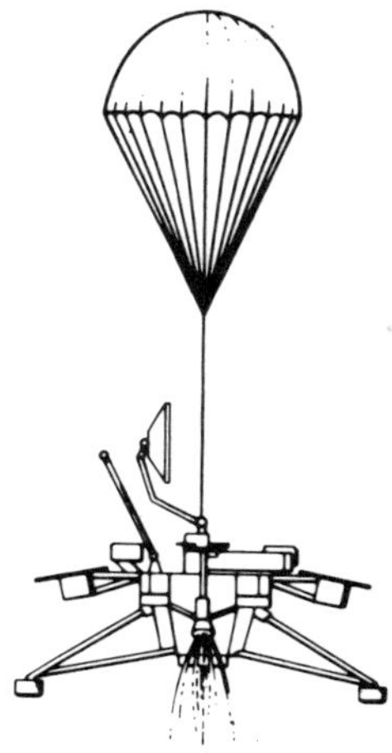



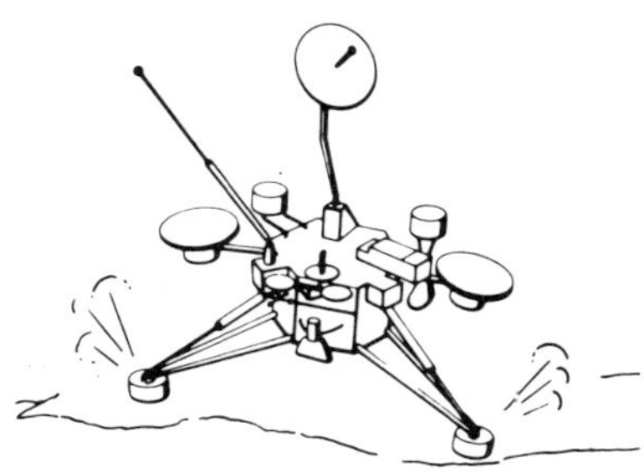

# P

PACE, N., AAS 61-115, Adv v10

PACE, R.E., JR, AAS 74-122, Adv v31 I

PACELA, A.F., AAS 65-154, S&T v6

PACKARD, R.F., AAS 67-009, S&T v16; 67-157, Mic v3

PACKER, A.H., AAS 69-159, Mic v16; 74-036, S&T v35

PAGE, T.L., AAS 70-035, Adv v27

PAGE, W.A., AAS 71-144, Adv v29II

PAHMEIER, R., AAS 67-495, Mic v4

PAIGE, H.W., AAS 66-034, S&T v10

PAINE, T.O., AAS 76-187, S&T v45

PAINTER, J.E., AAS 73-122, Adv v30

PALMER, C.E., AAS 68-055, Adv v24

PALMER, E.H., AAS 69-063, Mic v13

PALMITER, M.T., JAS v14-4 pp147-157 1967; v17-6 pp313-336 1970

PALMORE, J., AAS 68-085, Mic v7

PALUDAN, C.T.N., AAS 69-483, Adv v26

PALUSZEK, M.A., AAS 77-274, Adv v36I

PAN, C.H.T., AAS 73-064, S&T v34

PANCHENKOVA, E.F., JAS v19-1 pp77-81 1971

PANDE, K.C., JAS v22-1 pp1-20 1974; v22-1 pp36-54 1974; AAS 75-024, Adv v33

PANITZ, H.J., AAS 75-101, Mic v26; 78-039, Adv v37II

PANNETOV, R.J., AAS 69-609, S&T v23

PARANJAPE, S.V., AAS 66-225, Mic v9; 66-231, Mic v9

PARDOE, G.K.C., AAS 67-533A, Mic v6; 67-533B, Mic v6; 73-119, Adv v30

PARENTE, W.E., AAS 67-273, Mic v10

PARK, A.B., AAS 67-108, Mic v3; 61-531 Mic v6; 69-061, Adv v26

PARK, A.C., AAS 67-414, Mic v4

PARK, R.A., AAS 65-022, Adv v19; 67-188, S&T v17; 71-121, Adv v29I; 71-165, Adv v29II

PARKER, E.N., AAS 65-003, Adv v19

PARKER, R., AAS 75-237, Mic v25, Adv v32II

PARKER, R.A., AAS 74-050, S&T v38; 74-117, Adv v31 I

PARKINSON, W.H., AAS 68-219, Adv v25

PARKS, D.K., AAS 67-231, Mic v10

PARKS, R.J., AAS 63-023, Adv v13

POTTER, A.E., AAS 74-141, Adv v31II

POTTS, B.C., AAS 61-071, Sp v4

POTTS, G.R., AAS 69-592, Mic v12

POTTSEPP, L., AAS 71-366, Mic v20

POUND, E.F., AAS 67-239, Mic v10

POWE, W.E., AAS 63-129, Adv v16I

POWERS, W.F., JAS v16-2 pp88-94 1969; 71-367, Mic v20; 73-224, Mic v21; 73-240, Mic v21; ed. Adv v33 1976; ed. JAS v25 1977; AAS 77-132, Mic v27; 77-136, Mic v27; JAS v26-1 pp47-68 1978; AAS 78-138, Mic v30 & JAS v27-1

PRAGLUSKI, W.J., AAS 68-110, Mic v7; AAS 75-199, Adv v32I

PRATT, W.P., AAS 70-033, Adv v27

PREHODA, R.W., AAS 77-244(Abstract), Adv v36II

PRESCOTT, E.J., AAS 65-081, S&T v5

PRESSBURG, B.S., AAS 66-257, Mic v9

PRESTON, G.W., AAS 72-006, S&T v28

PREW, H.E., AAS 56-009, Adv v1; 57-026, Adv v2; JAS v4-1 pp9-15 1957

PRICE, I., AAS 69-528, S&T v22

PRICE, P.B., AAS 74-163(Abstract), Adv v31 II

PRICE, R.W., AAS 67-226, Mic v10

PRIFTIS, G., AAS 70-089, S&T v26

PRIGGE, J.S., JR, AAS 55-020; JAS JAS v3-1 pp4-8,26 1956

PRIMEAUX, G.R., AAS 74-177, Adv v31 I

PRINGLE, J.K., AAS 65-122, Adv v20II

PRINGLE, R., JR, AAS 68-125, Mic v7

PRITCHARD, E.B., AAS 70-030, Adv v28; 70-047, Adv v28

PRITCHARD, R.E., JAS v15-2 pp85-91 19 JAS v10-2 pp41-54 1963

PRITCHARD, W., AAS 76-026, S&T v44

PROBST, S.E., AAS 76-029, S&T v44

PROCTOR, L.D., AAS 66-033, S&T v10; 67-125, Mic v3

PRODAN, J., AAS 63-134, Adv v16I; 65-082, S&T v5

PRUSSING, J.E., AAS 73-238, Mic v21; JAS v25-2 pp123-128 1977

PURCELL, J., AAS 75-004, S&T v40

PURDY, J.G., JAS v18-5 pp281-305 1971

PURDY, W., AAS 66-154, S&T v12

PURDY, W.G., AAS 66-036, S&T v10

PURVIS, C.K., AAS 77-243, Adv v36I

PUTEGNAT, J., AAS 67-036, S&T v15

# Q



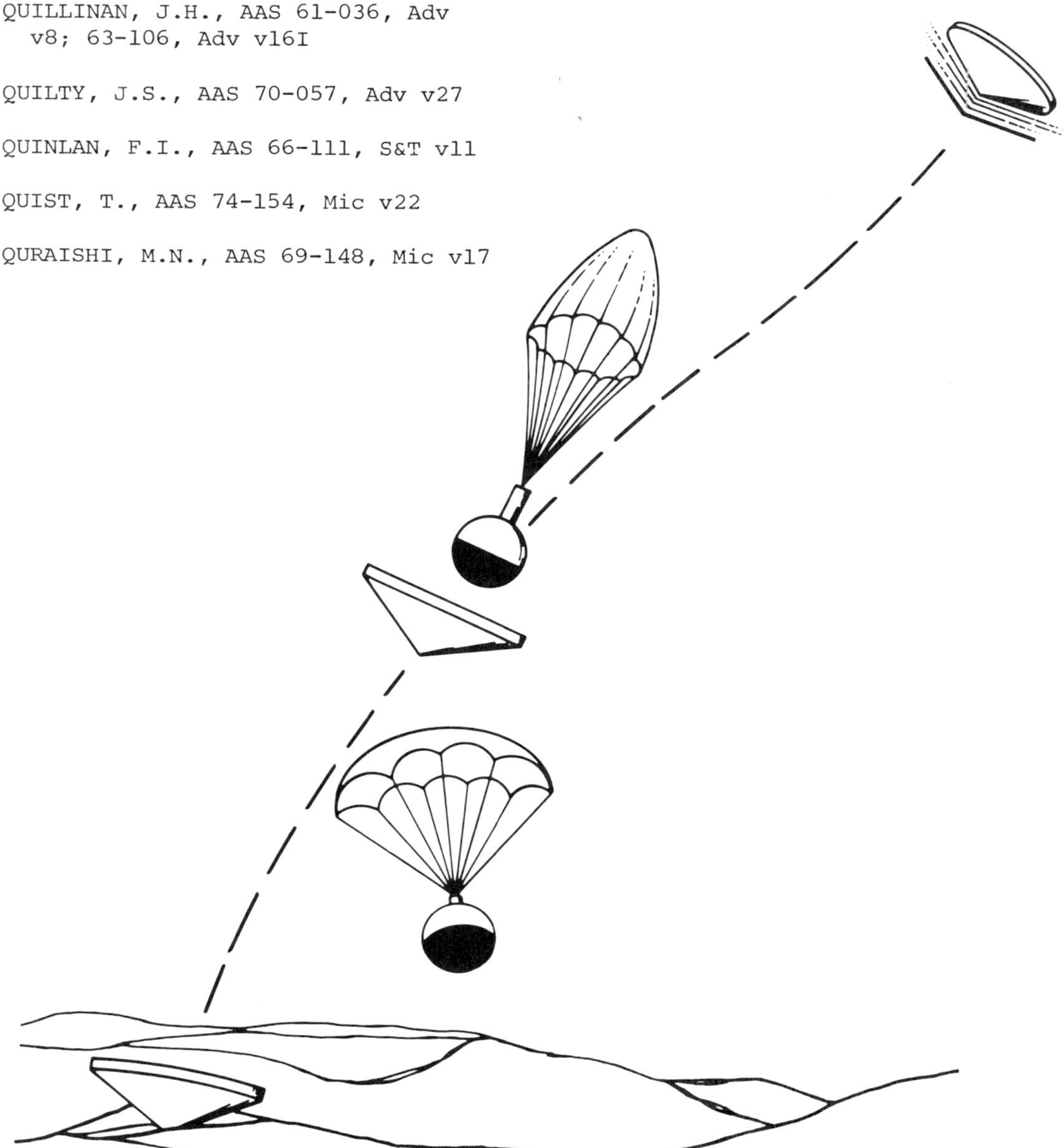

# R

RAYMES, F., AAS 69-027, S&T v21

RAYMOND, H., AAS 79-026, Adv v39

REA, D.G., AAS 68-192, Adv v25; 71-105, Adv v29I

REA, F.G., AAS 75-051, Mic v26

REBMAN, J.A., AAS 67-341, Mic v4

RECHTIN, E., AAS 63-081, Adv v15

REDISCH, W.N., AAS 76-046, Adv v35

REED, D.R., AAS 67-353, Mic v4

REED, J.F., AAS 65-156, S&T v6

REED, L., AAS 74-145, Mic v22

REED, R.R., AAS 67-385, Mic v4

REED, W.D., AAS 78-162, Adv v38II

REESE, D.E., JR, AAS 65-024, Adv v19

REEVES, E.M., AAS 68-219, Adv v25; 74-169, Adv v31 II

REGENBRECHT, D.E., AAS 69-174, Mic v18

REHDER, J.J., AAS 75-065, Mic v26

REHDER, J.R., AAS 75-054, Mic v26

REICHERT, R.J., AAS 71-318, Mic v20; 73-252, Mic v21

REICHMUTH, D.R., AAS 67-505, Mic v4

REID, J.H., AAS 65-165, S&T v6

REIFF, G.A., AAS 65-033, Adv v19; 69-066, Mic v19

REIGER, S.H., AAS 65-197, S&T v8

REILY, J.C., AAS 74-136, Adv v31 I

REINHEIMER, J., AAS 60-055, Adv v6

REISS, M.H., AAS 61-081, Adv v9

REITER, G.S., JAS v7-2 pp29-34 1960; AAS 62-042, Sp v5

REITZ, J.G., AAS 60-100, Sp v1

RENGER, W., DGLR 3.3 1976, Bonn Proc

RENNELS, D.A., AAS 69-368, Mic v19; 73-058, S&T v34

REPIC, E.M., AAS 70-028, Adv v28; 68-122, Mic v7

RESS, E.B., AAS 75-228, Mic v25, Adv v32I

RESWICK, J.B., AAS 65-019, Adv v19

RETTIE, R.S., AAS 67-535, Mic v6

REYN, J.W., JAS v12-2 pp46-50 1965

REYNOLDS, J.B., AAS 63-061, Adv v14

REYNOLDS, R.C., AAS 74-057, S&T v38

REYNOLDS, W.C., AAS 60-094, Sp v1



REZNICEK, P., AAS 71-143, Adv v29II

RHOTE, W.R., AAS 78-153, Adv v38I

RIBARICH, J.J., AAS 66-129, S&T v11

RICCIARDI, H.J., AAS 71-009, S&T v27

RICH, E., AAS 67-271, Mic v10

RICH, L.G., AAS 60-025, Adv v6

RICHARDS, A., AAS 65-078, S&T v5

RUEFLI, T.W., AAS 69-365, Mic v17

RUFF, G.E., AAS 57-021, Adv v2

RUGGLES, G.S., AAS 69-612, Mic v12

RULAND, W.C., AAS 66-251, Mic v9

RUMMEL, W.D., AAS 75-169, Adv v32II

RUMPEL, W.F., AAS 67-238, Mic v10; JAS v15-4 pp207-209 1968; AAS 68-268, Mic v8

RUNCON, S.R., AAS 68-302, S&T v25

RUNGE, F., AAS 75-166, Mic v25, Adv v32II

RUNGE, F.C., AAS 70-031, Adv v27; 78-052, Adv v37II

RUNKEL, H.C., AAS 64-013, Adv v18

RUPERT, P.R., AAS 71-120, Adv v29I

RUPP, C.C., JAS v26-1 pp1-17 1978; AAS 78-048, Adv v37II; 79-033, Adv v39

RUPPE, H.O., AAS 63-070, Adv v15

RUSSAK, S., AAS 68-266, Mic v8

RUSSAK, S.L., AAS 69-321, Mic v14

RUSSELL, A.S., AAS 74-008, S&T v35

RUSSELL, D.R., AAS 70-078, S&T v26

RUSSELL, J., AAS 58-058, Adv v4

RUSSELL, P.E., AAS 71-381, Mic v20

RUSSELL, R.K., AAS 71-385, Mic v20; 73-246, Mic v21

RUTHER, W., AAS 76-123, S&T v43

RUTLAND, C., AAS 69-468, Mic v19

RUTOWSKI, R.W., AAS 59-026, Adv v5

RUTTENBERG, S., AAS 69-120, Mic v19

RYABOVA, T.Y., JAS v20-5 pp304-309 1973

RYAN, I.J., AAS 66-198, S&T v14

RYAN, L.M., AAS 70-054, Adv v27

RYAN, R.E., JAS v23-4 pp287-324, 1975

RYAN, R.S., AAS 66-117, S&T v11

RYBAK, J.P., AAS 69-282, Mic v19

RYBAK, S.C., AAS 78-104, Mic v29

RYKEN, J.M., AAS 63-121, Adv v16I

RYKER, N.J., AAS 64-015, Adv v18; 70-065, Adv v27

RYSHKEWITCH, E., AAS 66-35C, S&T v24

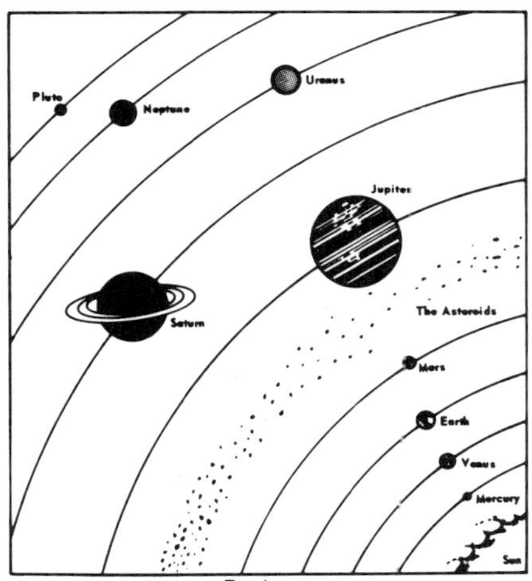

The solar system

# S

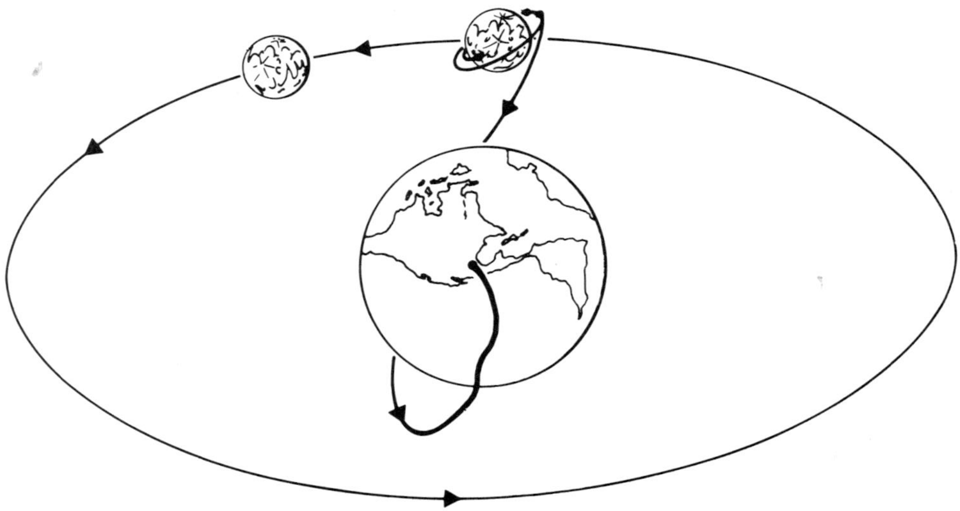

# T

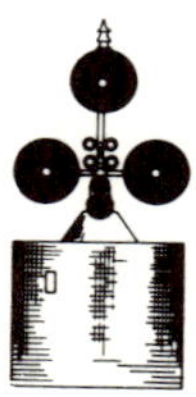

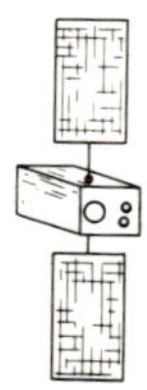

# U



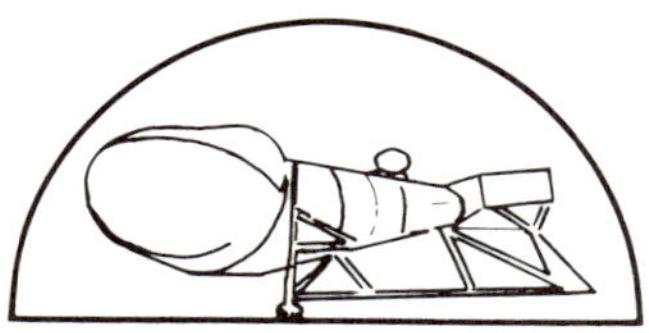

U

# V

VACHINO, R.F., AAS 58-037, Adv v3

VADEBONCOEUR, J.R., AAS 69-294, S&T v22

VAETH, J.G., JAS v1-1 pp1-6,30,32, 1954

VAGNERS, J., ed. Adv v29 1971

VAHLBERG, C.J., JR, AAS 71-368, Mic v20

VAJK, J.P., AAS 77-226, Adv v36II; 78-144, Adv v38II

VALI, V., AAS 58-007, Adv v3

VALLERIE, E.M., III, AAS 67-512, Mic v4

VAN ALLEN, J.A., JAS v3-3/4 pp59-60,80 1956; 58-042, Adv v4

VAN ALLEN, R.E., AAS 77-138, Mic v27

VAN AUSDAL, A.W., AAS 71-162, Adv v29II

VANDENBERG, F., AAS 75-070, Adv v33

VANDENKERCKHOVE, J.A., AAS 70-153, S&T v33; 73-120, Adv v30

VAN DER HA, J.C., JAS v25-4 pp283-306 1977;

VANDER JAGT, G., AAS 68-180, Adv v25

VANDER VEEN, A.A., AAS 66-125, S&T v11; 68-114, Mic v7

VANDER VOORT, R., AAS 79-014, Adv v39

VANDREY, J.F., JAS v15-4 pp177-182 1968; AAS 68-287, Mic v8; JAS v17-2 pp95-103 1969

VAN DRIEST, E.R., ed. S&T v2 1964

VANICA, D.F., AAS 66-285, Mic v9

VAN PATTEN, R.A., AAS 75-041, Adv v33; 75-042, Mic v26; JAS v24-2 pp137-146 1976; ed. Adv v36 1978

VAN SANT, C.T., AAS 61-029, Adv v8

VANSELOUS, T.M., AAS 73-152, Adv v30

VAN VOOREN, R.H., AAS 78-115, Mic v29

VAN WART, F.D., AAS 56-016, Adv v1; 57-019, Adv v2; JAS v4-3 pp52-53,59 1957; AAS 60-060, Adv v6

VAN WINKEL, D.D., AAS 68-148, Mic v7; 77-111, Mic v27

VAN ZYTVELD, P.J., AAS 71-300, Mic v20

VARGO, L.G., AAS 58-020, Adv v3; JAS v7-4 pp78-86 1960; AAS 60-048, Adv v6

VARNERS, J., AAS 66-094, S&T v11

VASIN, M.V., JAS v19-1 pp77-81 1971

VAUGHAN, D.H., AAS 69-152, Mic v18

VAUGHAN, O.H., AAS 67-326, Mic v4; 68-320, S&T v25

VAUGHAN, O.H.,JR., AAS 67-327, Mic v4

VON ROOS, O.H., JAS v25-3 pp227-250
1977

VON ROSENBERG, D.U., AAS 66-227,
Mic v9

VORIS, C.R., AAS 68-279, Mic v8

VOSS, R.G., AAS 60-057, Adv v6

# W

WINKLER, S.H., AAS 60-040, Adv v6

WINN, B.E., AAS 77-110, Mic v27; JAS v26-4 pp343-368 1978

WINN, C.B., AAS 66-119, S&T v11; 68-137, Mic v7; 68-234, Mic v8; 75-105, Adv v33

WINN, F.B., AAS 77-141, Mic v27

WINNEBERGER, R.A., AAS 71-301, Mic v20

WINSTON, G.C., AAS 78-118, Mic v29

WINTER, D.L., AAS 78-010, Adv v37I

WINTER, E.R.F., AAS 66-255, Mic v9; 67-445, Mic v4; 67-471, Mic v4; 69-395, Mic v18

WINTERNITZ, P.F., JAS v3-2 pp43-47, 1956; v3-3/4 pp65-71,80 1956

WINTERS, W.D., AAS 61-114, Adv v10

WIRTHS, G., DGLR 5.5-1976, Bonn Proc; AAS 78-011, Adv v37I

WISCHNIA, H.F., AAS 72-020, S&T v28

WISE, D.U., AAS 69-209, Mic v14

WISSINGER, A.B., AAS 75-188, Mic v25, Adv v32I

WITHBROE, G.L., AAS 68-219, Adv v25

WITHEE, W.W., AAS 70-034, Adv v27

WITHERSPOON, J.P., AAS 76-022, S&T v44

WITHOLDER, R., AAS 75-230, Adv v32I

WITT, A.F., AAS 74-053, S&T v38

WITTEBORN, F.C., AAS 75-234, Mic v25, Adv v32II

WITTEN, L., AAS 56-022, Adv v1; JAS v4-3 pp46-48 1957

WITTING, J., AAS 68-191, Adv v25

WOLBERS, H.L., AAS 68-059, Adv v24; 75-291, Mic v25, Adv v32II; 78-044, Adv v37II

WOLF, H., AAS 60-106, Sp v2

WOLF, H.D., AAS 74-182, S&T v37; IAF 75-013, S&T v41, JAS v24-1 pp1-34 1976

WOLFE, B., AAS 74-022, S&T v35

WOLFE, J.F., JAS v7-2 pp47-48 1960

WOLFE, J.H., AAS 71-124, Adv v29I; 76-112, S&T v43

WOLFF, E.A., AAS 77-256, Adv v36I

WOLLENHAUPT, W.R., AAS 66-105, S&T v11

WOLLENSAK, R.J., AAS 75-176, Mic v25, Adv v32I

WOLVERTON, B.C., AAS 70-071, S&T v26

WOMBLE, M.E., AAS 73-226, Mic v21

WONG, C.M., ed. S&T v6 1965

WONG, R.E., AAS 65-109, Adv v20I

WOO, K.E., AAS 75-253, Mic v25, Adv v32II

WOOD, D.B., AAS 78-165, Adv v38II

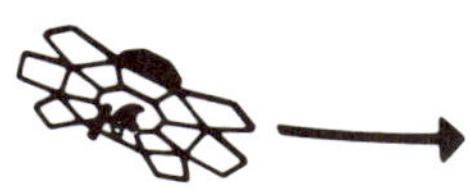

X

Y



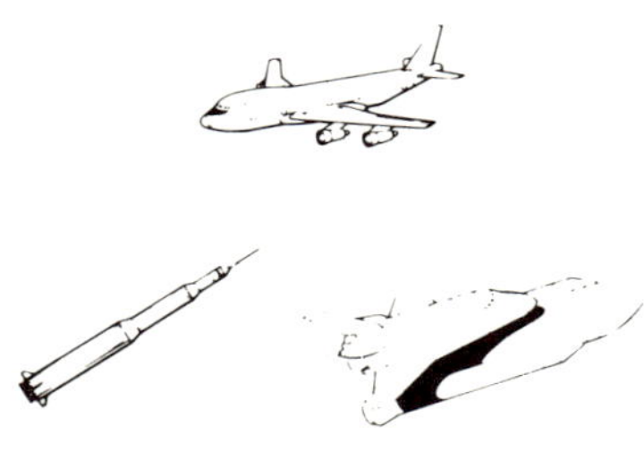

# Z

# APPENDICES

Meetings of the American Astronautical Society
(1954-1979)

Book and Microform Publications of the
American Astronautical Society
(1954-1978)

# APPENDICES

The following material is appended:

I. Meetings of the American Astronautical Society (1954-1979)--Appendix I. Generally only those meetings are listed for which the AAS was the principal sponsor. In each case, an indication of published proceedings is given when applicable. The precise citation for the proceedings is given in succeeding appendices.

II. All AAS meetings held in conjunction with the American Association for the Advancement of Science (AAAS), listed in Appendix I, are brought together.

III. All AAS Goddard Memorial Symposia, listed in Appendix I, are brought together.

IV. All books published by or for the American Astronautical Society are listed chronologically, with complete citation by series, as follows:

1. Advances in the Astronautical Sciences
2. Special AAS volumes not in AAS series
3. Science and Technology Series
4. AAS History Series
5. AAS Microfiche Series

# APPENDIX I

## MEETINGS OF THE AMERICAN ASTRONAUTICAL SOCIETY (1954-1979)

First Annual Meeting, December 3, 1954, New York, N.Y.
(One paper presented by W. von Braun - published in *Astronautics**)

Second Annual Meeting, December 1, 1955, New York, N.Y.
(Nine papers - mostly published in *Astronautics**)

Third Annual Meeting, December 6-7, 1956, New York, N.Y.
*(Volume 1, Advances in the Astronautical Sciences+)*

Fourth Annual Meeting, January 29-31, 1958, New York, N.Y.
*(Volume 2, Advances)*

First Western Regional Meeting, August 18-19, 1958, Palo Alto, California
*(Volume 3, Advances)*

Fifth Annual Meeting, December 27-31, 1958, Washington, D.C.
*(Volume 4, Advances)*

Second Western National Meeting, August 4-5, 1959, Los Angeles, California
*(Volume 5, Advances)*

Sixth Annual Meeting, January 18-21, 1960, New York, N.Y.
*(Volume 6, Advances)*

Symposium on Physical and Biological Phenomena under Zero G Conditions, July 1, 1960, Hawthorne, California
(AAS Special Volume 1)

Third Western National Meeting, August 4-5, 1960, Seattle, Washington
*(Volume 7, Advances)*

Symposium on Lunar Flight (AAS/AAAS Symposium), December 27, 1960, New York, N.Y.
(AAS Special Volume 2)

---

* *Now designated the JOURNAL OF THE ASTRONAUTICAL SCIENCES*

+ *Henceforth abbreviated as ADVANCES*

*Unless otherwise indicated, all proceedings are available from Univelt, Inc., P.O. Box 28130, San Diego, California 92128.*

Seventh Annual Meeting, January 16-18, 1961, Dallas, Texas
*(Volume 8, Advances)*

Interactions of Space Vehicles with an Ionized Atmosphere, First Goddard Memorial Symposium, March 17, 1961, Washington, D.C.
(Published by Pergamon Press, 1965)

Fourth Western Regional Meeting, August 1-3, 1961, San Francisco, California
*(Volume 9, Advances)*

Symposium on Manned Lunar Flight (AAS/AAAS Symposium), December 29, 1961, Denver, Colorado
*(Volume 10, Advances)*

Eighth Annual Meeting, January 16-18, 1962, Washington, D.C.
*(Volume 11, Advances)*

Torques and Attitude Sensing in Satellites, Second Goddard Memorial Symposium, March 16-17, 1962, Washington, D.C.
(Published by Academic Press)

Symposium on Scientific Satellites - Mission and Design (AAS/AAAS Symposium), December 27, 1962, Philadelphia, Pennsylvania
*(Volume 12, Advances)*

Interplanetary Missions, Ninth Annual Meeting, January 15-17, 1963, Los Angeles, California
*(Volume 13, Advances)*

Second AAS Symposium on Physical and Biological Phenomena under Zero G Conditions, January 18, 1963, Los Angeles, California
*(Volume 14, Advances)*

Symposium on the Exploration of Mars, June 6-7, 1963, Denver, Colorado
*(Volume 15, Advances)*

Symposium on Space Rendezvous, Rescue, and Recovery, September 10-12, 1963, Edwards, California
*(Volume 16-I and II, Advances)*

Symposium on Bioastronautics (AAS/AAAS Symposium), December 30, 1963, Cleveland, Ohio
*(Volume 17, Advances)*

Lunar Flight Programs, Tenth Annual Meeting, May 4-7, 1964, New York, N.Y.
*(Volume 18, Advances)*

Manned Space Reliability Symposium, June 9, 1964, Anaheim, California
*(Volume 1, Science and Technology)*

Towards Deeper Space Penetration (AAS/AAAS Symposium), December 29, 1964, Montreal, Canada
*(Volume 2, Science and Technology)*

Symposium on Unmanned Exploration of the Solar System, February 8-10, 1965, Denver, Colorado
*(Volume 19, Advances)*

Scientific Experiments for Manned Orbital Flight, Third Goddard Memorial Symposium, March 18-19, 1965, Washington, D.C.
*(Volume 4, Science and Technology)*

Symposium on Physiological and Performance Determinants in Manned Space Systems, (AAS-LA Section/HFS), April 14-15, 1965, Northridge, California
*(Volume 5, Science and Technology)*

Post Apollo Exploration, Eleventh Annual Meeting, May 3-6, 1965, Chicago, Illinois
*(Volume 20-I and II, Advances)*

Space Electronics Symposium (AAS/AES Meeting), May 25-27, 1965, Los Angeles, California
*(Volume 6, Science and Technology)*

Impact of Space Exploration on Society, August 18-20, 1965, San Francisco, California
*(Volume 8, Science and Technology)*

Second AAS Symposium on Interactions of Space Vehicles with an Ionized Atmosphere, November 26-27, 1965, Miami, Florida
(No proceedings published)

Recent Developments in Space Flight Mechanics (AAS/AAAS Symposium), December 29, 1965, Berkeley, California
*(Volume 9, Science and Technology)*

Practical Space Applications, February 21-23, 1966, San Diego, California
*(Volume 21, Advances)*

Space Age in Fiscal Year 2001, Fourth Goddard Memorial Symposium, March 15-16, 1966, Washington, D.C.
*(Volume 10, Science and Technology)*

The Search for Extraterrestrial Life, Twelfth Annual Meeting, May 23-25, 1966, Anaheim, California
*(Volume 22, Advances; Volume 1, AAS Microfiche)*

Space Flight Mechanics Symposium, July 6-8, 1966, Denver, Colorado
*(Volume 11, Science and Technology; Volume 2, AAS Microfiche)*

Interpretation of Lunar Probe Data, September 17, 1966, Huntington Beach, California
*(Volume 14, Science and Technology)*

National Conference on the Management of Aerospace Systems, November November 16-18, 1966, Columbia, Missouri
*(Volume 12, Science and Technology)*

Symposium on Missiles and Aerospace Vehicles Sciences, December 5-7, 1966, Huntsville, Alabama
*(Volume 9, AAS Microfiche)*

Symposium on the Physics of the Moon (AAS/AAS Symposium), December 29, 1966, Washington, D.C.
*(Volume 13, Science and Technology)*

Voyage to the Planets, Fifth Goddard Memorial Symposium, March 14-15, 1967, Washington, D.C.
*(Volume 16, Science and Technology)*

Symposium on Future Space Program and Impact on Range and Network Development, March 22-24, 1967, Las Cruces, New Mexico
*(Volume 15, Science and Technology)*

Commercial Utilization of Space, Thirteenth Annual Meeting, May 1-3, 1967, Dallas, Texas
*(Volume 23, Advances; Volume 3, AAS Microfiche)*

Symposium on Planetary Geology and Geophysics, May 25-27, 1967, Boston, Massachusetts
*(Volume 17, Science and Technology; Volume 5, AAS Microfiche)*

Saturn V / Apollo and Beyond, June 11-14, 1967, Huntsville, Alabama
*(Volume 4, AAS Microfiche)*

Rocky Mountain Resources for Aerospace Science and Technology, July 13-14, 1967, Denver, Colorado
*(Volume 10, AAS Microfiche)*

Astronautics International, November 27-29, 1967, New York, N.Y.
*(Volume 6, AAS Microfiche)*

Exobiology--the Search for Extraterrestrial Life (AAS/AAAS Symposium), December 30, 1967, New York, N.Y.
*(Volume 19, Science and Technology)*

National Specialists Symposium on Orbital Resonance, January 8-9, 1968, Redondo Beach, California
(No proceedings published)

Symposium on the Interpretation of Lunar Probe Data II, January 25-26, 1968, Huntington Beach, California
*(Volume 25, Science and Technology)*

Role of the Space Program in the Development of Modern Society (Sixth Goddard Memorial Symposium), March 12-13, 1968, Washington, D.C.
*(Volume 18, Science and Technology)*

Water Resources of the Inner Planets, April 10-12, 1968, Les Vegas, Nevada
*(Volume 25, Science and Technology)*

Exploitation of Space for Experiment Research, Fourteenth Annual Meeting, May 13-15, 1968, Dedham, Massachusetts
*(Volume 24, Advances)*

Space Projections, July 15-16, 1968, Denver, Colorado
*(Volume 8, AAS Microfiche)*

Astrodynamics Specialist Conference (AIAA/AAS), September 3-5, 1968, Grand Teton National Park, Wyoming
*(Volume 7, AAS Microfiche)*

Advanced Space Experiments, September 16-18, 1968, Ann Arbor, Michigan
*(Volume 25, Advances)*

Space as a National Resource, October 24, 1968, Los Angeles, California
*(Volume 11, AAS Microfiche)*

Symposium on Bioengineering and Cabin Ecology, (AAS/AAAS Symposium), December 30, 1968, Dallas, Texas
*(Volume 20, Science and Technology)*

Symposium on Lunar Transients, February 17-18, 1969, Huntington Beach, California
*(Volume 25, Science and Technology)*

Reducing the Cost of Space Transportation, Seventh Goddard Memorial Symposium, March 4-5, 1969, Washington, D.C.
*(Volume 21, Science and Technology)*

Planning Challenges of the 70s in Space and in the Public Domain, Fifteenth Annual Meeting, June 17-20, 1969, Denver, Colorado
*(Volume 26, Advances; Volume 22, Science and Technology; Volumes 13-19, AAS, Microfiche)*

Space Technology and Earth Problems, October 23-25, 1969, Las Cruces, New Mexico
*(Volume 23, Science and Technology; Volume 12, AAS Microfiche)*

Exploration of the Planets, (AAS/AAAS Symposium), December 26-27, 1969, Boston, Massachusetts
(No proceedings published)

Aerospace Systems Development--Implications for the Law, Eighth Goddard Memorial Symposium, March 1970, Washington, D.C.
(No proceedings published)

Space Stations and Manned Planetary Exploration, Sixteenth Annual Meeting, June 8-10, 1970, Anaheim, California
*(Volumes 27 and 28, Advances)*

Technology Utilization Ideas for the 70s and Beyond, October 30, 1970, Winrock, Arkansas
*(Volume 26, Science and Technology)*

The Surface Features of Mars, November 16, 1970, Huntington Beach, California
(No proceedings published)

Lunar Sample Analysis (AAS/AAAS Symposium) December 1970, Chicago, Illinois
(No proceedings published)

International Cooperation in Space Operations and Exploration, Ninth Goddard Memorial Symposium, March 11, 1971, Washington, D.C.
*(Volume 27, Science and Technology)*

Variable Geometry and Expandable Structures, (AAS/AIAA Meeting), April 21-23, 1971, Anaheim, California
(No proceedings published)

Capabilities for the Design of Manned and Unmanned Space Missions, May 26-27, 1971, Pasadena, California
(No proceedings published)

The Outer Solar System, Seventeenth Annual Meeting, June 28-30, 1971, Seattle, Washington
*(Volume 29-I and II, Advances)*

AAS/AIAA Astrodynamics Conference, August 17-19, 1971, Fort Lauderdale, Florida
*(Volume 20, AAS Microfiche)*

Astronomy from a Space Platform (AAS/AAAS Symposium), December 27-28, 1971, Philadelphia, Pennsylvania
*(Volume 28, Science and Technology)*

Transfer of Space Technology to Community and Industrial Activities, Tenth Goddard Memorial Symposium, Eighteenth Annual Meeting, March 13-14, 1972, Washington, D.C.
*(Volume 29, Science and Technology)*

Conference on Health Care Systems, November 21-22, 1972, Dallas, Texas
*(Volume 32, Science and Technology)*

Space Shuttle Payloads (AAS/AAAS Symposium), December 27-28, 1972, Washington, D.C.
*(Volume 30, Science and Technology)*

The Second Fifteen Years in Space, Eleventh Goddard Memorial Symposium, March 8-9, 1973, Washington, D.C.
*(Volume 31, Science and Technology)*

Management and Design of Long-Life Systems, April 24-26, 1973, Denver, Colorado
*(Volume 34, Science and Technology)*

International Congress of Space Benefits, Nineteenth Annual Meeting, June 19-21, 1973, Dallas, Texas
*(Volume 30, Advances)*

User/Developer Conference on Textiles and Apparel, June 26-28, 1973, Atlanta, Georgia
(No proceedings published)

AAS/AIAA Astrodynamics Specialist Conference, July 16-18, 1973, Vail, Colorado
*(Volume 21, AAS Microfiche)*

Energy Delta-Supply vs. Demand, (AAS/AAAS Symposium), February 25-27, 1974, San Francisco, California
*(Volume 35, Science and Technology)*

Skylab Science Experiments (AAS/AAAS Symposium), February 28, 1974, San Francisco, California
*(Volume 38, Science and Technology)*

Progress Report on Skylab and Pioneer Programs, Twelfth Goddard Memorial Symposium, March 8, 1974, Washington, D.C.
*(Volume 36, Science and Technology)*

The Skylab Results, Twentieth Annual Meeting, August 20-22, 1974, Los Angeles, California
*(Volume 31-I and II, Advances; Volume 22, AAS Microfiche)*

Future Space Activities, Thirteenth Goddard Memorial Symposium, April 11, 1975, Washington, D.C.
*(Volume 40, Science and Technology)*

AAS/AIAA Astrodynamics Conference, July 28-30, 1975, Nassau, Bahamas
*(Volume 33, Advances; Volume 26, AAS Microfiche)*

Space Shuttle Missions of the 80s, Twenty First Annual Meeting, August 26-28, 1975, Denver, Colorado
*(Volume 32-I and II, Advances; Volume 25, AAS Microfiche)*

Viking Mars Science Experiments (AAS/AAAS Symposium), February 19, 1976, Boston, Massachusetts
(No proceedings published)

Satellite Communications in the 80s, Fourteenth Goddard Memorial Symposium, March 12, 1976, Washington, D.C.
*(Volume 44, Science and Technology)*

The Bicentennial Space Symposium--New Themes for Space: Mankind's Future Needs and Aspirations, Twenty Second Annual Meeting, October 6-8, 1976, Washington, D.C.
*(Volume 35, Advances)*

Two Hundred Years of Flight in America, a Bicentennial Survey, November 4, 1976, Washington, D.C.
*(Volume 1, AAS History)*

Export of Aerospace Technology, Fifteenth Goddard Memorial Symposium, March 31-April 1, 1977, Washington, D.C.
*(Volume 46, Science and Technology)*

AAS/AIAA Astrodynamics Conference, September 7-9, 1977, Jackson, Wyoming
*(Volume 27, AAS Microfiche)*

The Industrialization of Space, Twenty Third Annual Meeting, October 18-20, 1977, San Francisco, California
*(Volume 36-I and II, Advances; Volume 28, AAS Microfiche)*

Space Shuttle and Spacelab Utilization, What are the Near-Term and Long-Term Benefits for Mankind? Sixteenth Goddard Memorial Symposium, Twenty Fourth Annual Meeting, March 8-10, 1978, Washington, D.C.
*(Volume 37-I and II, Advances)*

Annual Rocky Mountain Guidance and Control Conference, March 10-13, 1978, Keystone, Colorado
*(Volume 29, AAS Microfiche)*

The Future United States Space Program, Twenty Fifth Anniversary Conference, October 30-November 2, 1978, Houston, Texas
*(Volume 38, Advances)*

Annual Rocky Mountain Guidance and Control Conference, February 24-28, 1979, Keystone, Colorado
*(Volume 39, Advances; Volume 31, AAS Microfiche)*

Making Space Work, Seventeenth Goddard Memorial Symposium, March 28-30, 1979, Washington, D.C.
*(Science and Technology and AAS History)*

AAS/AIAA Astrodynamics Conference, June 25-27, 1979, Provincetown, Massachusetts
*(Advances and AAS Microfiche)*

Dawn of the Age of the Shuttle, Twenty Sixth Annual Meeting, October 29-November 1, 1979, Los Angeles, California
*(Advances and AAS Microfiche)*

# APPENDIX II

## *AAS MEETINGS IN CONJUNCTION WITH THE AMERICAN ASSOCIATION FOR THE ADVANCEMENT OF SCIENCE

As a member of the American Association for the Advancement of Science, the American Astronautical Society has regularly held symposia in conjunction with AAAS annual meetings. They are listed below:

Lunar Exploration and Spacecraft Systems, December 27, 1960, New York, N.Y.
(AAS Special Volume 2, 1962)

Manned Lunar Flight, December 29, 1961, Denver, Colorado
(*Volume 10, Advances in the Astronautical Sciences,* 1963)

Scientific Satellites--Mission and Design, December 27, 1962, Philadelphia, Pennsylvania
(*Volume 12, Advances in the Astronautical Sciences,* 1963)

Bioastronautics--Fundamental and Practical Problems, December 30, 1963, Cleveland, Ohio
(*Volume 17, Advances in the Astronautical Sciences,* 1964)

Towards Deeper Space Penetration, December 29, 1964, Montreal, Canada
(*Volume 2, Science and Technology,* 1965)

Recent Developments in Space Flight Mechanics, December 29, 1965, Berkeley, California
(*Volume 9, Science and Technology,* 1966)

Physics of the Moon, December 29, 1966, Washington, D. C.
(*Volume 13, Science and Technology,* 1967)

Exobiology--the Search for Extraterrestrial Life, December 30, 1967, New York, N.Y.
(*Volume 19, Science and Technology,* 1969)

Bioengineering and Cabin Ecology, December 30, 1968, Dallas, Texas
(*Volume 20, Science and Technology,* 1969)

Whither Lunar and Planetary Exploration in the 1970s?, December 26-27, 1969, Boston, Massachusetts
(No proceedings published)

---

* *Unless otherwise indicated all AAS/AAAS Symposium proceedings are available from Univelt, Inc., P.O. Box 28130, San Diego, CA 92128*

Lunar Sample Analysis, December 1970, Chicago, Illinois
(No proceedings published)

Astronomy from a Space Platform, December 27-28, 1971, Philadelphia, Pennsylvania
(*Volume 28, Science and Technology*, 1972)

Space Shuttle Payloads, December 27-28, 1972, Washington, D. C.
(*Volume 30, Science and Technology*, 1973)

Energy Delta, Supply vs. Demand, February 25-27, 1974, San Francisco, California
(*Volume 35, Science and Technology*, 1975)

Skylab Science Experiments, February 28, 1974, San Francisco, California
(*Volume 38, Science and Technology*, 1975)

Viking Mars Science Experiments, February 19, 1976, Boston, Massachusetts
(No proceedings published)

# APPENDIX III

## *AAS GODDARD MEMORIAL SYMPOSIA (1961-1979)

The American Astronautical Society has held each year since 1961--with few exceptions--a symposium in conjunction with the annual meeting of the National Space Club. These meetings are held in March or early April in Washington, D.C. Following is a list of these events:

First (1961), Interactions of Space Vehicles with an Ionized Atmosphere (*International Series of Monographs in Aeronautics and Astronautics, Division IX Symposia, Volume 18,* Pergamon Press, 1965)

Second (1962), Torques and Attitude Sensing in Satellites (*Mathematics and Mechanics, Volume 7,* Academic Press, New York, 1964)

Third (1965), Scientific Experiments for Manned Orbital Flight (*Volume 4, Advances in the Astronautical Sciences,* 1965)

Fourth (1966), Space Age in Fiscal Year 2001 (*Volume 10, Science and Technology,* 1967)

Fifth (1967), Voyage to the Planets (*Volume 16, Science and Technology,* 1968)

Sixth (1968), Role of the Space Program in the Development of Modern Society (*Technology and Social Progress,Synergism or Conflict? Volume 18, Science and Technology,* 1969)

Seventh (1969), Reducing the Cost of Space Transportation (*Volume 21, Science and Technology,* 1969)

Eighth (1970), Aerospace Systems Development: Implications for the Law (Not published)

Ninth (1971), International Cooperation in Space Operations (*Volume 27, Science and Technology,* 1971)

Tenth (1972), Transfer of Space Technology to Community and Industry (*Volume 29, Science and Technology,* 1972)

Eleventh (1973), The Second Fifteen Years in Space (*Volume 31, Science and Technology,* 1973)

---

* *Unless otherwise indicated all AAS Goddard Memorial Symposia Proceedings are available from Univelt,Inc.,P.O.Box 28130,San Diego,CA 92128*

Twelfth (1974), Progress Report on Skylab and Pioneer Programs (*Skylab and Pioneer Report, Volume 36, Science and Technology,* 1973)

Thirteenth (1975), Future Space Activities (*Volume 40, Science and Technology,* 1976)

Fourteenth (1976), Satellite Communications in the 80s (*Volume 44, Science and Technology,* 1977)

Fifteenth (1977), Export of Aerospace Technology (*Volume 46, Science and Technology,* 1978)

Sixteenth (1978), Space Shuttle and Spacelab Utilization (*Volume 37, Advances in the Astronautical Sciences,* 1978)

Seventeenth (1979), Making Space Work (*Volume 49, Science and Technology; Volumes 2 and 3, AAS History,* 1979-1980)

# APPENDIX IV

## *BOOKS PUBLISHED BY OR FOR THE AMERICAN ASTRONAUTICAL SOCIETY

+ADVANCES IN THE ASTRONAUTICAL SCIENCES
Series Editor: H. Jacobs

*Third Annual Meeting, Volume 1,* ed. N. V. Petersen, 1957, 184p

*Fourth Annual Meeting, Volume 2,* eds. N. V. Petersen, H. Jacobs, 1958, 440p

*First Western National Meeting, Volume 3,* eds. N. V. Petersen, H. Jacobs, 1958, 530p

*Fifth Annual Meeting, Volume 4,* ed. H. Jacobs, 1959, 462p

*Second Western National Meeting, Volume 5,* ed. H. Jacobs, 1960, 364p

*Sixth Annual Meeting, Volume 6,* eds. H. Jacobs, E. Burgess, 1961, 968p

*Third Western National Meeting, Volume 7,* eds. H. Jacobs, E. Burgess, 1961, 464p

*Seventh Annual Meeting, Volume 8,* ed. H. Jacobs, 1963, 602p

*Fourth Western Meeting, Volume 9,* ed. E. Burgess, 1963, 910p

*Manned Lunar Flight, Volume 10,* eds. G. W. Morgenthaler, H. Jacobs, 1963, 310p

*Eighth Annual Meeting, Volume 11,* ed. H. Jacobs, 1963, 808p

*Scientific Satellites, Volume 12,* ed. I. E. Jeter, 1963, 262p

*Ninth Annual Meeting, Interplanetary Missions, Volume 13,* ed. E. Burgess, 1963, 690p

*Physiological Phenomena in a Weightless State, Volume 14,* eds. E. T. Benedikt, R. W. Halliburton, 1963, 382p

---

* *Books of this series are available from Univelt, Inc., P.O. Box 28130, San Diego, California 92128*

\+ *Library of Congress number for this series: 57-43769*
*International Standard Book Number: 87703*

*Exploration of Mars, Volume 15,* ed. G. W. Morgenthaler, 1963, 634p

*Space Rendezvous, Rescue, and Recovery, Volume 16-I,* ed. N. V. Petersen, 1963, 1028p

*Space Rendezvous, Rescue, and Recovery, Volume 16-II,* ed. N. V. Petersen, 1963, 380p

*Bioastronautics--Fundamental and Practical Problems, Volume 17,* ed. W. C. Kaufman, 1964, 128p

*Lunar Flight Programs, Volume 18,* ed. R. Fleisig, 1964, 630p

*Unmanned Exploration of the Solar System, Volume 19,* eds. G. W. Morgenthaler, R. G. Morra, 1965, 1000p

*Post Apollo Space Exploration, Volume 20-I,* ed. F. Narin, 1966, 572p

*Post Apollo Space Exploration, Volume 20-II,* ed. F. Narin, 1966, 648p

*Practical Space Applications, Volume 21,* ed. L. L. Kavanau, 1967, 508p

*The Search for Extraterrestrial Life, Volume 22,* ed. J. S. Hanrahan, 1967, 388p and supplement of 4 microfiche

*The Commercial Utilization of Space, Volume 23,* eds. J. R. Gilmer, A. M. Mayo, R. C. Peavey, 1968, 512p and supplement of 24 microfiche

*Exploration of Space, Volume 24,* ed. H. Zuckerberg, 1968, 363p

*Advanced Space Experiments, Volume 25,* eds. O. L. Tiffany, E. Zaitzeff, 1969, 530p

*Planning Challenges of the 70s in Space, Volume 26,* eds. G. W. Morgenthaler, R. Morra, 1970, 470p and supplement of 13 microfiche

*Space Stations, Volume 27,* eds. L. Larmore, R. L. Gervais, 1970, 606p

*Space Shuttles and Interplanetary Missions, Volume 28,* eds. L. Larmore, R. L. Gervais, 1970, 488p

*The Outer Solar System, Volume 29-I,* ed. J. Vagners, 1971, 618p

*The Outer Solar System, Volume 29-II,* ed. J. Vagners, 1971, 740p

*International Congress of Space Benefits, Volume 30,* ed. F. S. Johnson, 1974, 528p

*The Skylab Results, Volume 31-I,* eds. W. C. Schneider, T. E. Hanes, 1975, 616p

*The Skylab Results, Volume 31-II,* eds. W. C. Schneider, T. E. Hanes, 1975, 558p and supplement of 8 microfiche

*Space Shuttle Missions of the 80s, Volume 32-I,* eds. W. J. Bursnall, G. W. Morgenthaler, G. E. Simonson, 1977, 598p

*Space Shuttle Missions of the 80s, Volume 32-II,* eds. W. J. Bursnall, G. W. Morgenthaler, G. E. Simonson, 1977, 766p and supplement of 65 microfiche

*Astrodynamics 1975, Volume 33,* eds. W. F. Powers, H. E. Rauch, B. D. Tapley, C. E. Velez, 1976, 390p and supplement of 36 microfiche

*Apollo Soyuz Mission Report, Volume 34,* ed. C. M. Lee, 1977, 336p

*Bicentennial Space Symposium, New Themes in Space: Mankind's Future Needs and Aspirations, Volume 35,* ed. W. C. Schneider, 1977, 242p

*The Industrialization of Space, Volume 36-I,* eds. R. A. Van Patten, P. Siegler, E. V. B. Stearns, 1978, 610p

*The Industrialization of Space, Volume 36-II,* eds. R. A. Van Patten, P. Siegler, E. V. B. Stearns, 1978, 550p and supplement of 9 microfiche

*Space Shuttle Spacelab Utilization, Volume 37-I,* eds. G. W. Morgenthaler, M. Hollstein, 1978, 400p

*Space Shuttle Spacelab Utilization, Volume 37-II,* eds. G. W. Morgenthaler, M. Hollstein, 1978, 465p

*The Future U. S. Space Program, Volume 38 I-II,* eds. R. S. Johnston, A. Naumann, Jr., C. W. G. Fulcher, 1979

*Guidance and Control 1979, Volume 39,* ed. R. D. Culp, 1979

## SPECIAL AAS VOLUMES NOT IN AAS SERIES

(1)* *Weightlessness - Physical Phenomena and Biological Effects,* ed. E. T. Benedikt, 1961, American Astronautical Society, 182p

(2)* *Lunar Exploration and Spacecraft Systems,* ed. R. Fleisig, E. A. Hine, G. J. Clark, 1962, American Astronautical Society, 214p

(4) *Interactions of Space Vehicles with an Ionized Atmosphere, International Series of Monographs in Aeronautics and Astronautics,* ed. S. F. Singer, Pergamon Press, 1965, 520p

(5) *Torques and Attitude Sensing in Earth Satellites, Applied Mathematics and Mechanics, Volume 7,* ed. S. F. Singer, Academic Press, 1964, 276p

---

* *Available from Univelt, Inc., P.O. Box 28130, San Diego, California 92128*

*SCIENCE AND TECHNOLOGY SERIES
Series Editors (Volumes 1-20): E. Burgess; (Volumes 21-46): H. Jacobs

*Manned Space Reliability Symposium, Volume 1,* ed. P. Horowitz, 1964, 112p

*Towards Deeper Space Penetration, Volume 2,* ed. E. R. van Driest, 1964, 182p

*Orbital Hodograph Analysis, Volume 3,* S. P. Altman, 1965, 150p

*Scientific Experiments for Manned Orbital Flight, Volume 4,* ed. P. C. Badgley, 1965, 372p

*Physiological and Performance Determinants in Manned Space Systems, Volume 5,* ed. P. Horowitz, 1965, 220p

*Space Electronics Symposium, Volume 6,* ed. C. M. Wong, 1965, 404p

*Theodore von Karman Memorial Symposium, Volume 7,* ed. S. Thomas, 1966, 140p

*Impact of Space Exploration on Society, Volume 8,* ed. W. E. Frye, 1966, 382p

*Recent Developments in Space Flight Mechanics, Volume 9,* ed. P. B. Richards, 1966, 280p

*Space in Fiscal Year 2001, Volume 10,* eds. E. Konecci, M. W. Hunter II, R. F. Trapp, 1967, 458p

*Space Flight Mechanics Specialist Conference, Volume 11,* ed. M. L. Anthony, 1967, 618p and supplement of 9 microfiche

*The Management of Aerospace Programs, Volume 12,* ed. W. L. Johnson, 1967, 392p

*Physics of the Moon, Volume 13,* ed. S. F. Singer, 1967, 260p

*Interpretation of Lunar Probe Data, Volume 14,* ed. J. Green, 1967, 270p

*Future Space Program and Impact on Range and Network Development, Volume 15,* ed. G. W. Morgenthaler, 1967, 583p

*The Voyage to the Planets, Volume 16,* ed. S. F. Singer, 1968, 184p

*Use of Space Systems for Planetary Geology and Geophysics, Volume 17,* ed. R. D. Enzmann, 1968, 623p and supplement of 6 microfiche

---

* *Books of this series are available from Univelt, Inc. P.O.Box 28130, San Diego, California 92128.*

*International Standard Book Number: 87703*

*Technology and Social Progress, Volume 18,* ed. P. K. Eckman, 1969, 170p

*Exobiology, The Search for Extraterrestrial Life, Volume 19,* eds. M. M. Freundlich, B. W. Wagner, 1969, 184p

*Bioengineering and Cabin Ecology, Volume 20,* ed. W. B. Cassidy, 1969, 162p

*Reducing the Cost of Space Transportation, Volume 21,* ed. G. K. Chacko, 1969, 264p

*Planning Challenges of the 70s in the Public Domain, Volume 22,* eds. W. J. Bursnall, G. K. Chacko, G. W. Morgenthaler, 1970, 504p and supplement of 15 microfiche

*Space Technology and Earth Problems, Volume 23,* ed. C. Q. Ford, 1970, 418p and supplement of 15 microfiche

*Aerospace Research and Development, Volume 24,* ed. E. A. Steinhoff, 1970, 500p

*Geological Problems in Lunar and Planetary Research, Volume 25,* ed. J. Green, 1971, 750p

*Technology Utilization Ideas for the 70s and Beyond, Volume 26,* eds. F. W. Forbes, P. Dergarabedian, 1971, 312p

*International Cooperation in Space Operations and Exploration, Volume 27,* ed. M. Cutler, 1971, 194p

*Astronomy from a Space Platform, Volume 28,* eds. G. W. Morgenthaler, H. D. Greyber, 1972, 416p

*Space Technology Transfer to Community and Industry, Volume 29,* eds. R. H. Tripp, J. K. Stotz, Jr., 1972, 196p

*Space Shuttle Payloads, Volume 30,* eds. G. W. Morgenthaler, W. J. Bursnall, 1973, 532p

*The Second Fifteen Years in Space, Volume 31,* ed. S. Ferdman, 1973, 212p

*Health Care Systems, Volume 32,* ed. E. B. Konecci, 1974, 264p

*Orbital International Laboratory, Third and Fourth OIL Symposia, Volume 33,* ed. E. A. Steinhoff, 1974, 322p

*Management and Design of Long-Life Systems, Volume 34,* ed. H. M. Schurmeier, 1974, 198p

*Energy Delta, Supply vs. Demand, Volume 35,* eds. G. W. Morgenthaler, A. N. Silver, 1975, second printing 1976, 604p

*Skylab and Pioneer Report, Volume 36,* eds. P. H. Bolger, P. B. Richards, 1975, 160p

*Space Rescue and Safety 1974, Volume 37,* ed. P. H. Bolger, 1975, 294p

*Skylab Science Experiments, Volume 38,* eds. G. W. Morgenthaler, G. E. Simonson, 1976, 182p

*Environmental Control and Agri-technology, Volume 39,* ed. E. B. Konecci, 1976, 346p

*Future Space Activities, 13th Goddard Memorial Symposium, Volume 40,* ed. C. H. Tross, 1976, 182p

*Space Rescue and Safety 1975, Proceedings of the Eighth International Symposium, Volume 41,* ed. P. H. Bolger, 1976, 230p

*The End of an Era in Space Exploration, From International Rivalry to International Cooperation,* J. C. D. Blaine, 1976, 216p

*The Eagle Has Returned, Volume 43,* ed. E. A. Steinhoff, 1976, 370p

*Satellite Communications in the Next Decade, 14th Goddard Memorial Symposium, Volume 44,* ed. L. Jaffe, 1977, 188p

*The Eagle Has Returned, Second Part, Volume 45,* ed. E. A. Steinhoff, 1977, 454p

*Export of Aerospace Technology, 15th Goddard Memorial Symposium, Volume 46,* ed. C. H. Tross, 1978, 174p

*AAS HISTORY SERIES
Series Editor: E. M. Emme

*Two Hundred Years of Flight in America, A Bicentennial Survey, Volume 1,* ed. E. M. Emme, 1977, 326p

---

** Available from Univelt, Inc., P.O.Box 28130, San Diego, California 92128*
*International Standard Book Number: 87703*

AAS MICROFICHE SERIES,
Series Editor: H. Jacobs

*The Search for Extraterrestrial Life, Volume 1,* (Supplement to Volume 22, Advances in the Astronautical Sciences), 1968, 4 papers, 4 microfiche

*Space Flight Mechanics Specialist Symposium, Volume 2,* (Supplement to Volume 11, Science and Technology), 1968, 13 papers, 9 microfiche

*Commercial Utilization of Space, Volume 3,* (Supplement to Volume 23, Advances in the Astronautical Sciences), 1968, 72 papers, 24 microfiche

*Saturn V/Apollo and Beyond, Volume 4,* ed. S. S. Hu, 1968, 147 papers, 62 microfiche

*Use of Space Systems for Planetary Geology and Geophysics, Volume 5,* (Supplement to Volume 17, Science and Technology), 1968, 8 papers, 6 microfiche

*Astronautics International, Volume 6,* 1968, 16 papers, 10 microfiche

*Astrodynamics Specialist Conference, Volume 7,* 1968, 69 papers, 21 microfiche

*Space Projections, Volume 8,* 1969, 63 papers, 28 microfiche

*Southeastern Symposium on Missiles and Aerospace Vehicle Sciences, Volume 9,* ed. S. S. Hu, 1969, 109 papers, 30 microfiche

*Rocky Mountain Resources for Aerospace Science and Technology, Volume 10,* 1969, 72 papers, 22 microfiche

*Space as a National Resource, Volume 11,* ed. P. Dergarabedian, 1969, 8 papers, 5 microfiche

*Space Technology and Earth Problems, Volume 12,* (Supplement to Volume 23, Science and Technology), 1970, 34 papers, 15 microfiche

*Planning Challenges of the 70s in the Public Domain, Volume 13,* (Supplement to Volume 22, Science and Technology), 1970, 32 papers, 15 microfiche

*Planning Challenges of the 70s in Space, Volume 14,* (Supplement to Volume 26, Advances in the Astronautical Sciences), 1970, 20 papers, 13 microfiche

*Operations Research - Management Applications, Volume 15,* 1970, 13 papers, 10 microfiche

*Operations Research - Urban Problems, Volume 16,* 1970, 18 papers, 13 microfiche

*Operations Research - Mathematical and Computer Methods, and Military Applications, Volume 17,* 1970, 37 papers, 23 microfiche

*Spacecraft and Space Technology, Volume 18,* 1970, 40 papers, 25 microfiche

*Space Exploration, Space Science, and Application, Volume 19,* 1970, 26 papers, 17 microfiche

*Astrodynamics Specialist Conference 1971, Volume 20,* 1971, 91 papers, 66 microfiche

*AAS/AIAA Astrodynamics Specialist Conference 1973, Volume 21,* 1973, 44 papers, 28 microfiche

*The Skylab Results, Volume 22,* (Supplement to Volume 31, Advances in the Astronautical Sciences), 1975, 7 papers, 9 microfiche

*International Space Rescue Symposia - First, Second, and Third, Volume 23,* (Symposia of the International Academy of Astronautics), 1975, 35 papers, 10 microfiche

*International Space Rescue Symposia - Fourth, Fifth, and Sixth, Volume 24,* (Symposia of the International Academy of Astronautics), 1975, 35 papers, 11 microfiche

*Space Shuttle Missions of the 80s, Volume 25,* (Supplement to Volume 32, Advances in the Astronautical Sciences), 1976, 120 papers, 65 microfiche

*AAS/AIAA Astrodynamics Conference 1975, Volume 26,* (Supplement to Volume 33, Advances in the Astronautical Sciences), 1976, 59 papers, 36 microfiche

*AAS/AIAA Astrodynamics Conference 1977, Volume 27,* 1977, 73 papers, 31 microfiche

*Industrialization of Space, Volume 28,* (Supplement to Volume 36, Advances in the Astronautical Sciences), 1978, 20 papers, 9 microfiche

*Guidance and Control Conference, Volume 29,* 1978, 22 papers, 9 microfiche

*The Future U. S. Space Program, Volume 30,* (Supplement to Volume 38, Advances in the Astronautical Sciences), 1979, 5 papers, 60 abstracts, 6 microfiche

*Guidance and Control 1979, Volume 31,* (Supplement to Volume 39, Advances in the Astronautical Sciences), 1979